ROUTLEDGE LIBRARY EDITIONS:
ENERGY ECONOMICS

Volume 15

ENERGY AND AGRICULTURE: THEIR INTERACTING FUTURES

ENERGY AND AGRICULTURE: THEIR INTERACTING FUTURES

Policy Implications of Global Models

Edited by
MAURICE LÉVY AND JOHN L. ROBINSON

Routledge
Taylor & Francis Group
LONDON AND NEW YORK

First published in 1984 by Harwood Academic Publishers Gmbh

This edition first published in 2018
by Routledge
2 Park Square, Milton Park, Abingdon, Oxon OX14 4RN

and by Routledge
711 Third Avenue, New York, NY 10017

Routledge is an imprint of the Taylor & Francis Group, an informa business

British Library Cataloguing in Publication Data
A catalogue record for this book is available from the British Library

ISBN: 978-1-138-10476-1 (Set)
ISBN: 978-1-315-14526-6 (Set) (ebk)
ISBN: 978-1-138-30688-2 (Volume 15) (hbk)
ISBN: 978-1-315-14210-4 (Volume 15) (ebk)

Proceedings of The United Nations University Symposium
on Energy and Agriculture Futures:
Policy Implications of Global Models,
held in Paris, February 22-25,1982

ENERGY AND AGRICULTURE: THEIR INTERACTING FUTURES
Policy Implications of Global Models

edited by

Maurice Lévy
Université Pierre et Marie Curie, Paris
Director of the Energy Sub-program,
The United Nations University, Tokyo

John L. Robinson
Coordinator of the Food-Energy Global Project
The United Nations University, Tokyo

harwood academic publishers
for The United Nations University

Published under license by OPA Ltd. for harwood academic publishers gmbh

harwood academic publishers

Postrasse 22
7000 Chur
Switzerland

42 William IV Street
London WC2N 4DE
England

58, rue Lhomond
75005 Paris
France

P.O. Box 15053
3501 BB Utrecht
The Netherlands

P.O. Box 786
Cooper Station
New York, NY 10276
United States of America

Published for The United Nations University
by harwood academic publishers

Library of Congress Cataloging in Publication Data
Main entry under title:

Energy and agriculture.

(Global modelling and applications, ISSN 0739-5671; v.1)
Papers presented at a symposium organized by the United Nations University, held in Paris, Feb. 1982.
1. Agriculture and state-Congresses. 2. Food supply—Government policy—Congresses. 3. Energy policy—Congresses. 4. Agriculture-Energy consumption—Congresses. I. Levy, Maurice, 1922-
II. Robinson, John L. III. United Nations University.
IV. Series.
HD1405.E53 1983 333.79 83-18363
ISBN 3-7186-0187-7

ISSN: 0739 5671
ISBN: 3 7186 0187 7

Printed in the United Kingdom by Bell and Bain Ltd., Glasgow.

Contents

Preface to the Series

Ever since the famous report of the Club of Rome entitled "Limits to Growth", global modelling has become an important activity throughout the world. Several other models, taking into account the criticism which had been raised against the Club of Rome model, have since been set up, and this has become a very active sector, both within the academic community and in international or national agencies. Of course, the limitation of global models are now well understood, as well as the necessity to improve the methodology.

Another important development which has taken place in the recent years has been connected with the applications of global models. In many cases, when dealing with a given planning problem at the national or regional level, it has been realized that an essential input is the set of constraints provided by the world situation, which can be taken into account by using the world models as input into the problem. In another approach, specialized global models, for example for agriculture or energy, have been built by connecting a reasonably representative set of national models (the so called "bottom up" approach).

This series intends to reflect the wide diversity of the work now being carried out throughout the world on global models: the work on the models themselves, as well as the methodological problems and the applications of the models to specialized situations. It will also try to reflect the different approaches which have been used in the various parts of the world, whether in industrialized or in developing countries.

Introduction

In the past thirty years, considerable progress has been accomplished in the agriculture of developing countries. The so called "green revolution" has been able to increase the food production in these countries. This helped greatly to avoid, but not eliminate, the recurrence of famine, which has been known to occur at regular intervals in the past, even though the population level had been increasing markedly during this time interval.

As is well known, however, the food problem of the developing countries is far from being solved, because it does not depend only on the improvement of the agricultural production, but more importantly on a number of complex socio-economic factors which keep the poorest fraction of the population from having access to food available through commercial channels.

In the future, more difficulties can be predicted for the developing countries, particularly those with a growing population, both rural and urban, as a consequence of the energy crisis:

1. The "green revolution" has been achieved by an appreciable increase in the energy consumption for agriculture (irrigation, mechanization, fertilizers). Even more energy is being consumed along the "food chain" from the field to the mouth, including food conservation, transformation and cooking.

2. The fuelwood crisis is now very much apparent in the rural areas of developing countries and stems from both rural consumption itself and from the growing needs of many cities. Through various commercial and noncommercial channels, the latter cause depopulation of forests in vast surrounding areas. This leads, among other consequences, to the deterioration of the environment and to the reduction of available land for agriculture, since many countries were accustomed to using a rotation cycle.

3. One now observes in several developing countries the temptation to use agriculture for the production of energy

itself: for example, liquid fuels of biological origin. As long as this usage is limited to agricultural residues, or to plants grown on lands inappropriate for food production, this "biomass' utilization is certainly a good means of alleviating the energy crisis in rural areas. However, the pressure of the world energy market can have a bad influence on the trade-off in land allocation between energy and agriculture. Such tendencies are already apparent in countries like Brazil.

4. Finally, for an increasing number of developing countries which are neither self-sufficient in food nor in energy supply, the increase in the price of energy creates dramatic trade-off problems between importations of food and of energy, in the face of a deterioration of their foreign trade balance.

The problem of the interaction between energy and agriculture is therefore a complex one. In many countries, agriculture is being developed as if energy were still a cheap commodity. In the same countries, or in others, energy planning is done independently of the other socio-economic sectors, without paying attention to negative evolution of the world food market. Looking into the long-term future, one sees that a new type of crisis is perhaps in the making for the cities of the developing countries: a "food crisis" which might become as difficult for them as the energy crisis is now. In any case, an increase in the price of food in the cities would certainly worsen the already dramatic situation of the poor, who find it difficult to feed themselves even under the condition of adequate production.

The United Nations University (UNU) has been aware for some time of this problem and decided that research was needed in order to make recommendations to decisionmakers. Even if it might take a long time for the problem to become very serious, these people should still take immediate steps to prevent a crisis.[1] It is also recognized that, in order to tackle this problem seriously, two types of approach are necessary:

[1] Soedjatmoko, Rector of the UNU: "Turning Point in Development: The Food Energy Pivot", Inter-governmental Meeting of Development Assistance Coordinators in Asia and the Pacific, Delhi, India, 23-28 February 1981.

1. A "local" approach, based on a series of monographic studies in different areas of the developing world, where the food-energy crisis has already surfaced. These local studies can enable us first of all to understand the full complexity of the problem, and also to experiment with solutions of balanced food-energy co-production.

2. A "global" approach, the necessity of which is clear when one realizes that any decision taken in one part of the world has immediate consequences in other regions, sometimes very far away.

It is with the second approach in mind that a synposium was organized by the UNU in Paris, in February 1982, bringing together experts in global modelling and specialists on food and energy problems, as well as decisionmakers of the developing countries. This was one of the first high-level discussions on the whole problem of the energy-agriculture interation, and the first to put an emphasis on the global nature of the potential crisis. However, it was also well recognized during the discussion – and through some of the presentations – that this global outlook had to be complemented by detailed analyses of a number of typical local situations, and a general scheme for these analyses was even presented and discussed in detail. In other words, although the emphasis was initially on global models, the presentations and discussions of the Paris Symposium covered a large fraction of the total spectrum and gave clear indications of the directions in which one had to look for solutions.

Another purpose of the Symposium was to formulate a research programme to be implemented not only by The United Nations University, but by other scientific groups or international organizations as well. The very high level of the participants and the abundant time available for discussions made the whole Symposium particularly successful. The publication of its Proceedings, including all the papers presented as well as the main part of the discussions, should therefore prove very useful to many researchers and decisionmakers throughout the world who feel that the "food-energy nexus" is becoming one of the crucial problems of the Third World.

We would like to express our thanks to all the persons who helped us make this Symposium a success, particularly to Mr. Jean-Pierre Chevènement, Minister of State, Minister of Research and Technology, who opened the Symposium with a very thoughtful and provocative presentation. Our gratitude is also due the Director of the "Institut International d'Administration Publique", who offered his generous hospitality to the Symposium, and to numerous members of his staff who helped us through many difficult organisational problems during the meeting. Finally, we are very indebted to Dr. John Robinson, who undertook the very difficult task of editing these Proceedings, firstly by reordering the papers into a logical order, and secondly by extracting the most relevant parts of the discussions and reformulating them again into a logical sequence.

MAURICE LÉVY

1 Opening Address

J.P. CHEVENEMENT[1]

Ladies and Gentlemen,

It is a great honour for France this week to host this important Symposium of the United Nations University, and it is a great pleasure for me to be with you this morning.

I would like to begin by welcoming on behalf of the French Government Mr. Soedjatmoko, Rector of the United Nations University, the vice-rectors, and the whole gathering of eminent participants, some of whom have come from far away to consider over these four days the future of a double theme which is at present so relevant for mankind: energy and agriculture.

For six years France has followed with increasing interest the activities of the United Nations University and at the same time we have been able to appreciate fully the importance of the work conducted within the framework of the three main programmes of this institution, concentrating on the central questions of world hunger, human and social development, and the use of natural resources. We expect much from the future topics which the Council has recently adopted, and which will allow the United Nations University, thanks to your new initiative Mr. Rector, to fulfil effectively the tasks with which it has been entrusted by its Charter. The Government of France is therefore pleased to have the opportunity this week to welcome the senior members of the United Nations University, and to hope that useful links will be forged at this meeting which will allow the strengthening of French participation in the present and future programmes of the University.

Ladies and gentlemen, as you know a month ago an event to which we attached great importance came to an end, a major national colloquy which began in the regions and then moved

[1] Minister of State, Minister of Research and Technology

on to the national level, which gathered tens of thousands of research workers, businessmen, trade unionists, representatives of all sorts, with the aim of redefining the foundations and the interests of our policy in respect to research and technology.

I think that one can say without exaggeration that a broad concensus has formed in our country to make scientific research and technological development a high priority as a guide to all other fields.

This Colloquy met not only to permit an exchange of views. It allowed the definition of a draft law which will be presented before the Parliament and whose purpose will be to define not only a schedule which will allow us to raise our expenditure on research to 2.5% of gross domestic product, but which should also, over and above this schedule, allow us to define the major concerns of research policy. This will take place, for example, at the level of regionalisation, or the priorities, the choices, of the yield of research, the diffusion of scientific and technical information which should permit France to catch up on a certain backwardness which has developed over the last dozen years and to raise it once again, as it was at the end of the nineteen sixties, to the forefront of the major scientific and technological countries in the world.

The international dimension was the object of special studies and in particular we turned our attention to the priority which is the concern of all our research ability as a function of requirements not only of our countries but equally of Third World countries. A special commission was entrusted to Professor Jacques Berque who is well-known to many of you as an Arabist.

In conjunction with several experts Professor Jacques Berque considered ways in which we could reorientate the whole of our research capacity, be it fundamental research, medical or agronomical research, educational, oceanological or space research, as a function of the needs which are making themselves felt in the countries of the Third World and to lay the foundations for a cooperation which would be of mutual benefit, because we ourselves are very concerned to understand the processes taking place in the vast countries of the Third World and which can equally enrich our ideas.

I myself was in India at the beginning of December 1981 and was able to admire the achievements of Indian scientific research, notably in respect to agronomy and the problem of integrating technology into the social context of the Indian village.

This is an extremely important point for all those who consider that the problems of peace in the world cannot be solved independently of the problems of development and that it is today towards the south, looking at the problems of Third World countries, that one can contribute most effectively to the solution of the problems of war and peace.

As you also know, the President of the French Republic, François Mitterand, has announced an important initiative concerning cooperation with the Third World, paying attention to the requirements, the needs, of the Third World countries.

For us this means contributing our resources to establish a new model of development in international relations, and it is in this spirit moreover that France would like to increase its participation in the activities of multilateral organisations.

Ladies and gentlemen, despite your extremely busy schedules elsewhere you have agreed to take part in this Symposium because you believe that the interaction of the energy and agricultural crises is one of the most pronounced problems of the end of the century. This week you will try to establish global models, approaches which will be able to guide the decision-makers in defining their policies.

I would like modestly to share with you some personal thoughts on these questions.

I personally tend to think that while the development of global models is very useful to guide thought, the value of these global models can only be relatively limited because they are by definition essentially economic, and no model can replace the so complex human reality and in particular no model can effectively take into account the ultimately decisive political factors. It is because of this that in the past the conclusions of certain models have frequently been invalidated by major political decisions which have not been foreseen in the considered economic scenarios.

Another factor has to be equally taken into account in order not to negate the interest of this thought, which on the contrary is extremely useful to illuminate the consideration of political men, who in the name of pragmatism often founder in the daily round of events. But it seems to me that important technological breakthroughs cannot be foreseen and they equally well overturn equations as correlations based on relatively constant technologies. Who could have guessed just a few short years ago the speed with which micro-precessors or biotechnology, to cite but two examples, would have become so important.

If I allow myself to put forward these few personal thoughts it is really not to deny the valuable help that global models can give, taking into account the complexity of the mechanisms that govern social relations.

If we want to be able to examine the major tendencies, evaluate the direct or indirect implications of possible choices, consider at the global level the consequences of sectional choices and thus to remedy the unfortunate effects of too narrow specialisations which all too often ignore the consequences of decisions on neighbouring fields, then the work of this Symposium is unquestionably very useful.

The role of these global models is therefore to aid leaders in decision-making, as instruments of forecasting allowing the risks raised by this or another policy to be better anticipated.

Nevertheless, it still remains that in the final analysis all choices are and will remain political. We must therefore dismiss the dream, which I am certain is not yours though embraced by some in earlier times, that a perfect model would definitively give the advantage to the economy or to economics over politics.

This is an old story and in France since Auguste Comte, and probably much earlier with the physiocrats, we had already held this always returning dream. Do not think, though, that I am defending some sectional interest in upholding the primacy of politics. It seems to me to be simply the lesson of experience.

We attach great importance to the two themes which you will be debating:

- agriculture and rural development are in our opinion a primary condition for the development and the "survival" of mankind.

One of the things that struck me most forcibly during my recent trip to India was the importance of research conducted in the field of agronomy and the positive effects that they had achieved in raising agricultural productivity and production. This has tripled since independence in 1947 whereas, over the same period, the population has barely doubled, and there are all the signs of a successful take-off, to which we have probably, paid insufficient attention. Hence the priority of agricultural development to secure self-sufficiency and alimentary independence for the many peoples of the Third World who are far from its achievement.

• energy is the motor of industrial expansion, and we can see throughout history the close links that have existed between energy consumption and the level of development.

That these two questions are taking on a cardinal importance in the North-South dialogue no one would dream of denying. There is great concern to promote real international cooperation on these subjects. Even though the idea of physical limits has somewhat passed out of fashion, the problems are no less important for the fact that resources are not evenly distributed throughout the world.

It is clear that the population of the world should increase between now and the end of the century to reach a figure of about 6 thousand millions, which is an increase of 50% over 25 years.

This demographic explosion allied to growth in the world economy, in all respects desirable, will inevitably be translated into a considerable rise in energy and agricultural consumption.

At present it is forecasted that the consumption of grains will nearly double in comparison to its 1975 level, surpassing 1.2 to 2.2 thousand million tons. At the same time energy consumption will at least double.

Certainly, the world has the physical and economic ability to produce the necessary energy and food to meet this increasing demand. The energy reserves and the development of new technologies of energy production are such that the challenge can be taken up. As for food, it also seems possible to produce the required quantities by the use of suitable technologies and the application of specific fertilisers.

But once again this excessively global view masks the profound disparities which can exist between different countries.

Energy resources are not distributed equally between nations but are, on the contrary, concentrated in certain regions, in the same way as food production is in surplus in certain countries which are glutted with surpluses, and which sometimes even destroy these surpluses. The scandal is all the greater while other countries live in poverty, and sometimes in famine.

Energy consumption by head varies considerably from one country to another, and at the same time shocking, even monstrous, inequalities appear at the level of food consumption between regions where endemic famine rages and those of countries threatened by over-eating.

This is why, if the world is able to respond to the major challenges facing it, there is good reason to fear that without a strengthening of international cooperation in these fields the solutions adopted will be extremely inadequate.

We must move towards planning at the regional level and also at the world level.

Taking into consideration the importance of the human and social stakes and the complexity of the present international political problems, we consider it essential that the foundations of a just cooperation in energy and agriculture are laid as soon as possible since it is not possible to delay and conduct a dialogue which only postpones the reckoning.

It is therefore important to analyse the inter-relations between agriculture and energy, and your Symposium is highly relevant from this point of view.

Energy consumption in the agricultural sector essentially arises from two types of activities: on the one hand, the mechanics of agriculture, that is to say the operations of ploughing, sowing, harvesting and transport of food products, and on the other hand, the chemical fertilisers that really have to be produced.

The introduction of fertilisers and mechanisation have made possible an increase by a factor of about five the yield per hectare of cereal crops, and a division by about the same amount of the number of workers required for their production.

At the world level the portion of agriculture in commercial energy consumption is 3.5 per cent, but will continue to increase from now to the end of the century, taking into account above all the need to modernise the agricultural sector of the developing countries. The increase in energy consumption of the agricultural sector will be greater than the expected increase in agricultural production because the annual rate of increase of agricultural energy consumption should, according to the forecasts of the Food and Agriculture Organisation, increase by 7.5% a year while at the same time agricultural production will increase by 3.8% from now to the year 2000.

It goes without saying that such an increase in agricultural production also presupposes an extension of farming to new areas. At the same time, as we have seen, the energy requirements will more than double. This sets the conditions for changes in energy sources, especially given the fact that the reserves of fossil fuels tend to become scarcer even though new reserves are constantly being discovered. If we look at today's petroleum market we see if anything a surplus, but this is a very short-term tendency which can be rapidly reversed. I am not in favour of basing one's plans of action on market trends since all of history indicates that they can be rapidly reversed.

In any case, this energy transition, which must be prepared for and which will be inadequate, will be peculiar in that it will take place in circumstances of a rise in real prices of energy whereas in the past energy transitions took place under conditions of a fall in energy prices. This was the case in the transition, for example, from wood to coal or coal to petroleum. This is a characteristic feature of the period in which we find ourselves. Therefore we shall have to resort to new forms of energy, not only fossil or nuclear energy, but equally renewable energy sources, and we are in the process of scheduling a very important research effort in this field.

One of the most promising technologies is the production of combustible gases or liquids from biomass.

But here we are faced with a new problem: if it is correct that a very important proportion of this energy production derived from biomass will come from waste products of vegetable or animal origin, it is not impossible that we shall see the growth of competition between agricultural production for food purposes and agricultural production for energy purposes.

The implications of such a competition are still not clear, but many people are afraid that it might manifest itself by a diminuition in food resources, properly understood. The poorest populations would above all suffer from this because of a combination of a decrease in cultivated area devoted to food and the transformation into combustibles, if the prices justify it, of a proportion of cereal production.

You shall be considering these very important questions and I shall not go any further for fear of presumption. Possible competition between various usages of agricultural production must be studied, and since the problems that you will be tackling are extremely complex and that few studies have so far been carried out in this field, it seems a matter of urgency to identify the multi-disciplinery research programmes which will allow the best evaluation of the importance of all the factors which have to be taken into account. Conclusions have to be drawn which could be implemented both by national development plans and also, in my opinion, through global planning for world development.

More than ever, the modern world is faced with global problems which present us with major challenges, and which require a vigorous mental response.

The investment that you are making in this absolutely crucial thinking will not be in vain, I am convinced, if you will be able to go forward with wisdom and foresight.

In any case, Your Honour the Rector, ladies and gentlemen, this gathering illustrates the urgency and the seriousness of the questions of energy and agriculture. These are perhaps the main questions facing mankind and that is why France is pleased to welcome your Symposium. I am certain that it will be marked by fruitful exchanges which will allow you to fulfil the expectations of the international community and to guide the decisions of politicians.

Thank you.

1 Discours d'ouverture

J.P. CHEVENEMENT[1]

Monsieur le Recteur, Mesdames, Messieurs,

C'est pour la France un grand honneur d'accueillir cette semaine cet important Symposium de l'Université des Nations Unies, et c'est pour moi un très grand plaisir d'être ici parmi vous ce matin.

Je voudrais donc tout d'abord, au nom du Gouvernement Français, souhaiter la bienvenue à M. Soedjatmoko, Recteur

[1] Ministre d'Etat, Ministre de la Recherche et de la Technologie

de l'Université des Nations Unies, à Messieurs les Vice-Recteurs et à l'ensemble des éminents participants qui sont venus parfois de très loin pour réfléchir durant ces quatre jours sur le futur d'un double thème déjà si actuel pour l'humanité: l'énergie et l'agriculture.

C'est avec un intérêt croissant que la France a suivi les activités de l'Université des Nations Unies depuis six ans et c'est ainsi que nous avons pu apprécier pleinement le sérieux des travaux entrepris dans le cadre des trois programmes principaux de cette institution, orientés autour des questions prioritaires que constituent la faim dans le monde, le développement humain et social et l'utilisation des ressources naturelles. Nous attendons beaucoup des thèmes futurs que le Conseil vient de retenir, et qui permettront à l'Université des Nations Unies, grâce à votre impulsion nouvelle, Monsieur le Recteur, de remplir avec efficacité les missions qui lui sont confiées par sa Charte. Le Gouvernement Français est donc heureux de pouvoir cette semaine accueillir les hauts responsables de l'Université des Nations Unies, et il espère qu'à cette occasion seront noués des liens utiles qui permettront un renforcement de la participation française aux programmes présents et futurs de l'Université.

Vous le savez Mesdames et Messieurs, il y a un mois s'est terminée une manifestation à laquelle nous avons attaché beaucoup d'importance, un grand Colloque National qui s'est d'abord déroulé dans les régions puis au niveau national, qui a rassemblé des dizaines de milliers de chercheurs, d'industriels, de syndicalistes, de représentants de toutes sortes et dont le but était de redéfinir les bases et les axes de notre politique en matière de Recherche et de Technologie.

Je crois qu'on peut dire sans exagération, qu'un très large consensus s'est dégagé dans notre pays pour faire de la Recherche Scientifique et du développement technologique une haute priorité qui éclairera toutes les autres.

Ce Colloque ne s'est pas réuni uniquement pour permettre un échange de vues. Il a permis la définition d'un projet de loi qui va être soumis au Parlement et dont le but sera de définir non seulement la programmation de l'effort qui doit nous permettre de porter notre dépense de Recherche à 2,5% de la production intérieure brute, mais au-delà de la programmation, ce projet de loi doit nous permettre de définir les grandes orientations en matière de politique de recherche, au niveau par exemple, de la

Régionalisation ou bien des priorités, des choix, de la valorisation de la recherche, de la diffusion de l'information scientifique et technique qui devrait permettre à la France de rattrapper un certain retard qui a été pris pendant une dizaine d'années et de la porter à nouveau, comme c'était le cas à la fin des années 60, à la tête des grands pays scientifiques et technologiques dans le monde.

La dimension internationale a fait l'objet de travaux particuliers et en particulier, nous nous sommes attachés à réfléchir à cette priorité qui est l'orientation de tout notre potentiel de recherche en fonction des besoins qui sont ceux, non pas seulement de nos pays mais également des pays du Tiers Monde. Une mission prioritaire a été confiée au Professeur Jacques Berque que beaucoup d'entre vous connaissent comme un Arabisant célèbre.

Le Professeur Jacques Berque réfléchit avec un certain nombre d'experts aux moyens qui nous permettraient de réorienter l'ensemble de notre potentiel de recherche qu'il s'agisse de Recherche Fondamentale, de Recherche Médicale, de Recherche Agronomique, de Recherche en matière d'Education, en matière d'Océanologie, d'Espace, en fonction des besoins qui se manifestent dans les pays du Tiers Monde et pour jeter les bases d'une coopération qui sera une coopération d'intérêt mutuel, car nous sommes nous-mêmes très intéressés à connaître les expériences qui se déroulent dans de vastes pays du Tiers Monde et qui peuvent également enrichir notre réflexion. J'étais moi-même en Inde au début du mois de décembre et j'ai pu admirer les réalisations de la Recherche Scientifique Indienne notamment en matière d'agronomie ou en matière d'intégration de la technologie dans le contexte social du village indien.

Il y a là une direction extrêmement importante pour tous ceux qui pensent que les problèmes de la paix dans le monde ne peuvent pas être résolus indépendamment des problèmes du développement et que c'est aujourd'hui vers le sud, en regardant vers les problèmes qui sont ceux des pays du Tiers Monde, qu'on peut contribuer le plus efficacement à la solution des problèmes de la guerre et de la paix.

Vous le savez aussi, le Président de la République Française, François Mitterand, a annoncé un effort très important en matière de coopération avec le Tiers Monde, en étant attentif à ce qu'est la demande, à ce que sont les besoins des pays du Tiers Monde.

Il s'agit pour nous de contribuer avec les moyens qui sont les nôtres, à fonder un nouveau modèle de développement dans les relations internationales et c'est dans cet esprit d'ailleurs que la France souhaiterait accroître sa participation aux activites des Organisations Multilatérales.

Vous avez, Mesdames, Messieurs, malgré vos emplois du temps par ailleurs si chargés, vous avez accepté de participer à ce Symposium car vous êtes convaincus que l'interaction des crises de l'énergie et de l'agriculture constitue un des problèmes les plus aigus de la fin du siècle. Vous chercherez donc cette semaine à dégager des modèles globaux, des orientations qui puissent guider les décideurs dans la définition de leurs politiques.

Je ne voudrais modestement que vous faire part de quelques réflexions personnelles.

J'incline personnellement à penser que si l'élaboration de modèles globaux est très utile pour guider la réflexion, la portée de ces modèles globaux ne peut rester que relativement limitée car ils sont par nature essentiellement économiques et aucun modèle ne peut remplacer la réalité humaine si complexe et en particulier, aucun modèle ne peut réellement prendre en compte les facteurs politiques qui sont décisifs.

C'est ainsi que dans le passé, les conclusions de certains modèles ont souvent été invalidées par quelques décisions politiques majeures qui n'avaient pas été prévues dans les scénarios économiques retenus.

Un autre facteur doit également être pris en compte pour non pas nier l'intérêt de cette réflexion qui est au contraire très utile pour éclairer la réflexion d'hommes politiques, qui au nom du pragmatisme, sombrent souvent dans l'action au jour le jour. Mais il me semble que des percées technologiques importantes ne peuvent pas être prévues et elles bouleversent également bien des équations, bien des corrélations qui sont établies sur la base de technologies relativement constantes.

Qui aurait pu il y a quelques années à peine deviner l'importance que prendraient si rapidement les micro-processeurs ou les biotechnologies, pour ne citer que ces exemples?

Si je me permets d'avancer ces quelques réflexions personnelles, ce n'est vraiment pas pour dénier l'aide précieuse que peuvent apporter les modèles globaux, compte tenu de la complexité de mécanismes qui régissent les comportements sociaux.

Si nous voulons pouvoir examiner les grandes tendances, estimer les implications directes ou indirectes des choix qui peuvent être faits, évaluer au niveau global les conséquences des choix sectoriels et remédier ainsi aux effets néfastes des spécialisations trop poussées qui trop souvent ignorent les implications des décisions dans les domaines connexes, les travaux de ce Symposium sont à coup sûr très utiles.

Le rôle de ces modèles globaux est donc bien d'être une aide à la décision au service des dirigeants, un instrument de prévision permettant au mieux d'anticiper les risques que pourraient présenter telle ou telle politique.

Il n'en reste pas moins qu'en dernier ressort tout choix est et reste politique. Nous devons donc écarter le rêve mais je suis persuadé qu'il n'est pas le vôtre, caressé autrefois par certains, qu'un modèle parfait donnerait définitivement à l'économie ou à l'économique le pas sur la politique.

C'est une vieille affaire et en France, depuis Auguste Comte et même probablement bien avant avec les physiocrates, nous avons déjà caressé ce rêve toujours renaissant. Croyez bien cependant, que je ne défends aucun intérêt de boutique en défendant le primat du politique mais c'est, me semble-t-il, simplement la leçon de l'expérience.

Nous attachons une grande importance aux deux thèmes dont vous allez débattre:

- l'agriculture et le développement rural constituent selon nous une condition première du développement et de la "survie" de l'humanité.

Une des choses qui m'a le plus frappé lors de ce récent voyage en Inde, est l'importance des recherches faites dans le domaine agronomique et les conséquences positives qu'elles ont pu avoir sur l'accroissement de la productivité agricole et de la production qui a triplé depuis 1947, depuis la date de l'indépendance alors que, dans le même temps, la population ne faisait que doubler et il y a là les éléments d'un décollage réussi sur lequel probablement, on n'a pas assez attiré l'attention. Donc priorité au développement agricole pour assurer l'autosuffisance et l'autonomie alimentaire de nombreux peuples du Tiers Monde qui sont loin d'y parvenir.

• l'énergie constitue le moteur du développement industriel, comme en témoigne l'étroite liaison qui a existé tout au long de notre histoire entre la consommation d'énergie et le niveau de développement.

Que ces deux questions revêtent une importance capitale dans le dialogue Nord Sud, nul ne songe à le nier. Mais il y a un intérêt important à promouvoir une véritable coopération internationale sur ces sujets. En effet, si je crois que le thème des limites physiques est quelque peu passé de mode, des problèmes importants ne résultent pas moins du fait que les ressources ne sont pas également réparties dans le monde.

Bien sûr la population mondiale devra croître pour atteindre un chiffre voisin de 6 milliards d'individus d'ici la fin du siècle, ce qui correspond, en 25 ans, à une augmentation de 50%.

Cette explosion démographique alliée à une croissance économique du monde, à tous égards souhaitable, se traduira inévitablement par une augmentation considérable des consommations énergétiques et agricoles.

On prévoit actuellement que la consommation de grains doublera presque par rapport à son niveau de 1975, passant de 1,2 à 2,2 milliards de tonnes. Dans le même temps, la consommation d'énergie devrait au moins doubler.

Certes, le monde a la capacité physique et économique de produire l'énergie et les aliments nécessaires afin de répondre à cette demande croissante. Les réserves d'énergie et le développement des nouvelles technologies de production énergétique sont tels que le défi peut être relevé. En matière alimentaire, il semble aussi possible de produire les quantités requises grâce à des technologies adaptées et à l'emploi de fertilisants spécifiques.

Mais encore une fois cette vision trop globale masque la disparité profonde qui peut exister entre les différents pays.

Les sources d'énergie ne sont pas distribuées également entre les nations, mais sont au contraire concentrées dans certaines régions, tout comme la production alimentaire est excédentaire dans certains pays qui regorgent d'excédents, qui détruisent certaines fois ces excédents et le scandale apparait d'autant plus grand que dans d'autres, c'est la misère, quelquefois, la famine.

La consommation d'énergie par tête varie considérablement d'un pays à l'autre, et de mêmes des inégalités choquantes, quelquefois monstrueuses apparaissent au niveau de la consommation alimentaire, entre des régions où sévit une famine endémique et celle des pays qui sont menacés par la suralimentation.

C'est pourquoi, si le monde a la capacité de répondre aux grands défis qui lui sont jetés, il est fort à craindre que, sans un renforcement de la coopération internationale dans ces secteurs, la réponse adoptée ne soit profondément inadéquate.

Il faut que nous allions vers une planification au niveau régional mais aussi au niveau mondial.

Compte tenu de l'importance des enjeux humains et sociaux et de la complexité de la problématique politique internationale actuelle, nous considérons comme essentiel que les bases d'une coopération juste en matière énergétique et agricole soient jetées aussitôt que possible, car il n'est pas possible d'attendre et de tenir un discours qui ne fait que reculer les échéances.

Il est donc important d'analyser les interactions qui existent entre l'agriculture et l'énergie, et votre Symposium s'inscrit avec pertinence dans cette perspective.

La consommation d'énergie du secteur agricole provient essentiellement de deux types d'activités: d'une part le machinisme agricole, c'est-à-dire les opérations de labour, de semis et de récolte et le transport des produits alimentaires, et d'autre part les engrais chimiques, qu'il faut bien produire.

L'introduction des engrais et de la mécanisation a permis de multiplier par un facteur voisin de cinq le rendement à l'hectare des cultures céréalières, et de diviser d'à peu près autant la main d'oeuvre requise pour leur production.

Au niveau mondial, la part de l'agriculture dans la consommation d'énergie commerciale est égale à 3,5 pour cent, mais devrait continuer à augmenter d'ici la fin du siècle, compte tenu notamment de la nécessité de moderniser le secteur agricole des pays en voie de développement, et la croissance de la consommation énergétique du secteur agricole sera plus rapide que la croissance de la production agricole qu'on peut atteindre parallèlement puisque le taux d'accroissement annuel de la consommation d'énergie pour l'agriculture devrait, selon les prévisions de l'Organisation pour l'Alimentation et l'Agriculture augmenter de 7,5% par an pour que dans le même temps la production agricole croisse de 3,8% d'ici l'an 2000.

Il va sans dire qu'une telle augmentation de la production

agricole suppose aussi une augmentation de la mise en culture de nouvelles terres. Et dans le même temps, nous l'avons déjà vu, la demande énergétique va plus que doubler, ce qui pose les conditions de la transition énergétique étant donné que les réserves d'énergie fossile tendent, semble-t-il à se raréfier bien que de nouvelles découvertes interviennent constamment. Si on regarde aujourd'hui ce qu'est le marché du pétrole, il se caractériserait plutôt par la surabondance mais c'est une tendance à très court terme qui peut se renverser rapidement, et je ne suis pas partisan de fonder des perspectives d'action sur des tendances du marché dont toutes les expériences historiques montrent qu'elles peuvent se renverser très facilement.

En tous cas, cette transition énergétique qui doit être préparée et qui l'est insuffisamment, aura ceci de particulier qu'elle se fera dans des conditions de renchérissement des coûts réels de l'énergie alors que traditionnellement, les transitions énergétiques du passé se faisaient dans des conditions de baisse du prix de l'énergie qu'il s'agisse du passage, par exemple, du bois au charbon ou du charbon au pétrole. C'est la caractéristique de la période dans laquelle nous sommes. Il faudra donc recourrir à de nouvelles formes d'énergie non seulement les énergies fossiles, l'énergie nucléaire mais également les énergies renouvelables et nous sommes en train de programmer un très important effort de recherche dans ce domaine.

Parmi les technologies les plus prometteuses, apparait la production de combustibles gazeux ou liquides à partir de la biomasse.

Mais nous sommes renvoyés là à un nouveau problème: c'est que s'il est vrai qu'une partie très importante de cette production énergétique dérivée de la biomasse proviendra des déchets d'origine végétale ou animale, il n'est pas exclu de voir surgir une compétition entre une production agricole à des fins alimentaires et une production agricole à des fins énergétiques.

Les implications d'une telle compétition ne sont pas encore claires, mais beaucoup craignent qu'elles ne se traduisent par une diminution des ressources alimentaires propement dites dont souffriront en priorité les populations les plus pauvres, du fait du jeu conjugué de la diminution de la part des surfaces cultivées produisant de la nourriture et de la transformation en combustible, si les prix le justifient, d'une partie de la production céréalière.

Vous allez réfléchir à ces questions très importantes et je ne saurais m'avancer plus loin sans tomber dans le travers d'immodestie. Il faut étudier la compétitivité potentielle entre les divers usages de la production agricole et par ailleurs, comme les problèmes que vous allez aborder sont extrêmement complexes, que peu de travaux ont été entrepris jusqu'ici dans ce domaine, il semble urgent d'identifier des programmes de recherche pluridisciplinaires qui permettront de mieux évaluer l'importance de tous les facteurs qui entrent en compte et de dégager des conclusions qui pourront se concrétiser à travers les planifications nationales du développement mais aussi, je le crois une planification même très globale du développement mondial.

Plus que jamais, le monde moderne se trouve confronté à des problèmes globaux qui nous posent des défis majeurs, et qui impliquent un vigoureux effort de réflexion.

L'investissement que vous faites dans une réflexion absolument cruciale ne sera pas perdu, j'en suis sûr, si vous savez avancer avec sagesse et clairvoyance.

En tous cas, Monsieur le Recteur, Mesdames, Messieurs, votre présence témoigne de l'urgence et de la gravité des questions énergétiques et agricoles. Ce sont peut-être les principales questions auxquelles se trouve confrontée l'humanité et c'est la raison pour laquelle la France est heureuse d'accueillir votre Symposium et je suis sûr qu'il sera l'occasion d'échanges fructueux qui vous permettront de répondre aux attentes de la communauté internationale et d'éclairer les décisions des politiques.

Je vous remercie.

2 Keynote Address: Food and Energy Demands on the Global Research Agenda

SOEDJATMOKO

Your excellency, Minister of State, Mr. Chairman, Ladies and Gentlemen,

It gives me great pleasure to welcome you to this first major scientific meeting to be held in France by the United Nations University since I first took office as Rector. France, with its brilliant scientific and humanistic traditions, is very important to the intellectual development of the University and I am certain that this is but the first of many meetings we will have together in this country.

We are met here, from many parts of the world, to explore how global modelling can be used in improving our ability to understand and deal with the agricultural and energy systems which are so important to the daily lives of people everywhere in the world. The essential core question before this symposium is: will we be able to provide sufficient food and energy to the globe's swelling billions as we move towards an increasingly crowded, competitive, and perilous 21st century world?

The food and energy problems are deserving of the label "global" in four major respects:

1) the regular and continuing availability of sufficient food and energy is a condition *sine qua non* of human survival, posing a formidable challenge of supply and demand that must be met simultaneously;

2) the future course of human progress and quality of life will be determined largely by the availability of food and energy;

3) food and energy loom large in the north-south confrontation as potential weapons and tools of domination;

4) Fourth, food and energy production both affect and are affected by the state of the environment. The patterns of energy, land and water use and manufacturing industry, for

example, will increasingly influence the climate in ways that could sharply alter global agricultural systems. All these four will be affected as world population grows from 4 billion to at least 6 billion by the year 2000.

Both agriculture and energy problems are thus tightly interwoven and they interact at many different levels.

One such level, for example, would be instances where food and energy production either compete for the same land or seek to ensure its more meaningful use through their complementarity. Under certain conditions, scarce crop land may be shifted to the production of biomass energy – for instance, if production of food proves less attractive to a market-controlled economy than production of automobile fuel.

At another level is the question of food and energy production competing for the same scarce resources of capital and foreign exchange. This will be particularly true in the third world, posing difficult policy choices for national planners and aid agencies alike with regard to allocation of resources between these competing yet essentially interrelated claims.

There is also the relationship between food and energy at the input level. Food prices have become highly sensitive to oil prices because of fertilizers and other related fossil fuel inputs. There are also obvious links between oil prices and attempts to increase agricultural production through desalinisation and use of hydroponics.

Another food-energy interaction results from the rapid exploitation of fuel wood by the poor, who increasingly lack the income to buy kerosene. This makes this energy source more scarce and food production more expensive, difficult and ecologically hazardous.

There is also linkage at the global level. Rising fuel prices have dramatically shifted the comparative advantage in favour of food-production technologies which are less dependent upon fossil energy in the developing countries. Over time, this could produce a major shift in food production and trade patterns. Also on a global level is the difficulty of establishing a pattern for food and energy security as any solution must take into account both key factors.

There has been a tendency so far to search for separate solutions to food and energy problems, but clearly this no longer makes sense. Such an approach may, at best, offer partial solutions or, more likely, no solution at all because of the failure to understand the essential linkages between these two inseparable factors.

We thus need to explore systematically and scientifically these linkages between food and energy issues in order to clarify policy options at national, regional and global levels and to strengthen institutional capabilities to deal in more multidisciplinary fashion with these problems.

The growing interactions between food and energy problems – which are already beginning to lock some of the poorer and most populous countries into seemingly implacable loops of negative synergy – would seem to call for a research agenda for the global scientific community, on which the following items should have high priority:

First, improving our understanding of the linkages between food and energy at various levels. In particular, co-production systems for food and biomass energy need to be developed to transform a potentially competitive relationship into a positive synergy.

Second, an analysis of the policy trade-offs of pursuing in isolation solutions for increasing food and energy production, with particular emphasis on the impact on the poor.

Third, identify the policy implications for national planning, if optimal solutions are to be found simultaneously to the food and energy crises.

Fourth, underline the implications of these analyses for global negotiations of food and energy.

Global models, I believe, could play a very valuable role in helping to tackle the immensely complicated interplay of elements on this research agenda. The theory and practice of modelling constitutes a highly useful analytical tool of modern science and scholarship. If properly applied, it can do much to help the decision-maker and the policy-planner see more clearly through the swirling mists of interweaving concerns whose urgencies and dimensions seem to be constantly shifting and changing. I look forward very much to the presentations and discussions of various modelling experiences over the next few days which I am sure will be greatly enlightening.

However, if global models are going to be truly useful to us, I think we need to be frankly realistic about their possible limitations and recognize that there is certain fine tuning which we simply cannot expect of them. We must not let global models become an end in themselves; what interests us here is in what ways and to what extent they might serve national planners in various countries and particularly in the third world.

In this connexion, it is important to keep in mind that modelling has generally developed in the industrialized world whereas the starkest dimensions of the food and energy problems are now, and will continue to be, in the poorest and most populous countries of the developing world. Modelling is based on assumptions, and the validity of those assumptions frequently hinges on the data we have available. It is precisely in the poorer, more populated countries that reliable statistics can be the hardest to come by.

Because of this dearth of reliable statistics in so much of the third world, the assumptions and aggregations of global models can mask vitally significant regional variations, local dynamics, structural disparities, value shifts, and changing power configurations that are intimately linked to the process of development – or lack of development.

We need to recognize, therefore, that modelling assumptions run the risk of ignoring or obscuring factors of great relevance in the third world. Much larger populations, even cities of 15 to 30 Million, vast population movements, urbanization and modernization, much lower levels of income all affect production and consumption patterns and expectations and can lead to different systems.

A further complication is that all of these factors come into play in different ways and to differing degrees in the individual countries of the third world – which are at widely varying stages of development. This speaks forcefully to the need to develop ways to build in existing inequalities. Which, in turn, recognizes the importance of disaggregating data – however difficult a modelling task that may be.

A linked concern is the fact that too many models and scenarios, tend to assume stability in institutions and social units – when their erosion and breakdown is a more normal modern reality. All developing countries must take into account the likelihood of major structural and systemic discontinuities. These stem in large part from the inabilities of present institutions and strategies to answer the needs of the poor and marginalized who are now rightfully demanding to be heard. The developing countries cannot escape the fact that experience has shown how ineffective their national policies have been in reaching the poor. Development strategies must be rethought to find ways to stress development from below. The modeller could be of great service to the national planner in helping him to better understand the dynamics of such development.

It is clear, therefore, that the models that we create for the developing countries, will need to be quite different from those used for industrialized nations in many respects. Consider, to take just one example, the different variables at work in models of electrical energy use in the industrialized and the developing world. In the former, one could assume a normal increase in population and total access to electricity. In the developing country, however, we would need to build in a larger than normal population increase. Another important variable would be potential new customers for electricity either through expansion of the electrical grid or through migration from areas without electricity into areas offering this service. One can see how quickly complicated and uncertain such a modelling attempt becomes.

Another important variable in the developing countries is expectation which, fed by advances in communications and transport, is increasing rapidly throughout the third world. Total expectations are often much higher than any possibility of delivery, and a model should take this into account.

Yet another important difference between industrialized and developing world models concerns reliability of food and energy systems. In the industrialized world, a street lighting system that works only 90 per cent of the time would be considered totally unacceptable, but such a level of reliability could be a vast improvement in many parts of the third world.

Food and energy models will further need to build in the political consideration that these two vital resources will increasing invoke. Food and energy have already become political weapons – their future use in this capacity, unfortunate as it may be, will have to be taken into account in modelling efforts.

A tragically important concern is that the factors being calculated in models involve the lives of people living on the bare edge of subsistence; for whom a sudden crop failure, drop in income, or loss of resource can literally mean the difference between life and death. They have no margin for error in their lives.

This lends particular urgency to the search for modelling concepts and theories which will give us a more precise picture of the fine structure of the future – few advances in science could be more important to the planner. We now have models in which we can build in *what* might happen if war, famine, crop failure or other catastrophe strikes. What are needed, however, are models that enable us to have a far better sense of *when* things might break apart.

We intend at this symposium to give particular attention to the bottom-up approach in modelling, which may help us in our search for more relevant and more effective models for understanding the interactions of food and energy problems, particularly as they impact on the third world. Most models to date have tended to start from a global view and as they get down toward the national level become less meaningful and useful to planners.

We want to explore the implications of starting at lower levels – national or even sub-national – and from there beginning to piece things together, linking models to other models and progressively introducing wider linkages among them. On a country-by-country basis, of course, the data will vary widely – in some instances, we might have to work with only 50 per cent of what we would ideally like to have, in others far less. Even with these and other shortcomings, however, we believe that the bottom-up approach could have great value for planners and others who need far more information about the shape of the future.

The bottom-up approach is important, I believe, if for no other reason that it puts the emphasis where it must be – at the local level. For all their global dimensions and implications, the problems of the converging paths of food and energy supply and demand will ultimately have to be solved at the national, and indeed the local, level. Whatever insights we might gain from models, the information will only be meaningful if we are capable of translating it into positive, concrete action that villagers and small farmers can take themselves, within the context of their own cultural values and perspectives, to improve their own lives.

In the long run, new patterns of food distribution and

consumption and new allocations of energy will depend on countless millions of decisions by individuals around the globe. It will be these small day-to-day decisions by individuals that will ultimately count the most – whether another tree must be cut down for cooking or heating purposes. We must find ways to incorporate the reality of such decisions – ones that touch intimately on daily human existence – in our scientific and technological planning. Any discussion of improving our understanding of food and energy interactions must inevitably be concerned with better understanding of the problems of the poor.

We have made this very much our angle of vision at the United Nations University. The food and energy nexus is an important new area of concern to us, although it is obviously linked to past work we have done in the field of world hunger, energy and resources systems, and on appropriate use of technology. What is new is our attempt to study these and other related concerns together and explore their interplay. I am certain that the discussions here this week can shed much valuable light on the directions that our future work in this area might take.

Clearly a concerted and collaborative global effort is needed, drawing on the best of our scientific creativity and thought, if we are to hope to come to terms with the centrality in all our future lives of the interactions between the agricultural and energy sectors. The "crisis management" approach that has brought us such unsatisfactory piecemeal solutions in the past will no longer do.

I have put before you what seem to me to be some of the most important challenges we face in taking on the enormously complicated task of relating the global food and energy problem to its manifestations and impacts at the national and local level – where the problem ultimately must be solved. In attempting to meet these challenges, we will no doubt find that a combination of approaches will serve us best – drawing on models, scenarios, and country-specific studies. Such a combination, I believe, could be the basis for a creative, innovative, and realistic response to a problem now threatening the very survival of many millions of our fellow human beings.

It may therefore be useful to emphasize at this point that we have not gathered here to discuss how to improve global modelling in terms of its own stated objectives or in terms of long waves or the Kondratiev cycle. Nor to discuss the relative merits or demerits of the various global modelling approaches.

Our purpose at this seminar, it would seem to me, is to see to what extent and how existing global models might help us to illuminate the policy options planners in third world countries have to face, especially in the food/energy nexus. Also what kinds of linkages they have to bear in mind in planning towards greater local, national and regional food and energy security systems or schemes.

Focussing on these questions may help to bring out more sharply the adequacy or inadequacy of the economic, social and political assumptions on which these models are based, and at the same time force us to identify more sharply the assumptions that underlie the modelling efforts at the national level in developing countries that have to do with the different conditions, dynamics and different perceptions of desirable societies, but also with the effects of the political instability and the high incidence of discontinuities of various kinds that are inherent in societies in rapid and profound social, cultural and economic change, when established institutions, as well as society's capacity for community formation and cohesion, are seriously eroded by the pervasive impact of modern communications.

While the need for more sophisticated assumptions on which to base the various scenarios at the global and national level is obvious, I hope that our discussions will also bring out how these assumptions at the global and national – as well as subnational level – relate to each other.

And even if this symposium will not provide us with clear answers to these perplexing questions, I do hope that the discussions will at least suggest the directions in which we should continue our search for models capable of accomodating both the global and the widely varying national dimensions of models that will provide us with better conceptual and operational tools to deal with the food/energy nexus. Or, failing that, that our discussions may suggest at least a tentative research agenda that might bring us closer to that goal. Our analytical tools seem to be incapable of keeping up with the pace, urgency and complexity of our problems. Let us not leave from here without making the next concrete step.

3 The Food–Energy Problem

3.1 The Food–Energy Nexus: Seeking Local Solutions to Global Problems

I. SACHS

The food-energy nexus has been singled out by the UNU as a possible theme for an interdisciplinary, co-operative, international research sub-programme. The present paper explores the possible aims, scope and design of such a project building from the programmes currently implemented by UNU in the realm of food, energy, appropriate technologies, natural resources and human development. Though the author bears the entire responsibility for its shortcomings and the views presented here do not engage the UNU, he wishes to thank for the assistance provided in shaping its ideas by the Rector Soedjatmoko, vice-Rectors Kwapong and Mushakoji, the UNU staff, the participants of the meetings held in CNP9 (Brasilia) and COPPE (Federal University of Rio de Janeiro) in July 1981, as well as those who attended two workshops organized to discuss the project by the Maison des Sciences de l'Homme (Paris, September 1981) and BOSTID, National Academy of Sciences (Washington, October 1981).

Food and energy are global problems at least in four ways.

1) Their regular and continuing availability is a condition *sine qua non* of human survival, posing a formidable challenge that must be tackled simultaneously from the supply and from the demand side; piling up of food and fuel stocks will be of little help to starving people unless they can afford buying them, or else produce them for self-povisioning so long as they have access to sufficient land.

2) Assuming optimistically that mankind will manage to solve the problems posed by its bare survival, the future course of human progress and quality of life will still depend to a great extent on the increased supplies of both food and energy; their central role in a need-oriented development strategy is only too obvious.

3) Today, both food and energy loom large in the North-South confrontation as potential weapons and tools of domination; hence the importance of a global negotiation to modify the present gloomy picture and bring about some constructive international cooperation in both fields. Global food security and global energy security are at stake in the North-South negotiation about to be engaged at the United Nations. As we shall argue throughout this paper, a joint consideration of both the issues will be at some point necessary as any meaningful solution must cover both these key and closely interwoven factors.

4) The food and energy production affect and are affected by the state of the environment: the patterns of energy, land and water use will increasingly influence the climates with far-reaching consequences for the long term prospects of biomass production.

These two global problems are closely interwoven. According to an IIASA study, in the developing countries 30 to 70 per cent of the intermediate input costs of agricultural crop production are directly or indirectly related to energy. On the other hand, agriculture provides 20 to 90 per cent of primary energy through the supply of the so-called non commercial energy (wood, waste, dung, etc.).[1]

Yet in contrast with the considerable research effort expended on each of the two problems taken in separate,[2] little attention has been paid up to now to the food energy nexus, to a systematic exploration of the ways in which they are linked and, alas, reinforce each other in a negative cumulative causation process with a great potentiality for social and ecological disruption. The prospect for the next decades is dismal, indeed, judging from the projections of the *Global 2000 Report*:[3] while on a worldwide average there was about four-tenths of a hectare

[1] PARIKH, J., "Energy and agriculture interactions" in: *Food for All in a Sustainable World.* The IIASA Food and Agriculture Program, K. PARIKH and F. RABAR (Editors), August 1981 SR-81-2, IIASA.

[2] It is impossible to assess and comment in a paper like this the voluminous literature produced by FAO, IIASA, the various international research bodies, and still less by the national centers of learning.

[3] *The Global 2000 Report to the President.* – Entering the Twenty-First Century. A report prepared by the Council on Environment Quality and the Department of State. Volume one. Washington, 1980, 47 p.

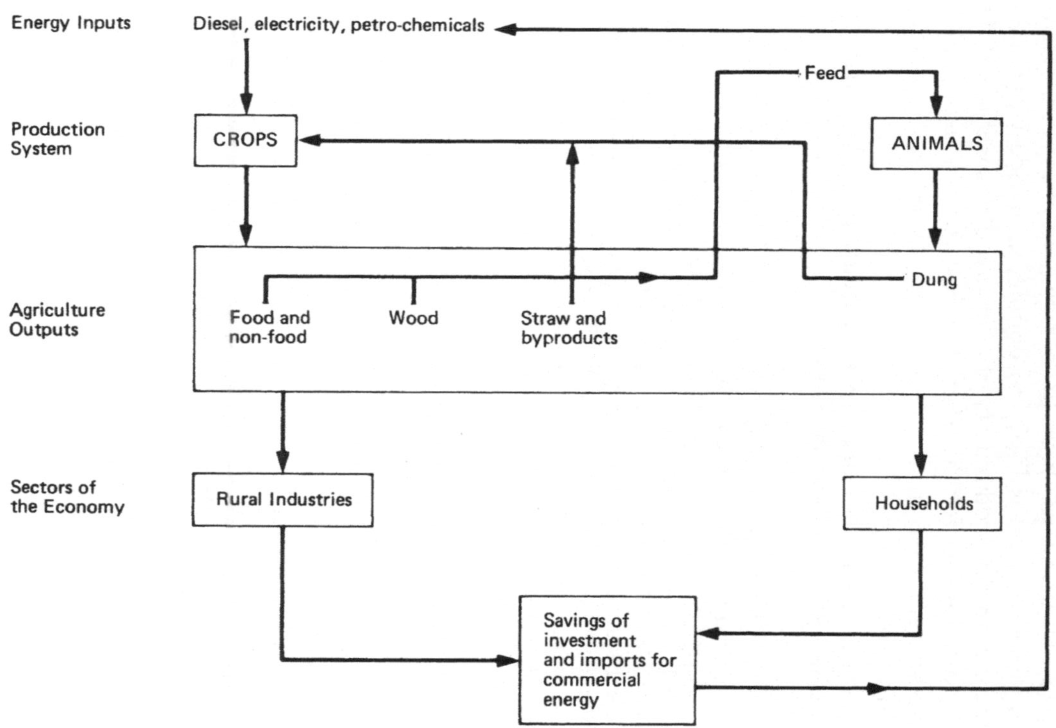

FIGURE I Energy inputs to and outputs from agriculture (taken from J. Parikh, 1981).

of arable land per person in 1975, there will be only about one-quarter hectare per person in 2000. By that year the world's per capita growing stock of wood will be 47 per cent lower than in 1978 and 40 per cent of the forests still remaining in the less developed countries will have been razed. In these conditions the real price of food is likely to double, alongside with long-term declines in the productivity of over-taxed renewable nature resource systems. As for the energy prices, the prospect is also for a steady increase in real terms, and, therefore, for more hardship for the poor.[1]

The vicious circle that must be broken has been described by Soedjatmoko in the following terms:

> Rising fuel prices, boosting transportation and agricultural costs, will inevitably push food prices beyond the reach of hundreds of millions of already hungry people. Rising populations, despite the best efforts to reduce fertility rates, will continue to increase the demand for both food and energy. The developing countries will not be able to solve their food problem without solving their energy problem and, without a satisfactory solution to both, their economic growth will be severely constrained. The centrality of the food and energy nexus calls for a comprehensive policy approach. Only through a clear understanding of this food-energy pivot can the situation be turned around.[2]

The intellectual challenge is to go beyond the simple analysis of data and assist the governments and nations in transcending the crisis management approach by means of development planning conceived as a societal learning process. This calls for the enhancement of the capacity of social organizations to respond creatively to new situations, constraints and opportunities and to shape new trajectories of industrialization and rural modernization different from the models derived from the historical experiences of the industrialized nations of the North. Specific local solutions must be found to the global problems of food and energy responding to the variety of ecological, cultural socio-economic and political contexts. Consideration of the

[1] See also the well documented article of Suzan GEORGE, "Le Commerce des céreales", *Le Monde diplomatique,* novembre 1981.

[2] SOEDJATMOKO, UNCNRSE, "First step towards a broader vision?" *ASSET* 3, (7) July/August 1981, UNU, Tokyo, Japan. See also for more details: SOEDJATMOKO, *Turning Point in Development: The Food-Energy Pivot,* Inter-governmental Meeting of Development Assistance Co-ordinators in Asia and the Pacific, Delhi, India, 23-28 February 1981.

food-energy nexus provides a convenient entry-point into these two crucial domains of development planning. The *differentia specifica* of the food-energy nexus sub-programme consists precisely in its concentration on planning, designing, implementing and evaluating of ecosystem-specific, culture-specific and even site-specific solutions to global problems, downstream from the ongoing UNU work on food and energy macro-models. Useful as the latter may be to sensitize policy-makers about the long-range implications of their decisions, or to test the coherence of the proposed policies, they do not yield themselves to a straightforward translation into planning guidelines at regional, micro-regional or project level. Macro-models must per force assume away the diversity of local conditions and, therefore, cannot do justice to peoples' resourcefulness, a key concept for successful ecodevelopment[1] planning. The food-energy nexus sub-programme by its "diagonal" character, cutting across the three main UNU programmes, could serve an integrative purpose within the UNU itself. By its nature it yields itself to be carried out through a co-operative network of national institutions pursuing the double and interrelated objectives of advancing the frontiers of knowledge and providing concrete inputs to actual development planning.

THE RESEARCH PROBLEMATIQUE

Competition for resources between biomass energy and food

Given the expected rise of real energy prices, biomass energy holds a considerable promise, if properly managed. An array of technologies of different degrees of sophistication, capital intensity and scale of production is already available. However, sound ecological principles are required to ensure the sustainability of a considerably increased production. No resource is renewable where the conditions of its renewability are not respected: a felled down forest where reafforestation is not undertaken is but a mine of timber and/or fuelwood. Stepping up biomass energy production with no regard to its immediate and long term, local and distant, environmental impacts might

[1] Ecodevelopment stands for a socially desirable, economically viable and ecologically sustainable resource use pattern.

bring about an ecological crisis.

The social consequences at stake are also very high indeed. Poor people all over the world need badly additional energy. But rich people can afford to pay more for it, so that biomass energy may end up fueling private cars rather than helping to cook the dinner of the poor. Moreover, in situation of scarcity of good agricultural land, a competition may engage between production of food and biomass energy, adding a new dimension to the traditional conflict between production of food for subsistence, industrial crops and agricultural commodities for exports.

In matter of fact, the competition is also likely to affect the allocation of capital and other scarce resources, as shown on Figure II.

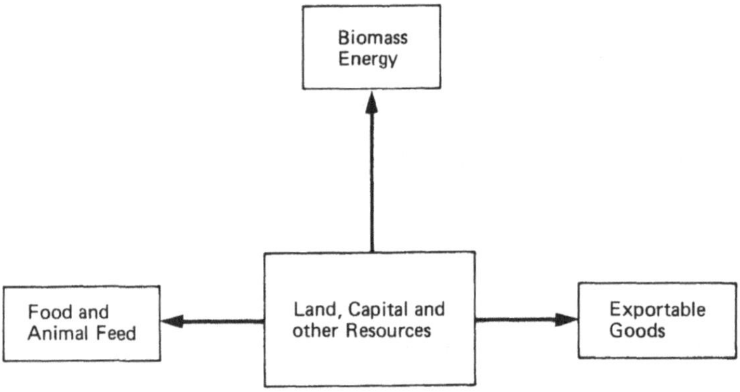

FIGURE II The food-biomass energy competition.

With a view at reducing the oil import bill, prime agricultural land, government capital funds and subsidies, as well as private savings, may e.g. go to the production of ethanol from sugar cane; at the expenses of the production of food for the internal market. Or else, forest land may be cleared to plant sugar cane.

Other less spectacular, but socially and ecologically costly, shifts in the allocation of resources, motivated by the shortage of fuel and/or by the rising prices of energy, may entail:

- predatory collection of fuelwood and the ensuing deforestation, loss of fruit trees, etc.;
- intensified use of animal droppings as fuel with loss of manure and the consequent decrease of agricultural production;[1]
- use of fuel of agricultural waste (e.g. rice straw) with negative impacts as livestock production and/or depletion of soil nutrients;[2]
- resorting to more time-consuming fire-wood collection at the expenses of work time devoted to food production and/or additional burden for women and children.

Another case of competition between energy, agriculture and forest production is the conversion of large stretches of land into man-made lakes in order to build reservoirs for hydroelectric power stations.

The social and ecological costs of all such shifts in the land, water and other resources use patterns, prompted by the energy shortages and/or the increases in the real prices of energy, must be carefully assessed and compared against alternative solutions.

The price spiral

While the effects of expansive energy on food prices are relatively well known, more investigation is necessary to document the regressive social impact of the increases in real prices of energy and food in terms of income distribution and shifts in the consumption patterns.[3]

---

[1] According to Nigel SMITH ("Wood: an ancient fuel with a new future", *Worldwatch Paper*, Washington, no. 42, January 1981, p. 14) worldwide, the use of livestock droppings as fuel is thought to lower grain production by some 20 million tons, enough food to nourish 100 million people.

[2] As pointed out by D. PIMENTEL ("Biomass energy: bonanza of boondoggle?" *Soft Energy Notes*, volume 3, no. 6, Dec-Jan. 1981), there exists a severe ecological constraint to the use of agricultural waste normally returned to the soil.

[3] In USA, the poorest decile of the population spends 34.1 percent of their before-tax income on energy consumption, while the richest decile spends only 5.2 percent (Data quoted by Thurow L.C.: *The Zero Sum Society*, 1980, Basic Books, New York). Comparable data for Third World countries are lacking, but energy and transportation costs constitute a heavy burden for the urban poor.

An additional complication may arise if real prices of energy continue to grow: the biomass energy prices may push up the food prices over and above their increase due to higher prices of agricultural inputs, adding dramatically to the plight of the urban poor, while intensifying the pressures for the concentration of land at the expenses of small landowners.[1]

Food and energy co-production systems

The chances of competition between food and biomass energy for land, but also capital and other scarce resources, will be greatly enhanced by the juxtaposition of single-purpose agricultural projects evaluated on a micro-economic basis with little or no consideration for the externalized social and ecological costs.

These fears should not distract us from seeing the importance of biomass energy both in Third World and industrialized countries. The more so that biomass energy has a far greater employment potential than any other form of energy production. But they force us to spell out and consider carefully the food-energy nexus in all its complexity and at the same time to build into all the projects the condition of sustainability.

Under circumstances, it is submitted that food and energy production should be considered jointly. We ought to develop food and energy "co-production systems" based on a careful analysis of the biomass potential of each ecosystem. In matter of fact, such an analysis should be also culture specific and, in order to become operational, even site specific. The best way of resolving the conflict over land, capital and skills between food and energy production is to deal with it *ex ante*.

We are thus referred to the ecodevelopment approach referred to in the introductory part of this paper. The aim is to look for socially desirable, economically viable and ecologically sustainable man-made production systems, paradigmatically inspired by the concept of ecosystem, and capable of jointly supplying

[1] See i.a. BROWN, Lester R., *Food or Fuel: New Competition for the World's Cropland,* Worldwatch Institute, Washington, March, 1980, *Energy Cropping Versus Agricultural Production,* Sixteenth FAO Regional Conference for Latin America. Havana, Cuba, 26 August-6 September 1980. FAO, LARC/80/8, June 1980, and PLOTKIN, Steven E., "Energy from biomass", *Environment,* November 1980, 22 (9): pp. 6-40.

food for people, animal feed and biomass energy. Environment appears in this perspective as a resource potential to be harnessed on a sustainable basis and, as much as possible, in an ecologically benign form. The design of such co-production systems implies, on the one hand, comprehensive land-use planning and, on the other, a careful consideration of the ecological complementarities between different activities, with special emphasis on the use of forest, agricultural and human waste.

Transforming a waste into a resource proves in many cases the best way of dealing with environmental disruption. All the social implications of the proposed co-production systems must be equally spelled out. The advantage of co-production systems over juxtaposed monocultures could well reside in their greater ability to use in a fuller and more equitable manner the available working force. The choice of appropriate technology mix is, of course, of paramount importance here.

Food and biomass energy co-production systems should meet the following five conditions:

firstly, be sustainable in the long run which entails an integrated land and water resource management;[1] mismanagement of soil, water and forest resources may transform potentially renewable resources into non-renewable ones;[2]

secondly, be primarily oriented towards improving the livelihood of local populations through a combination of self-provisioning and market operations, the increased production being both the cause and the consequences but not the solely aim of the planned development;[3]

thirdly, be energy, capital and resource efficient and, at the same time, science-intensive. As PEARSE (*op. cit.* p. 245) rightly observes, tropical production systems are capable of achieving a very delicate balance in complementaries and

[1] See VOHRA, *A Policy for Land and Water*, Sardar Patel Memorial Lectures 1980, Department of Environment, Government of India, New Delhi.

[2] See on this point i.a. PIMENTEL (*op. cit.*) and also on the tropical forest GOMEZ POMPA, A., VALQUEZ YANES, C. and GUEVARA, S., "The tropical rain forest: A nonrenewable resource", *Science*, 177, pp. 762-765.

[3] PEARSE, A., *Seeds of Plenty, Seeds of Want, Social and Economic Implications of the Green Revolution.* Clarendom Press, Oxford University Press, Oxford, 1980, p.245.

ecological adaptation. Such systems may be now rendered more effective by low-cost or no-cost contribution from contemporary science;

fourthly, be as much as possible oil saving, as commercial energy, and more especially oil-intensive food production and handling systems are beyond the reach of poor people; hence the importance of closely scrutinizing traditional systems and past practices with a view at upgrading them by low-energy scientific inputs;

fifthly, be based on a participatory approach to the planning and management of local environmental resources and development.

Energy efficient food systems

Whatever the future evolution of real energy prices, the highest priority should be attached to the design and implementation of energy efficient food systems, encompassing all stages i.e. crop and livestock production, storage and conservation, transport and distribution, processing and preserving as well as kitchen practices. In other words, the energy profile of the food systems, defined along the same lines as in the joint UNDP/UNRISD/IFPRI project, should be analysed with a view at identifying all the energy inputs as well as energy outputs and opening up the range of alternatives in terms of product mix, agricultural techniques (including the search for alternative sources of plant nutrient), product and process technologies (with particular reference to dehydration technologies), scale and location of production (with reference to the degree of local self-sufficiency in food and the consequent reduction of transportation flows), scope for energy conservation and use of wastes, methods of cooking and kitchen technology as well as changes in food habits. (See Figure III).

Rising fuel prices have dramatically shifted the comparative advantage in favour of food-production technologies which are less dependent upon fossil energy in developing countries and may over time lead to a major shift in food production and trade patterns.

An important input into the design of energy efficient food systems might come from a survey of traditional food-energy

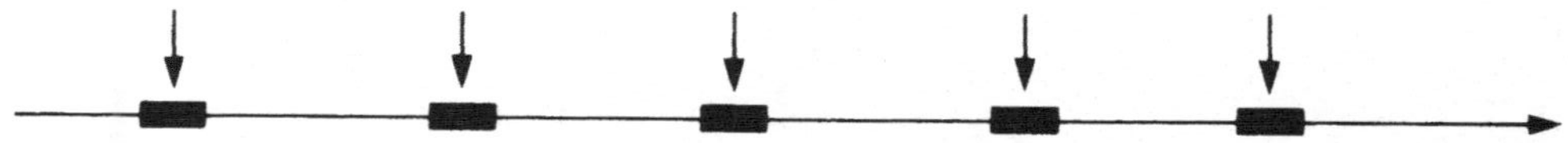

	Crop and Livestock Production	Storage and Conservation	Transportation and Distribution	Processing and Preserving	Kitchen Practices
Variables affecting the energy demand					
Opportunities for energy conservation and use of dissipated heat and other wastes					

FIGURE III The energy profile of food systems.

systems, cast in an ecosystem culture matrix. Such an approach would allow for two sorts of comparisons: how a given culture has adapted itself to the constraints of different natural environments or else, how diverse cultures faced the challenges and explored the opportunities of a given ecosystem.[1]

PROJECT DESIGN

There has been a tendency so far to search for separate solutions to food and energy problems. Such an approach may at best offer partial solutions or no solutions at all because of the failure to understand the essential linkage between the two inseparable factors. We thus need to explore systematically and scientifically these linkages between food and energy issues in order to clarify policy options at national, regional and global levels and strengthen institutional capabilities through research, training and dissemination of knowledge. UNU, neither being bound by the restrictions of a conventional UN body nor by those of a traditional university, would be able to look at the food energy problem more objectively than the more conventional approaches to policies and programmes. It could develop the fresh approach that is so essential for the creation of a dynamic programme of action and could draw upon its various disciplines and programmes in the execution of this project. It can also involve the concerned research institutions throughout the world several of which are already associated with UNU. The interdisciplinary nature of the project would require such approach for its successful execution.

The research agenda

The research agenda will have to cover at least the following issues:

a) Systematically explore the nature of the linkages between food and energy at the four levels outlined in the section A of

[1] SACHS, I., "Ecologie et développement dans les tropiques" in: *L'Ecologie en action: recherche d'une base scientifique pour l'aménagement du territoire*. Conférence organisée par l'Organisation des Nations Unies pour l'éducation, la science et la culture, Programme sur l'homme et la biosphère (MAB), Paris, 22-29 septembre 1981, MAB/CONF-81/1/8).

this paper and evolve a conceptual framework and the necessary analytical tools to deal with them.

b) Assess the long term social and ecological costs of alternative food and energy production systems, the policy trade-offs and price effects of pursuing in isolation solutions for increasing food and energy production, with particular emphasis on the impact on the poor.

c) Highlight the policy implications for national, regional and micro-regional planning if joint solutions are to be found simultaneously to the food and energy crises.

d) Underline the implications of these analyses for global negotiations on food and energy security.

The training agenda

Keeping with the UNU overall objectives, the project should have a strong training component, aimed at building competencies, especially in the third world, for undertaking multidisciplinary work of the kind postulated in this paper. In this process, vertical linkages will have to be built that would include policy makers, planners, researchers and the users of the results, i.e. peasants.

Attention should be also paid to the formulation of guidelines for planners and the introduction of the food-energy linkages into the curricula of higher learning institutions training development planners and policy makers as well as food and energy specialists.

Proposed activities

Workshops, literature surveys and informal consultations conducted in several developing and industrialized countries have led to the identification of three prioritary areas for policy-oriented inter-disciplinary activities.

a) Improving the energy efficiency of food production, conservation, distribution and utilization systems along the lines discussed in the section A of this paper.

b) Assisting in the design and testing of food-energy co-production systems, on the understanding that full-size local or micro-regional development projects implementing the new food-energy systems should find an independent funding from national, bilateral or international sources. The UNU project might be instrumental in generating such development projects, in providing the necessary research inputs, in collecting and disseminating comparative analytical data, in training planners and project managers for them, but should not aim at substituting itself for the national and international development planning agencies.

c) Assisting planners and research and development workers in dealing with the food-energy nexus. This task can be carried out by a number of means such as co-ordination of the "status of the art reports" being prepared by different academies, research councils and international bodies, surveying the ongoing research and other efforts by different institutions on food-energy problems; linking the national efforts and creating networks of institutions to stimulate each other and have a multiplier effect on the work to be done, while minimizing duplication and strengthening the inter and intra-institution research and development capacity; supporting research on developing analytical tools needed to assess social and ecological costs and trade-offs between food and energy production and their effect on price and income relationships.

The aim is thus to organize within a five-year time-frame open-ended self contained activities and fund them case by case as part of a dynamic global network. Research should be carried out primarily through projects to be implemented by national research institutions and as necessary by a network of institutes, in collaboration with planning agencies and international organizations. Activities would thus involve initiating projects, building networks along South-South lines (the TCDC component) or North-South lines, providing methodological assistance, circulating the relevant information and formulating guidelines (on the basis of comparative studies) for planners, policy-makers and project designers in third world countries and international development organizations. Several developing countries such as Brazil,[1] Mexico, India, have the necessary

[1] CNPq – the national research council – and FINEP – Project financing agency – are about to start two interesting programmes on "agro-energy" and on "agro-energetic communities".

research and development capabiltiies, are already working in the field of food-energy linkages and have responded favourably at the level of research institutions to the idea of a UNU project on the food-energy nexus. A multidisciplinary co-operative system can be established among them to support efforts of other developing countries and, as necessary, linkages can be created with appropriate institutions in industralized countries.

DISCUSSION POINTS

1) Dr. Hrabovszky was of the opinion that a study such as that suggested would require a very detailed knowledge of technologies which are available or forecast, as well as of a multiplicity of other factors from the physical, social and cultural environment. They had found that once an effort is made to consider more than a limited area, it is necessary to introduce considerable simplifications.

Dr. Norse on the other hand felt that some of the MAB models suffer from excessive complexity.

2) Dr. Norse referred to the fascinating ecosystem culture aspect of the problem, and to the existence of some very interesting anthropological studies, such as those of shifting cultures in Tanzania and transhumance in the Sahel. This is a very complex area.

Dr. Sachs emphasised that he believes in a socio-cultural perspective and that he was deeply indebted to anthropology. It is not possible to give justice to the diversity of real situations, to the mix of cultural, institutional and ecological conditions, other than by tackling the problems at a local level.

The question is then whether it is necessary to aggregate to a global level. This is not required in order to formulate policies, to define policy options or to assist local people to take care of themselves. Thus macro-economic and macro-social decisions must be related to this fantastic complexity at the local level. The central issue of the political sciences is how to formulate policies that reflect that diversity.

3) Professor Meadows agreed that the integrating overview of the large macrosystem must be balanced with the micro particular approach for an awareness of and respect for the small

economic activities that are very hard to integrate on the macro level. Linking the global and local views is very difficult and often confusing, yet it is important to hold both views simultaneously and to try to integrate them.

People with micro and macro views must communicate more, in order to make macro models more accurate and to allow micro views to be integrated into a total system. Micro decisions must not be permitted to cause macro instabilities.

Professor Richardson added that top-down and bottom-up approaches must be considered as complementary, not as dichotomic. A useful rule of thumb in modelling is that it is useful to have some representation of the model running within 1-3 months, even if this is so limited as to be almost a caricature of the overall vision. There have been almost tragic instances of projects which have started out with the image of building up from the local level to ultimately get to the macro level. Modellers are congenital optimists, and they underestimate the effort needed to move from what is believed to be possible to what is possible in practice. One reality is that funds, time and energies run out, and if this happens in mid-project then the overall vision, however imperfect, may never be realised. It is then very valuable to use that vision as a means of defining the issues and concerns that are to be considered at the local level, having always the vision in front of you, being constantly aware of its incompleteness, and using it as a guide rather than on a straight jacket.

4) Professor Malu wa Kalenga enquired how the problem of the complementarity of the production of energy and food could be regulated in practice (by authoritarian price controls or market forces, for example) and how should this be controlled.

Dr. Sachs did not believe that the problems of development could be resolved within a market economy. Some form of state intervention in the economy is necessary. It is not by simply playing with prices that a developing economy can be guided.

5) Professor Meadows felt that global models capture about 60% of the economic system of developing countries, while the other 40% in the informal sector never appears in the model national accounts.

Dr. Sachs added that in highly industrialised societies more time is spent in non-market activities than in market activities, and that perhaps macroeconomists should consider the hidden economy of the domestic sector.

3.2 Energy and the Global Predicament: Some Elements of a Sensible Strategy

J.P. HOLDREN

ENERGY AND HUMAN WELL-BEING

What are the problems which threaten human well-being in the world to-day? This is the context in which we should place deliberations about energy. The list is, in no particular order of importance:

– the poverty of people, i.e., the maldistribution of basic physical and social ingredients of human well-being around the world,

– the impoverishment of the physical and biological environment, which, as much as economic systems do, provides the underpinnings for human well-being,

– the probability of war, i.e., the probability that the ultimate social and political and environmental problem will engulf us all, rich and poor alike, and

– population growth, which tends to aggravate the other three problems.

Focusing on human well-being as the proper goal of our efforts, and pondering the ways in which energy and well being are related, leads to the view that the heart of the energy dilemma is the two-sidedness of that relation.

On the one hand, the use of energy makes essential positive contributions to well-being, both in the form of direct consumer services (such as heating, lighting, and motive power for transportation) and as an input to economic production (for example, fuel and electricity for agriculture, mining, construction, and manufacturing).

On the other hand, harvesting, processing, moving, and using energy produce a set of costs that subtract from well-being. Most obviously, these costs include the diversion of capital, labor, land, water, and other human and natural resources away from the provision of other compelling human needs. Less obvious to many people but no less important to human well-being are the environmental and sociopolitical costs of getting and using energy: injury and disease suffered by workers and by the public as a result of accidents and routine effluents; damage to buildings, crops, and other economic goods by air and water pollution; disruption of ecological and geophysical conditions and processes associated with such ""environmental services" as climate regulation and nutrient cycling; and disruption of sociopolitical conditions and processes with consequences from psychological distress to increased likelihood of war.

This symmetric view, cognizant of both the positive and the negative links between energy supply and human well-being, implies that energy problems may be of two different kinds. Most obviously, people may suffer from *insufficiency of energy supplies,* creating shortages of consumer energy services and contributing to reduced economic output, increased unemployment, and frustration of the aspirations of the poor. This is the traditional view of "the energy problem" – the perils of not having enough. But the other side of the coin is that it is possible to suffer from *excessive energy costs* – costs measured in dollars, in death and disease among workers and the public, in the loss of essential environmental services, in decreased security for people and for nations. This version of "the energy problem" can fairly be described as the hazards of too much – paying more than necessary to get the energy services we want, or even paying more than the benefits are worth.

Clearly, if maximizing human well-being is our goal, we must avoid not only the economic perils of too little energy but also the economic, environmental, and sociopolitical perils of too much energy that is too costly. The real task of energy policy must be to steer a prudent course between these twin perils – an exquisitely difficult task made even more so by the sizable uncertainties that often obscure the hazards on both sides.

HOW MUCH IS ENOUGH?

The task of finding the optimum level of energy supply can be illuminated with the aid of the schematic diagram in Figure I. The diagram requires us to imagine that we can tabulate all the costs and benefits of energy supply – monetary, environmental, sociopolitical – in a single currency. While this cannot actually be done, there is much reason to believe that, if it could be, the curves of costs and benefits versus rate of energy use would have qualitative shapes roughly as shown.

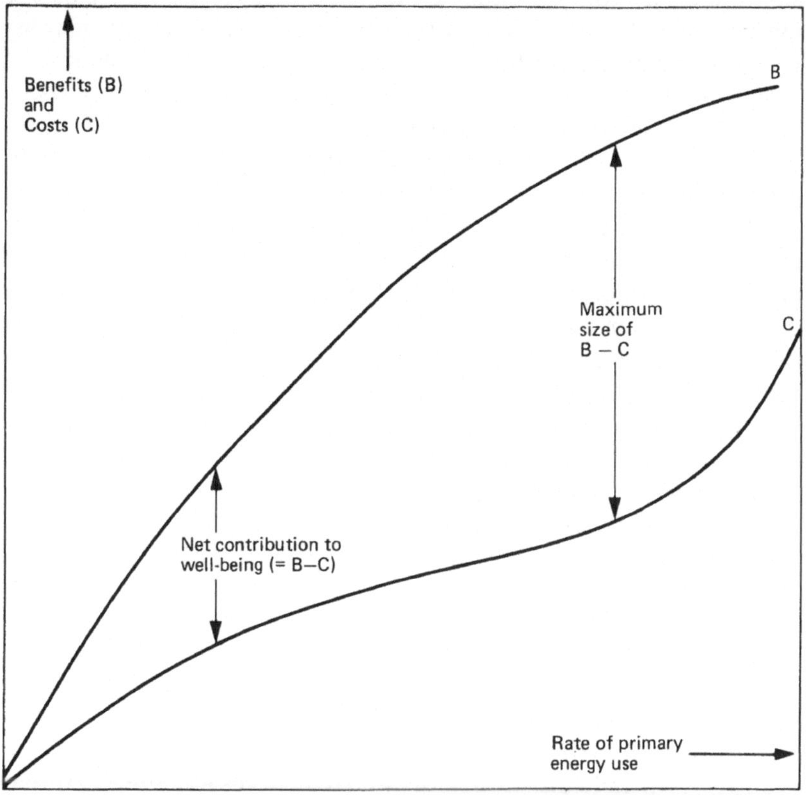

FIGURE I Idealized relation of energy benefits and costs to rate of primary energy use.

The initial flattening of the cost curve at modest rates of energy use corresponds to the realization of economies of scale, which permit, up to a point, the costs per additional unit of energy flow to fall as the size of the energy system grows. The upturn of the cost curve at higher rates of energy supply corresponds to rising marginal costs. This phenomenon results because a society tends to use its best resources (that is, the richest, most accessible, hence most cheaply exploitable ones) first. As the best resources are depleted (or, in the case of renewable ones, harnessed to their full capacity), the society must resort to leaner, more remote, less tractable resources, driving up costs. This trend is reinforced by the cost curve's environmental and sociopolitical components: at high enough levels of stress, the absorptive capacity of environmental and sociopolitical systems tends to saturate, with the result that further increments of stress produce disproportionate damages – the fabric of these systems ceases to stretch and starts to tear. (See, e.g., Holdren and Ehrlich, 1974; Ehrlich *et al.,* 1977; Ehrlich and Ehrlich, 1981.)

The curve of benefits versus the rate of energy use rises steeply at first, because there are so many important needs to be met when little energy is available. The flattening of the benefits curve at higher use rates reflects diminishing marginal benefits of increased energy use, the result of a rational society's having met the most compelling needs first. This syndrome is exemplified by some simple propositions: the first automobile per family makes a greater absolute contribution to real well-being than does the second; the first home per family a greater contribution than the second home; the first two thousand kilo-calories of food per person per day a greater contribution than the second; and so on.

Now it should be evident, looking at Figure I, that the *net* contribution to well-being associated with a specified rate of energy use is the *difference* between the two curves at that point – the total benefits minus the total costs. This net contribution to well-being has its maximum at the value of the energy use rate at which the slopes of the two curves are equal. Above that use rate, the costs of additional energy supply are increasing more rapidly than the benefits, so the net contribution of energy supply to well-being shrinks. A rational society, if it could identify this optimum point beyond which further growth in the energy use rate no longer pays, would not go beyond it.

Of course, a diagram such as that shown in Figure I, if we really could draw it at all, would apply only for the case that the resources and technologies for energy supply, as well as the applications for energy and the technologies for applying it, were specified and fixed. Technological improvements can change the picture in two general ways, as shown in Figure II.

First, such improvements can reduce the costs (monetary, environmental, sociopolitical) of energy supply. Second, they can increase the benefits derivable from any given level of energy

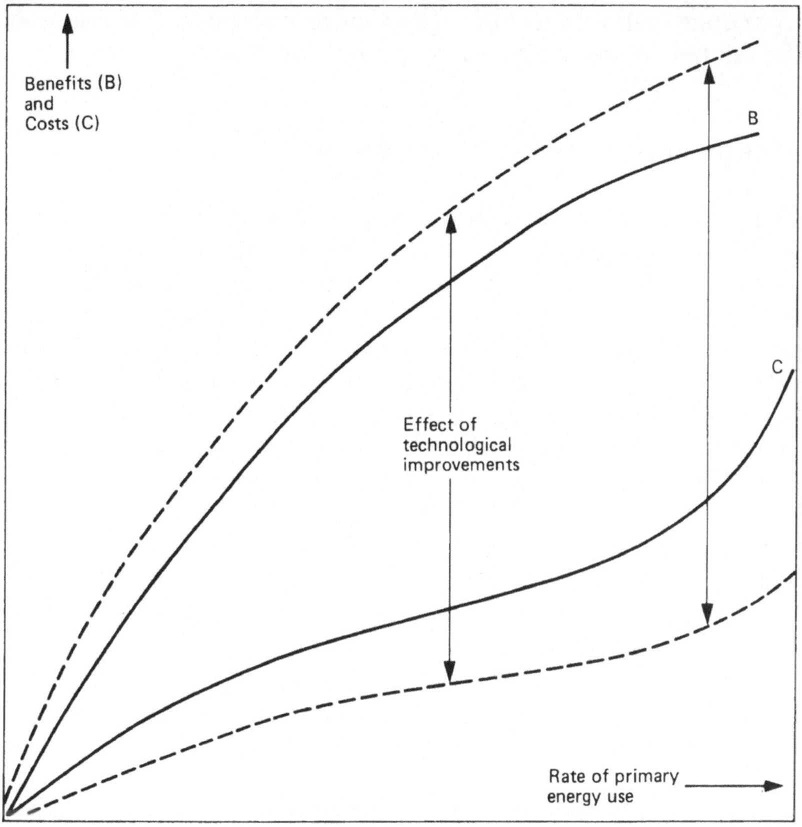

FIGURE II Idealized effects of technological improvements on benefits and costs of energy supply.

supply, either by increasing the efficiency of energy conversion and end-use or through development of beneficial new applications of energy. While such changes certainly can shift the cost and benefit curves – and hence can change the optimum energy use rate and the size of the net contribution to well-being associated with it – the phenomena underlying the general shapes of the curves are basic and pervasive. Thus the key ideas – that for *any* set of energy resources, technologies, and applications there exists an optimum rate of energy use beyond which further additions do more harm than good – can be expected to remain valid over time. (For more extensive discussion of the related economic issues, see, e.g., Daly, 1977.)

ENERGY AND ECONOMIC DEVELOPMENT

The foregoing considerations supply insight into the relative size of the *incentives* for growth in the rate of energy use at various levels of use rate already achieved. If we consider also the correlation between energy use rate and level of economic development (a correlation rather flexible in its details but robust enough across large income differences to be usable for the present purpose), we can also draw some conclusions about the relative *capabilities* of societies at different levels of energy use rate to increase that rate. Combining these considerations suggests that an idealized trajectory of a country's rate of energy use over time would have the form shown in Figure III, resembling the "logistic" or "S" curve familiar to biologists as describing in an idealized way the growth of an organism or a population of organisms.

The process of development, as loosely correlated with growth in the rate of energy use, can be characterized in the framework of Figure III as consisting of three phases.

- Phase I, in which the world's poorest countries today still find themselves, is associated with large incentives to increase the rate of energy use but small capabilities for doing so. The incentives are large because, at low rates of use, the net benefits of energy supply grow very rapidly as the use rate is expanded. But these countries' capabilities to expand the rate of energy use – capabilities measured in terms of access to capital, technology, and managerial talent – are extremely limited. Countries such as Bangladesh, Tanzania, and Haiti most starkly exemplify this phase.
- Phase II, exemplified by such economically "intermediate" countries as Brazil, South Korea, and Turkey, combines large

incentives for increasing the rate of energy use with substantial capabilities for doing so. The result is the "take-off" or exponential-growth phase of the curve shown in Figure III.

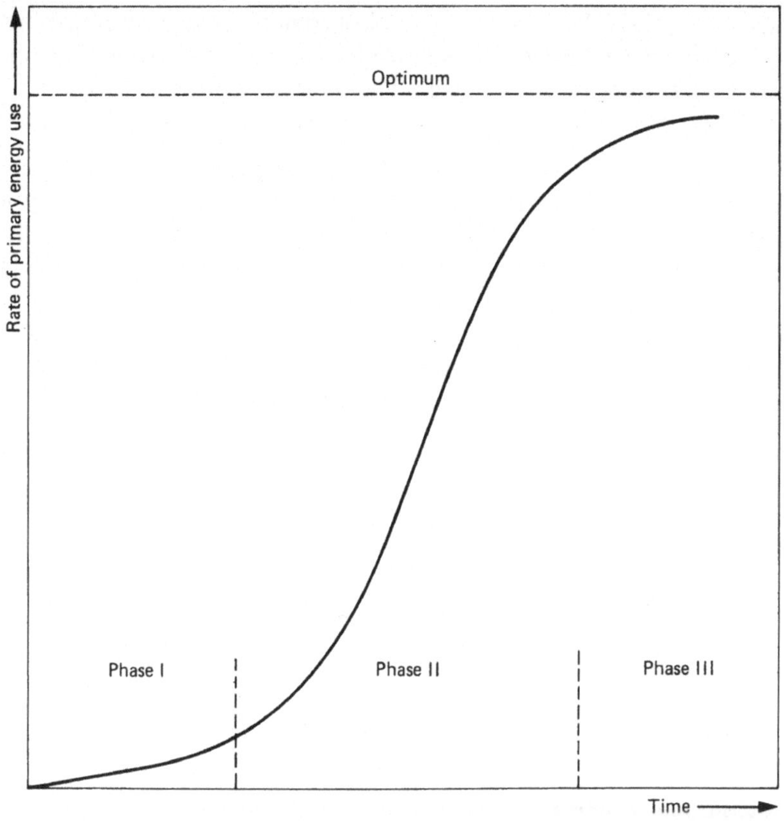

Phase I (poorest countries): big incentives for growth, but small possibilities
Phase II (intermediate countries): big incentives, big possibilities
Phase III (richest countries): small incentives

FIGURE III Idealized growth curve for national energy supply.

• Phase III, where the richest countries now reside, is characterized by large capabilities to further increase the rate of energy use, but small incentives. The incentives are small because, at the high levels of energy use these countries have already achieved, the marginal benefits of further increases are dwindling while the marginal costs tend to be rising sharply. These societies are, in short, getting closer to the "optimum" level of energy use for today's resources, technologies, and opportunities, beyond which further growth no longer will pay. (Indeed, some may have overshot this level.) Prominent examples of countries in Phase III are Sweden, the United States, and West Germany.

It follows that the nature of the energy problem looks different – and *is* different – at different stages of economic development. Both the appropriate energy goals and, accordingly, the dominant criteria for choosing among energy options, change as one moves from societies in Phase I, to those in Phase II, to those in phase III. Not surprisingly, the main obstacles to achieving a society's energy goals differ from phase to phase, as well. These differences are summarized in Table I.

TABLE I
The energy problem at different stages of economic development.

Appropriate energy goals	Criteria for energy choices	Obstacles to achieving goals
Poorest countries		
providing energy for basic human needs	low cost, simplicity, versatility	lack of capital and technical and management skills
Intermediate countries		
rapid growth in energy availability to fuel high economic growth	expandability, reliability	bottlenecks in mobilizing people and equipment; hasty choices leading to excessive costs
Richest countries		
avoiding overshoot of well-being optimum; building a reliable and sustainable energy system	environmental/ social compatibility, inexhaustibility	difficulty of integrating environmental and social characteristics into market and nonmarket decision mechanisms

HOW THINGS FELL APART IN THE SEVENTIES

The foregoing discussion falls short of explaining just how and why it was in the 1970s – and not either sooner or later – that energy emerged so suddenly from the bevy of human problems to assume a place of prominence on the technical, economic, and political agendas of rich countries and poor countries alike. The essence of the answer is that a number of disparate factors converged in the 1970s to push most countries rather rapidly onto the steeply rising part of the cost curve of Figure I.

For the rich countries, the immediate problem that resulted was to reduce rapidly their dependence on the particular energy source whose increased costs seemed most threatening of all – namely Mideast oil – without paying unnecessarily high costs for the replacements. Their trauma (and in some cases even paralysis) over energy choices was due not to any technical inability to expand energy supply, but rather to the growing perception in many constituencies that the marginal costs – monetary, environmental, sociopolitical – of the available ways to do so were becoming perilously close to the marginal benefits. As is typical when the spread between benefits and costs shrinks, moreover, the deliberations of the rich countries were complicated by increased concerns with distribution, equity, and accountability: Who will benefit, who will pay, and who will get to decide?

The poor countries found themselves in a vise (excepting the major oil exporters, who rapidly became "rich" while still "underdeveloped"). Locked, by the nature of the world market, onto the rising cost curve for oil that the cumulative and current consumption of the rich countries had brought about, the oil-importing less developed countries (hereinafter LDCs) had the worst of both worlds: the compelling requirements of the poor for energy to meet basic human needs, the poor's small ability to pay, and the same soaring oil costs faced by the rich.

This plight of the urban/industrial sector of the oil-importing LDCs, which threatened financial insolvency if the oil flow was maintained and massive deprivation if it was not, was paralleled by a substantially independent energy crisis in the rural sector – the fuelwood crisis. The most basic energy need – cooking – is met for nearly half the LDCs' 3,300 million people by fuelwood. (Another third of this nearly three quarters of the world's population use crop residues and dung.) But in the 1970s, over an increasing fraction of the world's surface, the regrowth of forests was no longer keeping pace with the growing rural population's depredations for fuelwood. The result was not only spreading deforestation and erosion (contributed to also by commercial logging and clearing of forest for agriculture), but a rural energy crisis manifested as punishing expenditures of time and cash for fuel acquisition by many of the world's poorest people (Eckholm, 1976; Hughart, 1979).

It is worth recalling that most of the economic development of the present industrial nations occurred in a period of sustained growth in the availability of convenient and versatile energy forms at steadily falling real monetary costs. (For most of this period, moreover, few people saw reason to think that the environmental and social costs of the expanded energy supply were anything but small compared to the net economic benefits.) These circumstances were able to be maintained for so long a time – roughly, the century between 1870 and 1970 – by a combination of economies of scale, successes in exploration for large and accessible deposits of the particularly convenient and versatile fluid fossil fuels, and technological advances in energy conversion and transmission (see, e.g., Cook, 1976).

The seventies marked the time when these favorable developments were no longer frequent enough, pervasive enough, or powerful enough to continue to offset worldwide the combined effects of two countervailing factors: the cumulative depletion of the most accessible and inexpensive nonrenewable resources and the sheer magnitude the instantaneous rate of energy use had attained. A century of unprecedented growth both in world population and in the use of industrial energy per head had taken its toll. Table II conveys quantitatively the essence of the matter. (It is a common mistake to attribute the turning point of the 1970s to less fundamental factors, notably the success of the OPEC cartel in "artificially" boosting the price of oil far above its production cost from Mideast wells. In reality, OPEC has been able to set the oil price where it has essentially *because* most of the rest of the world no longer can produce energy adequate to its needs, either from its own oil or from othe sources, at costs below the OPEC price.)

This is not to say that the world was running out of energy in

TABLE II

A century of growth in population and energy.

Year	World population (millions)	Population increase in previous 20 years (millions)	World use rate of industrial energy (terawatts)[a]	Cumulative energy use in previous 20 years (terawatt-years)[b]	Cumulative energy use since 1850 (terawatt-years)[b]
1870	1300	170	0.2	3	3
1890	1500	200	0.5	7	10
1910	1700	200	1.1	15	25
1930	2000	300	2.0	30	55
1950	2500	500	2.9	45	100
1970	3600	1100	7.1	100	200

[a] 1 terawatt (TW) = 10^{12} watts = 31.5×10^{18} joules/year = 14×10^6 barrels/day of oil

[b] 1 terawatt-year (TWy) = 31.5×10^{18} joules = 10^9 tonnes bituminous coal

Sources: Bogue (1969), Haefele and Sassin (1977)

the seventies. But it was running out of *cheap* energy – in particular, out of the one-time windfall of cheap and versatile oil and gas on which much of the prosperity of industrial society had been built. Even the cheap oil and gas was not – and is not – completely gone, but most of the world's consumption was in countries that could not produce any faster from their own resources and that were already heavily dependent on imports from the few (mostly OPEC) countries with production capacity greatly exceeding their own needs. In 1980, oil still supplied forty-two percent of the world's supply of industrial energy, and natural gas another eighteen percent. (See Table III). Half this oil and a tenth of the gas were sold in international trade.

TABLE III
Sources of world energy use, 1980 (terawatts).

Industrial energy forms[a]	
petroleum liquids	4.2 TW
coal	3.0 TW
natural gas	1.8 TW
hydropower[b]	0.8 TW
nuclear energy[b]	0.2 TW
Industrial Total	10.0 TW
Traditional energy forms[a]	
fuelwood	0.6 TW
dung	0.6 TW
crop wastes	0.2 TW
Traditional Total	1.0 TW

a Data for industrial energy forms are accurate to ± 10% or better; estimates for traditional energy forms are ± 50%.
b Hydropower and nuclear energy converted from electricity at 10 MJ/kWh (approximately the amount of fossil fuel "displaced").
Sources: British Petroleum (1981), World Bank (1980), 1981).

Many of the oil-exporting countries – and in particular those with the largest remaining supplies – are in politically unstable regions and/or have ideological conflicts with their customers. Thus the costs of imported oil must be reckoned not only in monetary terms, but also in vulnerability to interruption beyond the control of importer or exporter, in the actual and potential uses of oil as a political weapon, and indeed in the chance that military intervention by powers seeking to protect or deny access to oil will spark a regional or global conflagration.

Leaving aside the chance of world war sparked by oil, the most damaging consequences of the energy developments of the seventies surely will be borne by the oil-importing LDCs. In the long run, countries that were already rich in 1970 can adjust to expensive energy – with some disruption, to be sure, but at bearable cost. Where it proves necessary, they can *reduce* their use of energy sources whose economic, environmental, and/or sociopolitical costs (like those of Mideast oil) threaten to become intolerable. But the poor countries have seen the pathway to industrial prosperity followed by the countries now rich – industrialization fueled by cheap energy – blocked off before their eyes. Even what little industrial machinery they now possess was designed to use the cheap energy that no longer exists; thus they can hardly afford to operate it, but they can ill afford to replace it, either. The dimensions of the gulf that separates them from the rich are suggested by the data in Table IV.

EVALUATING THE OPTIONS

What are the energy options open to rich countries and poor ones, and what criteria should be used in choosing from among them?

The variety of energy resources and existing and possible technologies for harnessing them is large. Beyond conventional oil and natural gas, there are: heavy crude oils and tar sands; oil shades; unconventional gas of several types; coal, burned directly or transformed into fluid fuels; uranium, use in conventional or breeder reactors; lithium and deuterium from seawater, if fusion reactors succeed; geothermal energy in half a dozen forms; and renewables (mainly biomass, hydropower, solar heat, wind, solar electricity, and ocean heat). Even that is only a partial list, the most important omission from which is increased efficiency of energy *use* – as much an energy "source" as oil or gas, in that a kilojoule saved is fully as useful elsewhere in the economy as a kilojoule pumped out of the ground.

On what basis will a sensible society make its own energy

TABLE IV
Distribution of energy use and gross national product, 1980.

Country Group	Population (billions)	Average GNP per person (1980 $)	Industrial energy/person (kilowatts)	Traditional energy/person (kilowatts)	Total energy (terawatts)	Total GNP (billions of 1980 $)
Poorest	2.5	400	0.3	0.3	1.5	1,000
Poor	0.8	1,250	1.3	0.2	1.2	1,000
Rich	0.6	4,500	4.4	0.1	2.7	2,700
Richest	0.6	10,700	9.2	0.1	5.6	6,400
Total or average	4.5	2,500	2.2	0.2	11.0	11,100

Sources: Author calculations and estimates based on World Bank (1980, 81), British Petroleum (1981), Hughart (1980).

choices from the menu of options just outlined? Among the most important considerations are: fuel supply, timing, economics, reliability, versatility, and compatibility. These criteria are elaborated in Table V and taken up in more detail in the subsections that follow.

TABLE V
Criteria for choosing energy options

Fuel supply	
	non-renewable: size of stocks, geographical distribution, ownership
	renewable: size of flows, intermittency, geographical distribution, ownership
Timing	
	How soon can the option be deployed? (stage of feasibility)
	How rapidly can its use be expanded? (nature of constraints)
Economics	
	construction cost, capacity factor, system lifetime
	fuel cost and other operation and maintenance costs
	cost of back-up for low-reliability options
Reliability	
	degree of uncertainty about key characteristics
	vulnerability to large-scale disruptions
	frequency and predictability of other interruptions
Versatility	
	applicability to diverse end-use needs
	usability in diverse geographic, economic and cultural contexts
Compatibility	
	compatibility with co-existing and future energy technologies
	compatibility with nonenergy values and goals
	compatibility with choices and aspirations of other regions and nations

Fuel supply

With respect to magnitude of energy resources, Table VI underlines the point that the world is in no meaningful sense running out. Confusion on this point is sometimes engendered by preoccupation with *reserves,* which refer only to those resources whose location has been established and which can be profitably extracted with present technology at present prices. That reserves often stand at only a few decades' consumption at current rates is no indication that the fuel will all be gone at the end of that period; there simply is little incentive for firms or governments to spend money to establish the location of reserves more than a few decades in advance of the need for the material. More meaningful as indices of long-term adequacy of supply are estimates of *remaining ultimately recoverable resources,* which refer to all the remaining material that will ever be economically recoverable. Making such estimates requires assumptions about how good the technologies of extraction will ever become and about how high the prices of the fuels will ever go, and it also usually requires rather heroic extrapolations of limited geologic evidence, so the figures must be regarded as highly uncertain.

Expressing the estimates in terawatt-years facilitates translating resource magnitudes into time periods. Thus, an estimated 800 terawatt-years of remaining ultimately recoverable oil and gas would last something over 130 years at today's global use rate of about 6 terawatts for these fuels. At twice today's rate, they would last 60 or 70 years. This is not so reassuring, and it becomes even less so when the uneven distribution and likely high cost of the remaining oil and gas are taken into account. But, clearly, the resources of the less convenient fossil fuels – coal, heavy oils and tar sands, and oil shales – are far larger. If coal were meeting *all* today's fossil fuel consumption, it would last 500 years, and the oil shale, if one were willing to exploit material as lean as 40 litres of oil per tonne of rock, represents ten times more energy still. Uranium, if used in today's fuel-wasteful reactors, represents an energy source only about as large as oil and gas; but, if breeder reactors were perfected and accepted, even uranium in seawater would suffice as fuel, and a world more energy-hungry than today's could be run for a hundred thousand years. As Table VI shows, moreover, the lithium and deuterium in seawater become energy resources of staggering dimensions if the controlled fusion technology to harness them can be mastered. And, with no "ifs" whatever, energy from the sun will continue to reach earth's surface, at a rate of nearly 10,000 times the present energy use rate of civilization, for the few billion more years our star will live. Sheer adequacy of energy resources, in short, is not the problem.

TABLE VI
World fuel reserves and resources (TWy)[a].

	Reserves[b]	Remaining ultimately recoverable resources[b]
Petroleum	120	400-800
Natural gas	80	300-500
Coal	600	5,000-10,000
Heavy oils, tar sands	2	500?[c]
Oil shale	40	30,000?[d]
Uranium, in conventional reactors	30	400-1,000[e]
Uranium, in breeder reactors	3,000	> 1,000,000
Fusion fuels (lithium & deuterium)		> 100,000,000[f]
Sunlight reaching Earth's surface	88,000 TWy/year for the lifetime of the sun	

[a] 1 TWy (terawatt-year) = 31.5×10^{18} joules = 10^9 tonnes bituminous coal.

[b] Except as noted; sources for data are IIASA (1981) and Linden (1980).

[c] Assumes about half of the maximum estimate of identified "resources in place" as of 1979 (Meyers and Dietzmann, 1979) ultimately will be recoverable. The true value could be much larger or much smaller.

[d] Assumes about half of an estimate of total "oil equivalent in deposits" richer than 40 litres of oil per tonne (Hubbert, 1969) ultimately will be recoverable. The true value could be somewhat larger or much smaller.

[e] Higher figure assumes rest of world land area eventually will yield at least half as much uranium per 10^6 km^2 as U.S. Department of Energy's 1978 best estimate of "potential supply" (Harris, 1979) for the U.S. at \$50/lb (U_3O_8) forward cost. If extraction of 50% of uranium from seawater proves feasible, add 3,000 TWy (Holdren, 1978).

[f] If only "easiest" fusion reaction succeeds commercially, lithium is limiting fuel and stated figure corresponds to extracting 50% of lithium from seawater. If more difficult reaction using deuterium alone succeeds, 50% extraction of deuterium from seawater will yield 250,000,000,000 TWy (Holdren, 1978).

Some of the problems posed by the uneven geographic distribution of oil and gas already have been mentioned. Table VII depicts the situation quantitatively for the fossil fuels. The position of many of the oil-importing LDCs may not be quite so bad as one's first glance at the table's figures for oil and gas suggest, however, because even a very small share of these resources represents a large amount of energy compared to the still modest annual demands in the poorest countries. Deposits of oil and gas too small to have interested the multinational oil companies, who have had access to the much larger fields of the Mideast and the continental shelves, could be very important energy sources indeed to the LDCs that possess them. Most LDCs, moreover, have been less thoroughly explored for energy resources than have the industrial nations, and it seems likely that more exploration will yield at least modest additional discoveries of oil, gas, and coal.

TABLE VII
Distribution of conventional fossil fuels.

Percentage preceding slash is group's share of world reserves. Percentage following slash is group's percentage of remaining ultimately recoverable resources.

	Petroleum	Natural Gas	Coal
OECD[a]	9%/15%	17%/26%	50%/41%
Warsaw pact[b]	11%/18%	36%/28%	23%/37%
China	3%/5%	1%/6%	16%/14%
Other OELDC[c]	76%/34%	43%/25%	1%/1%
OILDC[d]	1%/28%	3%/15%	10%/7%

[a] OECD = Organization for Economic Cooperation and Development (market industrial economies)
[b] Warsaw Pact = USSR and Eastern Europe, less Yugoslavia and Albania
[c] OELDC = Oil exporting less developed countries
[d] OILDC = Oil importing less developed countries
Sources: World Bank (1980), IIASA (1981)

Of the 88,000 terawatts of energy continuously reaching earth's surface from the sun, about 40,000 evaporate water and thus are embodied in the hydrologic cycle. Only a tiny fraction of this amount – 2 to 3 terawatts – ends up as potentially harnessable hydropower (Hubbert, 1969; Deudney, 1981), but that is still a very large energy flow. As shown in Table VIII, its geographic distribution is quite favourable to the LDCs. The distribution of the most readily harnessed of the other renewable energy flows – the insolation itself, wind, and production of biomass – is uneven in each case, but most regions possess at least one of the three in some abundance: the desert and semi-arid regions have intense and reliable sunlight; the tropics and temperate zones, often too cloudy for ideal, solar-collector performance, enjoy high biomass production; and the high latitudes tend to be windy (Hayes, 1977; Sorensen, 1979). The total energy flow driving the winds is 1,000 to 2,000 terawatts (Sorensen, 1979; Flavin, 1981), and the flow captured by photosynthesis is about 100 terawatts (net, after subtracting the metabolic requirements of the plants themselves) (Ehrlich *et al.*, 1977). The fractions of these flows that will prove harnessable to human energy needs are highly uncertain.

TABLE VIII
Hydropower potential and installed capacity (electrical gigawatts).

	Exploitable potential	Installed capacity, 1980
Asia, excluding USSR	610	53
South America	430	34
Africa	360	17
North America	360	130
USSR	250	30
Europe	160	96
Oceania	45	7
World Total	2200	360

Source: Rounded to two significant figures from Duedney (1981)

Geothermal energy may be renewable or nonrenewable, depending on the kind of occurrence and the rate of exploitation. The total resource is very large: the geothermal heat stored in the first 10 kilometers of the earth's crust has been estimated at 40,000,000 terawatt-years, and the rate at which this heat theoretically could be tapped on a sustainable basis has been estimated at 80 terawatts (Ellis, 1981). In practice, geothermal energy is only being harnessed today in isolated, highly favorable circumstances – the total in 1980 amounted to perhaps 0.02 terawatts. A very ambitious scenario for the year 2000 envisions 0.35 terawatts (Ellis, 1981) and the IIASA energy study argues that one cannot count on a sustainable geothermal contribution much above 3 terawatts even in the very long term (IIASA, 1981). The latter is, however, a very substantial energy flow – equal to world coal use today – and there are attractive geothermal sites in many LDCs.

The geographic distribution of known high-grade resources of uranium is as uneven as that of oil, but there is some reason to think that this situation is more indicative of uneven exploration and imperfect understanding than of geologic reality. In any case, if nuclear technology is pushed to the level of breeder reactors and fusion, the raw fuel will be available to anyone with access to seawater.

Timing

Relevant to the question of *how soon* an energy option can be deployed are four degrees of feasibility:

1) *scientific feasibility,* meaning the scientific principles underlying the option are known and favorable;

2) *technological feasibility,* meaning a device has been built that successfully exploits these principles to yield a net energy gain in usable form;

3) *commercial feasibility,* meaning that the technology can be operated under conditions (including cost) attractive to paying users;

4) *social feasibility,* meaning that the observed consequences of using the technology are accepted by enough of those affected so that its use gains and retains political approval.

Commercial and social feasibility of course depend in substantial part on characteristics discussed here under other headings, notably economics, reliability, versatility, and compatibility.

A common error that leads to understating the difficulty of the energy predicament is to underestimate the time it takes for a new energy source to pass these stages of feasibility and become a major contributor to national or global energy supply. The scientific feasibility of nuclear fission as an energy source was demonstrated in 1942, for example, but it did not achieve commercial feasibility until around 1960, and in 1980 it produced only two percent of total global energy supply and less than ten percent of the world's electricity. Its social feasibility is still in doubt in some countries. Fusion research probably will achieve a scientific-feasibility demonstration before 1985, but technological feasibility is unlikely to come before 1995 and commercial feasibility could easily take until 2010 or later (Holdren, 1978). It will be a remarkable achievement if by 2025 fusion is contributing 10 percent of world electricity supply.

Relevant to this problem, of course, is not only how soon an energy option can first be deployed commercially, but also how rapidly its use can be expanded thereafter. This depends strongly on the presence or absence of bottlenecks in the supply of the necessary ingredients: capital, trained workers of specific kinds (e.g., exploration geologists, coal miners, welders), special materials (e.g., high-alloy steel, silicon crystals), fabricated equipment (e.g., drilling rigs), transport systems (e.g., railroads, pipelines), water, and so on. Shortages of such inputs occasionally have constrained the expansion of one energy source or another in industrial nations, but the situation in LDCs is much worse: shortages in most of these categories are endemic.

Economics

I believe the central economic reality about future energy supply is that *energy will never be cheap again,* in the sense of the three-dollar-per-barrel oil of a decade ago. This is so despite the potential availability of energy resources of enormous abundance, as discussed above, and despite the circumstance that, in the case of some of the most abundant of these (notably solar energy, fusion fuels, and fission fuels for breeders), the raw fuel is free or nearly so. (Uranium oxide at $1000 per kilogram – more than ten times today's price – would contribute only 0.02 cents to the cost of a kilowatt-hour of electricity from a breeder reactor; and deuterium, producible already from seawater at $400 per kilogram, would contribute about 0.001 cent to the cost of a kilowatt-hour from a deuterium-burning fusion reactor. For comparison, the cost of coal at $40/tonne contributes about 1.5 cents per coal-generated electrical kilowatt-hour.)

The reason cheap and abundant fuel do not mean cheap energy is the high cost of building and/or running the facilities needed to convert the raw fuel into the energy forms people use (for example, electricity and fluid fuels) and to distribute them when and where they are needed. It is the perverse nature of all the superabundant energy sources that the technologies needed to make them useful tend to be technically complex, or materials intensive, or labor intensive, or some combination of these – all of which translate into high costs. No one can be certain, of course, that some unforeseeable breakthrough will not someday restore an era of cheap energy; but the odds are so strongly against it that prudent individuals, firms, and governments must now base their planning on energy that is permanently expensive.

As to the *relative* monetary costs of the various energy options, one compared to another, almost any assertion is subject to argument – partly because the conditions governing costs vary so much from one region to another, and partly because of genuine uncertainties about the future cost of practically everything. With respect to electricity generation, coal and conventional nuclear fission invariably are cheaper than oil and usually are roughly comparable to one another, the advantage varying with location. (The disclaimer must be added that the real costs of fission are clouded by subsidies and uncertainties about long-run performance, and those of coal and fission alike are clouded by unresolved environmental issues that might require controls costlier than today's.) Breeder reactors and their fuel cycles are enough more complex than conventional fission that their electricity costs are virtually sure to be higher despite lower fuel costs. Fusion technology promises to be even more complex than breeders and can hardly be cheaper.

Among renewable sources of electricity, hydropower at good sites is roughly competitive with coal and fission, and wind at the best sites seems to be on the verge of becoming competitive. Use of biomass wastes for cogeneration of electricity and steam or other process heat is economical where large sources of these wastes are at hand. Electricity from solar-thermal-electric systems reamins two to five times as expensive as from conventional sources, and electricity from photovoltaics is more than ten times as expensive. (Both may be useful for specialized applications remote from electricity grids. Many proponents of photovoltaics consider the prospects for significant cost reductions – and hence much wider applications – to be promising.)

Fluid fuels derived from coal and oil shale remain more expensive than oil and gas from conventional sources. Although this eventually must change as oil and gas becomes costlier, the many past projections of an imminent crossover point all have proven to be premature. Alcohols made from biomass are perhaps closer to economic competitiveness, but they are being extensively used for fuel today only where heavily subsidized. Direct combustion of biomass – crop and animal wastes, wood, and wood wastes – for cooking and domestic and industrial heat is the dominant energy source in many of the poorest countries and is surprisingly important in some industrial nations as well. Clearly, such use of biomass is sometimes the most economical available energy source, but its opportunity costs (in time spent gathering it and in the diversion of the biomass from other uses such as animal feed and structural material) are rising in many circumstances.

Simple solar collectors for heating domestic water and swimming pools are competitive with the use of gas, oil, and electricity for these purposes in sunny climates where conventional energy is expensive. Passive solar space heating (relying on window size and placement, building orientation, judicious use of thermal mass, and designed-in pathways for natural convection, conduction and radiation) can also be an economically attractive alternative in many climates, but active solar heating (relying on more elaborate collectors and mechnical circulation) rarely pays if the less expensive passive measures and energy-conservation steps are taken first (Anderson, 1981). Solar-pond collectors may be economically attractive for low-temperature industrial process heat and even for driving heat engines (Taylor, 1978), but such systems have yet to prove themselves in large-scale and prolonged operation.

Reliability

I here intend reliability to cover a variety of characteristics related to the question, "Can we count on it?" One of these characteristics is our degree of certainty about the answers to questions addressed above: the quantity and location of the fuel, the timing of the energy source's availability, the affordability of the usable output. A second aspect of reliability is an energy option's freedom from or resistance to large-scale disruptions of supply by human action (embargoes, strikes, sabotage) or natural events (floods, droughts, earthquakes). A third aspect of reliability relates to the frequency and predictability of smaller interruptions, such as those associated with scheduled and unscheduled maintenance and with the intermittency inherent in such energy sources as sunlight and wind. (Interruptions in this last cateogry can be countered, at a cost, by means of energy storage.) The impact of interruptions of any kind may depend not only on the nature of the interrupted energy source but also on the characteristics of other energy sources in use in the same region – that is, reliability must be considered a characteristic of energy *systems,* not just of individual energy sources.

Versatility

Versatility means the ability to meet a variety of kinds of energy needs in a variety of geographic, economic, and cultural contexts. Key factors influencing versatility are: (a) the variety of useful energy forms the system can produce, (b) the degree of technical and managerial sophistication needed to operate the system, and (c) the minimum size of practical units.

The enormous increase in the use of oil and gas in this century has been due in substantial part to their versatility, and one of the greatest drawbacks of the most counted-upon replacements – coal and nuclear power – is their relative lack of it. Nuclear power plants come only in very large sizes (a few hundred megawatts and more), make only electricity, and are extremely demanding of technical and managerial skills. Coal is less demanding, more flexible in the size of the units that can use it, and can be used for more than electricity generation; but matching the full versatility of oil and gas by converting coal into synfuels with the same properties is an expensive, high-technology enterprise.

Most of the renewable energy sources are far superior to nuclear power with respect to versatility, and many are superior to coal. Their greatest asset in this respect is the relative ease with which they can be matched to dispersed end-use needs down to very small sizes – a particularly crucial factor in the rural sectors of LDCs.

Compatibility

This heading covers a number of issues. One is compatibility of an option with other energy options in use or expected to be used in the future: Are the options complementary in the kinds of needs met? Are there symbioses? Does the implementation of one option facilitate evolution or transition to better ones? No society's energy needs are likely to be met by a single energy source or technology, and sensible choices often require looking at the properties of different mixes of energy options as systems rather than just at the properties of the individual components.

A society's energy choices should be compatible also with the society's other values and goals, having to do with, for example: levels of employment, regional development, the maintenance of environmental conditions and processes supportive of human well-being, the distribution of wealth and income, patterns of decision making and the distribution of political power, human rights, and personal and national security. Of these links, those between energy and environment and between energy and national security have received the most attention, and they are treated in separate sections below. That decisions about energy also can profoundly affect other social conditions and processes has been recognized for a long time (see, e.g., Cottrell, 1955), but the idea that these social consequences of alternative approaches to energy supply ought to be a major determinant of which approaches we choose has become prominent in energy-policy discussions only recently (Lovins, 1976, 1977; Nader and Beckerman, 1978; Morrison and Lodwick, 1981).

Finally, *regional* compatibility among energy sources is important: choices made in one region should not impose undue burdens on – nor foreclose the options of – other regions. Difficult questions are raised by such considerations: What are the obligations of nations rich in energy resources to those that are energy poor? What are the obligations of the industrial nations to the LDCs? One of the most unfortunate aspects of twentieth century energy development is the degree to which the industrial nations – which have done almost all the world's energy research because they were the only ones who could afford it – have concentrated their efforts nearly entirely on large-scale, central-station electricity generation. The extent to which the most immediately threatening energy problems for the industrial nations in the 1970s had much less to do with electricity than with fluid fuels suggests that this focus on electricity was wrong even for the industrial nations themselves (National Academy of Sciences, 1979a). But, more importantly, the neglect of research on renewable sources of heat and fluid fuels further undermined the already dismal energy prospects of the LDCs, by ensuring the continued absence from the global menu of available energy options of some of the possibilities of greatest relevance to LDC needs.

ENERGY AND ENVIRONMENT

No energy option known or imagined – including solar energy, fusion, and increased efficiency of energy use – is completely free of environmental liabilities. These liabilities take the form of disruptions of a wide variety of environmental conditions and processes (including, when "environment" is given its most general interpretation, sociopolotical conditions and processes), and the resulting damages to human well-being cover a wide range of degrees of severity (Budnitz and Holdren, 1976; Holdren, 1981). Table IX lists, for a number of important or potentially important energy options, some of the environmental issues of greatest interest. The following subsections elaborate on three cases of particular importance: coal, nuclear fission, and renewables.

Coal and environment

The extraction, transport, and combustion of coal pose a number of direct and serious hazards to human health and safety, of which the most familiar are: aggravation or causation of public disease by air pollution from coal combustion, with particulate matter and oxides of sulfur thought to be the biggest dangers; accidents and black-lung disease befalling underground coal miners; and hazards to worker and public safety in the

TABLE IX
Some environmental liabilities of energy options.

Fossil fuels

Air pollution in the form of oxides of sulfur, oxides of nitrogen, photochemical oxidants, aromatic hydrocarbons, and particulate matter cause and/or aggravate disease.

Acid rain, in poorly buffered watersheds, can kill fish, accelerate leaching of nutrients, mobilize toxic trace metals, alter microbial activity.

Carbon dioxide from combustion, accumulating in atmosphere, may alter climate more rapidly than agriculture can adjust.

Accidents involve workers in coal mines and oil platforms; public in collisions with coal trains and trucks; public and workers in gas explosions.

Biomass

Deforestation from overharvesting fuelwood, with resulting erosion and long-term loss of carrying capacity.

Soil depletion and impaired water-flow control from removal of crop and forest residues.

Air pollution, particularly particulates, from combustion.

Accidents in harvesting and transport.

Hydropower

Destruction of river ecosystems and fertile bottom land.

Dam failure through miscalculation, earthquake, or sabotage.

Nuclear fission

Spread of nuclear weapons among nations and into subnational groups.

Sabotage and accidents at reactors and reprocessing plants, in waste transport and storage.

Routine emission and exposures from mining and milling, fuel reprocessing, reactors.

Geothermal energy

Water pollution by dissolved salts and toxic elements in geothermal water.

Air pollution by hydrogen sulfide gas (oxidized to $S0_2$ in atmosphere).

Solar space-heating and water heating

Falls from roofs during collector installation and maintenance.

Fires caused by collector overheating.

Leaks of collector fluid damaging building or contaminating household water supply.

TABLE IX (Continued)

Wind

Accidents: toppling towers, flying blades, and falls during construction and maintenance.

Aesthetic intrusion on mountain ridges and scenic coastlines.

Photovoltaics

Toxic substances as manufacturing hazards, or released from overheated or burning equipment, or entering environment when equipment is discarded.

Chemical, electrical, and explosive hazards of storage batteries.

Solar power satellites

Worker hazards in space: weightlessness, radiation, micrometeroroid.

Microwave exposures to people and other organisms by beam transmitting power to earth.

Upper-atmosphere pollution by launcher exhausts.

Military applications of satellites themselves or enhanced space capability.

Ocean thermal energy conversion

Marine ecosystem disruption by altered temperatures and currents, leaks of working fluids, anti-fouling chemicals, entrainment of organisms.

Worker hazards on floating plants, especially in storms.

Nuclear fusion

Tritium delivering radiation doses to workers in reactors and, through inevitable leaks and possible accidents, to members of the public.

Neutrons rendering reactor components radioactive, thus adding to worker hazards and requiring management as radioactive wastes.

Increased efficiency of energy use

Indoor air pollution by chemicals in insulation, or aggravated by reduced ventilation rates.

Carinogens and mutagens in exhausts of diesels.

Decreased crash safety in smaller automobiles.

---

form of accidents in the transport of coal by train and truck (Office of Technology Assessment, 1979). All these hazards are susceptible to being reduced substantially by means of available technologies and management practices, such as pollution control devices on power-plant stacks, better miner training and coal-dust abatement for underground mines, and overpasses and underpasses at train crossings. It must also be emphasized that the shocking incidence of disease often attributed to coal-produced air pollution in technical and popular articles alike – 100 to 200 deaths per large power plant per year – fall at the high end of a range of scientific uncertainty that extends down to zero on the low end (Lave and Seskin, 1977; Morgan *et al.*, 1978; U.S. National Academy of Sciences, 1978).

It may well turn out in the long run that the most serious environmental costs of coal – and of the other abundant fossil fuels such as oil shale – are not the direct impacts on health and safety but the damages to ecosystems and hence valuable environmental services (Holdren and Ehrlich, 1974; Harte and Jassby, 1978). These damages may arise from consumptive water use in the semi-arid regions where much coal and oil shale are found (Harte and El-Gasseir, 1978), from acid rain formed from sulfur oxides and nitrogen oxides released in combustion (Likens *et al.*, 1979), and from climate change induced by the build-up of fossil-fuel-produced carbon dioxide in the atmosphere (U.S. Council on Environmental Quality, 1981). For the last problem, no plausible cure has been proposed short of limiting the use of fossil fuels.

Nuclear fission and environment

Defining "environment" to include the sociopolitical issues of weapons proliferation among nations and access to nuclear weapons by subnational groups, I would rank the environmental liabilities of fission, from biggest to smallest, as follows: international proliferation, subnational nuclear weapons, accidents and sabotage, and routine emissions and exposures. International proliferation is taken up below under "Energy and War".

With respect to acquisition of nuclear weapons by subnational groups such as terrorists, there is no doubt that plutonium of the kind produced by presently commercial nuclear reactors can be used, with suitable sophistication in design and fabrication, to make devastating nuclear explosives (Willrich and Taylor, 1974; Gilinsky, 1977). There *is* uncertainty about: the likelihood that the necessary sophistication will be available to a subnational group; the number of subnational groups that may be motivated to make an attempt; the fraction of attempts to steal the material that will succeed; whether a group that manufactures a bomb will actually explode it (as opposed to merely threatening to do so for financial or political gain). None of these latter issues is susceptible to persuasive technical analysis. The question that society must face thus becomes: Is this uncertainty *itself* a reasonable trade for the benefits of nuclear power?

The possibilities with respect to accidents and sabotage are less threatening and more tractable, in my view, than the links between nuclear power and nuclear weapons, because the potential effectiveness of strictly technological safeguards is greater. Nevertheless, as the Three Mile Island incident emphasized (Kemeny, 1979), human frailties play important roles. These frailties, of course, in the future could include malevolence as well as carelessness and incompetence. Neither operating experience nor a combination of theoretical and experimental studies is adequate to the task of pinning down the risk of commercial nuclear operations (accident probability times consequences, summed over all possible events) to a narrow range. The widely publicized Reactor Safety Study of the U.S. Nuclear Regulatory Commission (1975) gave uncertainty ranges of factors of 3 to 6 up and down on both probability and consequences, but a variety of prestigious reviews have concluded that the real uncertainties are much larger, even without accounting for sabotage (Nuclear Energy Policy Study Group, 1977; Lewis *et al.,* 1978; U.S. National Academy of Sciences, 1979b). Accident possibilities at facilities other than reactors have been less thoroughly studied. In terms of actual risks, repositories for radioactive wastes probably are receiving more public attention than they deserve: the greater hazards occur *before* the material is placed in geologic storage (U.S. National Academy of Sciences, 1979b).

Routine emissions and exposures are better understood and generally more susceptible to control than are the problems discussed above. The uncertainty range on the health damages from ionizing radiation at low dose rates is not much more than a factor of 5 up and down from the "best estimates" used by standard-setting bodies, and these estimates indicate that routine emissions and exposures from the generation of a kilowatt-hour of electricity using nuclear fission probably do less damage than the routine emissions of air pollutants from generating a kilowatt-hour with coal or oil. The possible exception is the effects of radon gas from uranium-mill tailings and carbon-14 from reprocessing plants, integrated over the thousands of generations during which the emissions from a single year's nuclear power operations remain effective (U.S. National Academy of Sciences, 1979b). There is no accepted calculus for computing the "present value" of such drawn out effects, nor can one be confident that persistent contaminants from combustion of fossil fuels (specifically, trace metals such as mercuy and cadmium) do not produce similar consequences.

Renewables and environment

As Table IX suggests, renewable energy sources vary as widely in their environmental effects as in their technological and economic characteristics (Kendall and Nadis, 1980; El-Hinnawi and Biswas, 1981). Many make heavy demands on land, with potential damage to ecosystems. (Deforestation for fuelwood and the flooding of fertile valleys and wildlife habitat for hydropower reservoirs are very serious problems in this category today.) Some renewables make heavy demands on materials for constructing the equipment to harvest and process the energy, which leads to emissions of pollutatants and occupational injuries in materials acquisition and fabrication of components. Large-scale biomass plantations, if heavily irrigated, fertilized, and pesticided, could be very destructive ecologically (Harte and Jassby, 1978).

The available evidence suggests, however, that carefully selected and properly managed renewable energy options could make available large amounts of energy with considerably less troublesome environmental costs than either the solid fossil fuels or nuclear fission (Holdren *et al.,* 1980; Kendall and Nadis, 1980; Holdren, 1981a). This means not only that the environmental costs tend to be smaller per unit of energy delivered, but also that the biggest of these costs tend either to be relatively easy to bring into the economic balance sheets (e.g., compensation for occupational hazards being incorporated in the price of the energy) or at least to be borne by the beneficiaries of the energy use (e.g., the owner of a rooftop solar collector bearing the risk of damages if it leaks).

This last set of characteristics is in sharp contrast to those of the fossil fuels and nuclear fission, many of the most serious environmental costs of which are exported in space and time far beyond the users of the energy. If such exported costs are significant, the circumstance that the energy users neither bear them nor have to compensate the victims is both unjust and inefficient – the latter because users who do not pay the full costs of their consumption will not make socially efficient decisions about how much or what kinds of energy to use (Holdren, 1980). That carefully selected renewables permit a more equitable and efficient distribution of the environmental burdens of energy supply than do conventional energy options may prove, in the long run, to be a key advantage.

ENERGY AND WAR

The political, social, and environmental problem that dominates all others in its potential for damage to human well-being in the decades immediately ahead is thermonuclear war. The chances of avoiding such a war depend on progress in coping with a number of pernicious factors: ideological conflicts, religious hatreds, territorial disputes, destabilizing developments in armaments, and so on. Many of these factors have little or no direct connection with energy – that is, with how nations and regions choose to meet their energy needs. But there *are* links between energy and the prospects for war or peace – between energy and the roots of war, between energy and the proximate causes of war, between energy and the means of war. There are also ways in which the characteristics of a country's energy system – notably the degree of self-sufficiency, decentralization, redundancy, resilience of individual components, and back-up supplies – influence that country's *vulnerability* to military attack (see especially Clark and Page, 1981). The paragraphs that follow, however, focus only on the first set of linkages – the ways the energy situation affects (1) the likelihood of international conflict and (b) the chances that such a conflict will escalate toward thermonuclear war.

Energy and the roots of war

The rich-poor gap in energy mirrors and helps to maintain the gulf between rich and poor in nearly every measure of human well-being. Although, as emphasized above, the supply of energy itself should not be mistaken for an index of well-being, the poor countries cannot climb up from poverty without some amelioration of their energy woes. That amelioration is not a *sufficient* condition for eradicating the poverty that plagues nearly three quarters of the human race, but it is a *necessary* one: without it, the rich-poor gap will endure – a continuing source of frustration, despair, and, inevitably, social and political instability.

That the frustration and resentment of the disadvantaged is one of the roots of war can hardly be doubted. The most obvious connection is the historical one: regions of social and political instability have long proven irresistible targets of military intervention both by neighboring countries and by the superpowers; those interventions where superpower interests collide are sure to increase world tensions, and almost any one of them could, with bad luck, produce a miscalculation that leads to confrontation and war.

Until relatively recently, the second connection – the idea that the frustrated poor could themselves make war on the rich – was not taken seriously. But terrorist acts of increasing sophistication and audacity have begun to make plain the vulnerability of industrial society to attacks against which military might is no defense at all. Readers can spin out for themselves some of the scenarios by which concerted escalation of terrorist assaults – perhaps funded by governments and perhaps employing nuclear bombs – could lead to open war.

Energy and the proximate causes of war

The most obvious and (at least for the moment) most important way in which energy itself might be the immediate cause of war is military action to try to protect or deny access to Persian Gulf oil. (There is a large literature on this topic; for an introduction, see Deese and Nye, 1981). Such intervention might be stimulated, for example, by a revolution in a major supplier country or even by a local war that interrupts oil output to an extent importers find intolerable. While it seems bizarre, at best, that the United States might entertain the idea of going to war over the seven percent or so of its total energy supply that comes from the Persian Gulf, the far greater percentage dependence of Japan, France, and West Germany makes U.S. intervention on behalf of the Western industrial economies as a whole all too plausible. (No serious analyst seems to think that military action could possibly succeed in maintaining the oil flow if any significant adversary wants to cut it off, but that probably would not stop the attempt to do the impossible.)

In the future, regional energy resources other than Middle East oil may assume sufficient international importance to produce analogous problems. Also, there is the possibility of military conflict over energy resources of disputed ownership, notably oil in continental-shelf areas between nations and in deeper ocean waters subject to disputes about jurisdiction.

Finally, Israel's 1981 attack on an Iraqi nuclear reactor underlines the possibility that energy-related facilities with significant potential military applications may become targets of pre-emptive attack. Such an attack, of course, could set off a larger conflict. (That the Iraqi reactor was designed for research and training rather than actual power production does not blunt the general point.)

Energy and the means of war

Some energy technologies can provide the countries that develop and use them with increased capabilities for the development, production, or use of related weapons. The most widely recognized example of such a linkage is that between the spread of nuclear fission as an energy source and the spread of capabilities in nuclear weaponry. I treat this important case next, before touching more briefly on some other examples.

It is generally conceded that the knowledge needed to design and fabricate fission bombs is by now available to almost any nation (Gilinsky, 1977, Nuclear Energy Policy Study Group, 1977). Although the diffusion of knowledge that produced this situation was partly due to the promotion of fission power as a commercial energy source, the diffusion is already so complete that it is no longer given much weight as a liability of fission power. Lack of access to nuclear explosive materials, not lack of knowledge, has been the principal technical barrier to the spread of fission weapons capability for the last two decades. The continuing major threat to nuclear weapons proliferation posed by fission power is that it tends to provide this missing ingredient, either in the form of uranium enrichment capability or in the form of plutonium extractable from spent reactor fuel by means of chemical reprocessing.

Three main arguments have been used over the years by people contending that the link between fission power and fission power is not so serious. The first argument – that plutonium of the sort produced by power reactors is unsuitable for the fabrication of high-quality nuclear weapons – has been discredited publicly and repeatedly by professional weapons designers and high officials of the defense establishment. (See, e.g., Gilinsky, 1977; Nuclear Energy Policy Study Group, 1977. See Lovins, 1980, for a more technical review.)

The second argument is that there are more direct and cheaper ways for a country to acquire nuclear bombs than via its commercial nuclear energy facilities – centrifuges for uranium enrichment and special reactors dedicated to plutonium production are usually mentioned. This argument is technically correct but seriously misleading. Countries *can* get nuclear explosive materials by these other means, but doing so is made much easier if the requisite technical skills and infrastructure are already in place courtesy of a nuclear power program. The existence of commercial nuclear power in a country, moreover, provides a legitimating cover for nuclear activities which, without electricity generation as their manifest purpose, would be unambiguously weapons of oriented and thus subject both to internal dissent and external sanctions and countermeasures. Indeed, countries that initially have no intention of acquiring nuclear weapons might later find the built-in weapons capability that comes with nuclear power too tempting, particularly if their internal or external political circumstances change.

The third argument is that weapons proliferation is basically a *political* problem that must be handled – and is being handled – by such political measures as the Non-Proliferation Treaty (NPT), not by discouraging the use of nuclear power technology. That the problem is partly political is true, but the NPT offers little reassurance: many key countries have not ratified it; any of those that have are permitted to abrogate it with a few months' notice; and the safeguards that have been implemented to detect violations are hopelessly inadequate to the task (Moglewer, 1981; Richter, 1981).

It remains true that additional countries inevitably will acquire nuclear weapons, whether or not commercial nuclear power is available to facilitate the process. The key issue is the *rate* of spread of the weapons. If the proliferation problem is viewed as a race between the spread of nuclear weapons, on the one hand, and the growing (one hopes) effectiveness of political and moral barriers against the use of these weapons, on the other, then nuclear power's contribution to speeding up the spread of weaponry must be considered a very serious liability of this energy source.

Unfortunately, nuclear fission is not the only energy source with links to weapons capabilities. The inertial-confinement approach to harnessing nuclear fusion as an energy source is partly classified because of the principles it shares with fusion weaponry (Holdren, 1978); the spread of research on this technology is quite likely to accelerate the spread of fusion-weapons capability among nations already possessing the fission bombs needed to trigger thermonuclear weapons (Holdren, 1981b). Even solar energy may not be free of a potentially important weapons connection: the development and use of orbiting solar power satellites would enormously increase the scale of human operations in earth orbit, providing both the capability and the cover for correspondingly increased military uses of space.

Besides the links between new energy sources and new weaponry, there is a more pedestrian but perhaps equally threatening connection between contemporary energy circumstances and the means of war. I refer to a vicious circle associated with the very large money flows in the world oil market: the oil-exporting LDCs pour substantial amounts of their new-found revenue into conventional armaments, thus fueling regional arms races that increase both the chance of armed conflict and the chance that such conflict will expand in a way that draws in the superpowers; and the industrial oil-importers increase their exports of sophisticated armaments as rapidly as they can, driven in part by the need to offset the balance-of-payments deficits caused by their oil bills. Table X shows the phenomenon in quantitative terms. This situation is appalling enough in its contribution to the chances of regional and global war; but it is deplorable also for the diversion of desperately needed resources from the provision in support of human well-being.

TABLE X
Military expenditures and arms exports, 1971-1980.

All figures are in billions of 1980 U.S. dollars

	1971	1974	1977	1980
DC military expenditures	395	420	435	460
LDC military expenditures	94	112	128	150
DC arms exports	11	18	24	40
LDC arms imports	9	13	19	30

Source: U.S. Arms Control and Disarmament Agency (1980), Barnaby (1981).

ELEMENTS OF A STRATEGY

The main perils in today's energy predicament, as elaborated above, can be summarized rather concisely. The world is suffering simultaneously the consequence of having too little energy (these borne mainly at present by the poor, who have too little to meet basic human needs) and the consequences of having too much of especially costly or risky kinds (these borne by everyone in one form or another: excessive vulnerability to unpredictable loss of supply, increased chance of war arising from that vulnerability, increased chance that *any* war will be nuclear, and increasingly threatening disruptions of environmental conditions and processes crucial to human well-being).

What actions recommend themselves when, against the backdrop of this predicament, the available energy options are evaluated on the appropriate criteria? The answers will vary, of course, from one country and region to the next, depending on resources, pattern of needs, economic and social context, environmental conditions, and so on. But some elements of a sensible strategy apply nearly everywhere.

Increase efficiency

The cornerstone of a sensible energy policy in a world of costly energy must be increased efficiency of energy conversion and use. The term "energy conservation" is often used in this context, but to a physicist that term describes a law of nature that will be obeyed irrespective of policy, and to many members of the public it has a connotation of sacrifice – of belt-tightening – that is misleading and counterproductive. The needed program of increased efficiency entails not sacrifice but the use of human ingenuity to increase the contributions to well-being derivable from each litre of oil, each kilogram of coal, each kilowatt-hour of electricity.

How much energy can be saved, and how rapidly? Very detailed and rigorous studies of energy use in homes and offices, in industry, in agriculture, and in transport have shown for the United States and for many other nations that a suitable combination of efficiency measures could reduce the ratio of energy to real Gross National Produce by as much as half over the next twenty to thirty years. (See, e.g., Gibbons, 1978; Beijdorf, 1979; National Academy of Sciences, 1979c; Sant, 1979; Sawhill, 1979; Stobaugh *et al.*, 1979; Ross and Williams, 1981; Solar Energy Research Institute, 1981. Some of these analysts and others – see Lovins and Lovins, 1981 – now think even larger reductions are feasible.) A reduction of half in the ratio of energy to GNP means, of course, that a country can have twice the real GNP with the same energy use as before, or the same GNP with half the energy use.

One need not rely solely on projections for evidence of the potential of increased efficiency in response to higher energy costs, however. Schipper and Lichtenberg (1975) showed in a pioneering study how historically higher energy prices in Sweden as compared to the United States produced a ratio of energy to GNP some thirty percent lower in the former country than in the latter. Subsequent studies of other industrial nations revealed a similar pattern of flexibility in the ratio of energy to GNP, depending on energy price and other factors (Darmstadter *et al.*, 1978; Schipper, 1979). And data published by the U.S. Department of Energy (1981) show that, between the initial oil-price rise of 1973 and the first half of 1981, the ratio of energy to real GNP in the U.S. fell by more than 17 percent.

It is natural to suppose, of course, that the potential for gains from increases in efficiency is much smaller in the LDCs. The *absolute* potential savings are indeed smaller, because the total amount of energy used is so much less than in the industrial nations. But the *relative* savings – the reduction in the ratio of energy to service – may be as big or bigger than in the industrial nations. Nearly all the LDCs have industrial enterprises and commercial buildings (including, for example, tourist hotels), and the energy-efficiences in these sectors are at least as low as in the industrial-nation counterparts. Often, they are lower, because the chronic capital shortage in LDCs discourages the higher initial investments usually associated with the more energy-efficient technologies. (See, e.g., Howe and Tarrant, 1980; Schipper, 1980.) In the rural sectors of many LDCs, where the main energy sources are biomass materials (fuelwood, crop wastes, dung), the fraction of the fuel's energy actually applied to the task typically is only five to ten percent (Makhijani and Poole, 1974; Hughart, 1979; Newcombe *et al.*, 1979). Modest improvements in the way this fuel is burned could triple or quadruple the service derived from the biomass energy already being used in the rural sectors of the LDCs.

Like other energy options, of course, increasing the efficiency with which energy is converted is not a free of costs and difficulties – it costs money, takes time, in some cases poses new hazards to health and safety. (The emissions from fuel-efficient diesels may be more hazardous than those from spark-ignition engines, some insulation materials are highly toxic, reduced ventilation rates to decrease energy use for heating and cooling can increase indoor air pollution, and so on.) The implementation of programs to increase efficiency may pose difficult challenges to institutions. And, as with other energy sources, there is generally some uncertainty about the exact rate of return a given investment in increased efficiency will produce. The attractiveness of the efficiency option, then, is not that it is cost-free, risk-free, or uncertainty-free – it is simply that it is less costly, less risky, and less uncertain than any of the other options that are available. It is the rational response to the high cost of everything else.

Exploit local resources

The development and use of locally available energy resources makes particularly good sense in the LDCs, where transport systems and electric transmission grids generally are neither extensive enough nor reliable enough to permit the needs of most of the population to be met by imported or centralized energy supplies. Cost savings obtained by minimizing the transport of energy, moreover, can at least partly offset the higher costs that may be associated with using lower-quality local resources (e.g., smaller, leaner fuel deposits; less intense sunlight; more intermittent wind) as opposed to higher-quality remote ones.

Most LDCs have been far less thoroughly explored for oil and gas than have the major oil-producing regions, yet, as noted above, the quantities of these fuels already discovered there are by no means negligible when measured against the scale of *local* – as opposed to world – needs. At today's costs for other energy supplies, many of these small deposits – previously considered economically uninteresting – would repay the trouble and cost needed to exploit them. Effort spent looking for more also could yield considerable rewards (Howe and Tarrant, 1980; World Bank, 1980). Similarly, there is probably far more short-term potential in identifying and developing a myriad of small hydropower and geothermal sites than in focusing on the much less numerous gigawatt-scale possibilities.

In the industrial nations, too, there are far more opportunities than generally recognized for developing relatively small-scale energy supplies based on local resources and matched to local needs. Here the transport and transmission facilities to permit the use of more centralized supplies do exist, but transmission and transport costs are nevertheless high enough that avoiding them through the use of local supplies may pay off in some cases. Indeed, the whole question of economies versus dis-economics of scale and centralization – raised so provocatively and effectively by Lovins (1976, 1977; Lovins and Lovins, 1981) – deserves more systematic re-examination.

Promote diversity in energy supplies

A certain amount of diversity in energy supplies tends to emerge naturally from the search for the most economic mix of resources and technologies to meet an inevitably diverse array of energy end-use needs. But the nonmonetary costs and risks of energy supply provide incentives for countries to promote deliberately a greater degree of diversity of supplies than would result from economic considerations alone.

One such incentive is the vulnerability of oil importers to

cutoffs in that supply. Clearly, it is in the interest of oil-importing countries to spread that risk by buying their oil from a variety of exporters, preferably in a variety of different regions of the world, even when this means higher direct monetary costs, higher transactions costs, and so on. Similarly, the vulnerability of imports warrants larger investments in domestic energy sources than could be justified simply by comparing their monetary costs to the present cost of imported oil. Energy supplies other than oil are also subject to interruption, of course – coal by strikes, hydropower by drought, nuclear power by accidents or the fear of them – and not putting too many energy eggs in any one basket is a sensible safeguard against such vulnerabilities.

Indeed, diversity can be regarded as a form of insurance against uncertainty of all kinds. We do not know with much accuracy either the future monetary costs of most of our options, or their environmental costs, or their degree of "fit" with society's future needs and values. Yet the amount of time needed to change course if today's guesses prove to be wrong – the time needed to replace one major energy source with another – is dangerously long because of the inertia associated with the scale and complexity of the energy system. This is the strongest argument for developing capacity in alternative energy supplies before they are "competitive" in a narrow economic sense.

As with most insurance policies, diversity can be expensive, and we cannot afford an infinite amount of it. No society can afford to develop and deploy all the energy options theoretically available to it. Diversity is a reasonable argument for preserving a number of options, therefore, but it is not an argument for preserving all of them. As in nature, diversity in energy technologies is desirable in part because it allows for eventual rejection of those that prove least suitable.

Avoid or control weapons linkages

The link between the spread of nuclear power technology and the spread of the capability to build nuclear weapons is the strongest argument of those who contend that this particular energy option is so plainly unsuitable that it should be rejected "with all deliberate speed" (Lovins and Lovins, 1980). Even many of the fission option's most ardent supporters have admitted that international control over the most "sensitive" facilities is virtually a prerequisite if expanded global use of this energy option is to be seriously entertained. Yet many students of *political* reality think that the required degree of internationalization is unattainable in this century, if ever.

What, then, is one to conclude? It seems plain that the required degree of international control must be sought, notwithstanding the presently dismal prospects for attaining it. But, given the low probability of success and the quite possibly intolerable costs of failure, it is equally clear that no possibilities for reducing rather than increasing the world's dependence on nuclear fission should be overlooked. There is no doubt that "the genie is out of the bottle" to the extent that neither persuasion nor coercion alone could stop its spread – even in the unlikely event that the present main suppliers of nuclear technology were to agree on that goal. The best hope – the only hope – of reducing its use is the continued development and demonstration of alternatives (notably, in the near term, increased efficiency and diverse and largely decentralized renewable and fossil resources) that meet society's energy needs more efficiently and compatibly than fission can.

With respect to more embryonic energy technologies with deeply troubling links to weapons capabilities, most notably inertial confinement fusion but possibly also solar-power satellites, it can only be hoped that society finds a way to ask whether the benefits justify the risks *before* these genies, too, are irrevocably out of the bottle.

Cooperate to shrink the rich-poor gap

In response to suggestions that the United States and other industrial nations should assume greater responsibilities in helping the LDCs cope with their staggering problems, U.S. President Reagan has said, in essence, "The poor must solve their own problems." It is true that, in many important respects, the solutions to the problems of the LDCs must come from within. But to carry this notion to the extremes being pursued by the Reagan Administration – withdrawal of U.S. support for an energy-lending affiliate of the World Bank, refusal to support significant initiatives for North-South cooperation at the 1981 United Nations Conference on New and Renewable Energy

Sources, refusal to consider raising U.S. concessional aid to LDCs above its present 0.2 percent of GNP (one of the lowest figures in the industrial world) – is to ignore both the role of the industrial nations in creating the energy predicament of the poor and the mutual interests of rich and poor in ameliorating it.

That web of interests extends far beyond humanitarian concerns (although it would be nice if those were sufficient): it includes the links between poverty and the prospects for instability and war, dicussed above, it includes the existing and potential role of the LDCs as markets for the output of industrial economies; it includes the threat that LDC insolvency poses to the already overburdened and fragile world financial system. (A particularly useful exposition of the interlocking energy interests of industrial nations and oil-exporting and oil-importing LDCs has been provided by Hoffmann and Johnson, 1981.)

The list of opportunities for rich-country action in response to these interests is long. Direct transfers of capital are needed to finance the measurement and development of indigenous energy sources in the LDCs – oil, gas, geothermal, hydropower, wind. Management skills should be shared and transferred, to help extract more productivity from the human and natural resources available. Joint research projects involving industrial-nation and LDC scientists and technologists should be sponsored, to devise technologies for energy harvesting, conversion, and use matched to the needs and contexts of the LDCs.

More radical notions should also be entertained. One such would be a global compact restricting the export of armaments, thus reducing the opportunities for LDCs to spend precious resources in so counterproductive a way, while forcing the arms manufacturers to divert their talents to constructive rather than destructive pursuits. Another would be the adoption of the principle that any development of the resources of the global commons (the seabed, space, Antarctica), should be managed so as to give preferential treatment to the worst-off member in the commons – namely, the LDCs. (The Law of the Sea negotiations flirted with this concept, with very limited success. Yet, outside the international arena, the idea is not really so radical: a version of it is in force in nearly all professional sports, where the last-place teams get first choice of the new talent available in each year's "draft". How odd that athletic leagues should have recognized the merits of minimizing the gap between the best-off and worst-off groups, while the international community has not.)

Restrain population growth

The mid-twentieth-century revival of Malthus' insight that no combination of good technology and good management can cope with unlimited population growth on a finite planet (Osborne, 1948; Brown, 1954) is more relevant in the 1980s than ever. One sees, in the rising real costs of energy and the poor prospects for a reversal, the end of the illusion that technical fixes inevitably will outpace the effects of declining resource quality. And one sees, in the accumulating evidence of environmental disruption at continental and global scale – deforestation, erosion, salination, species extinctions, carbon dioxide buildup, intervention in nutrient cycles, groundwater depletion and contamination – the end of the companion illusion that human activities are but puny perturbations on the scale of natural processes.

Restraining to the greatest extent practical the further increases in world population will not by itself solve these problems, but it may make them solvable. The tired and discredited notion that the opposite is true – that further population growth will be a boon because "people are the ultimate resource" – is being promulgated again in the 1980s by a euphoric economist who writes that he rescued himself from a deep depression with this insight and wishes to share the good news (Simon, 1981); flying in the face of logic and the evidence, this view is dangerous nonsense.

The adverse effects of further population growth on the human prospect could hardly be plainer than they are in the energy area. Coping effectively with the energy problems of 1980's 11-terawatt world – 4.5 billion people at 2.4 kilowatts per person – manifestly has been beyond the technical, economic, and organizational capabilities available. Yet, for every billion people added to the world's population *at that same level of energy use per person,* new energy sources capable of sustaining an additional continuous drain of 2.4 terawatts must be mobilized and paid for – and their environmental impacts somehow absorbed.

This figure may well understate the problem. To narrow

the rich-poor gap in prosperity at a satisfactory rate probably will require some increase in energy use per person, even assuming that large gains in energy efficiency are rapidly achieved. Table XI illustrates the point with scenarios covering the time period 1980-2020. The "high population" and "low population" cases shown there probably span the realistic possibilities for population growth, barring catastrophe. (For a discussion of the factors that will prevent a more rapid slowing of world population growth, see Ehrlich *et al.,* 1977). The assumed rates of increase in the efficiency with which energy is used reduce the DC and LDC ratios of energy to real GNP to 45 and 49 percent of their 1980 values by 2020. The ratio of DC to LDC GNP per person falls from about 13 in 1980 to about 5 in 2020. Energy use per person rises from about 2.4 kilowatts in 1980 to about 2.6 kilowatts in 2020. If GNP per person grows any more rapidly than assumed or if energy efficiency grows any more slowly, the increase in energy use per person will be larger.

Under the stated assumptions, the *difference* in the 2020 rate of energy use between the high-population-growth and low-population-growth scenarios is 5 terawatts – equal to the total world rate of energy use for all purposes in 1960, and considerably larger than world petroleum use in 1980. Given, moreover, that the additional terawatts "on the margin" will be supplied from the most costly and probably most environmentally disruptive sources, the importance of the incremental contribution of extra population growth will be even greater than that suggested by simple addition. And, under less optimistic assumptions about energy efficiency, the energy consequences of the higher population growth rate would be still worse.

The world will need all the clever technology, management skill, and good luck it can muster if 7.3 billion people are to be provided with the energy they need in 2020 – not to mention the food, water, housing, health care, education, and so on – at tolerable monetary, environmental, and social costs. Anyone who claims that trying for 9.3 billion would do no great harm simply doesn't understand the situation.

D

TABLE XI

The effect of population in a high-efficiency energy future.

Assumptions: Average annual percentage growth of key factors.		
	1960-1980[a]	1980-2000
World population (DCs/LDCs)		
Low population scenario	1.4/2.3	0.4/1.7
High population scenario	1.4/2.3	0.8/2.5
Real GNP per person (DCs/LDCs)	4.1/2.3	2.0/4.0
Energy per real GNP (DCs/LDCs)	-1.0/-0.6	-2.0/-1.5

Outcomes: Mid-year values of key variables		
	1980	2000
World population, billions (DCs/LDCs/total)		
Low population scenario	1.2/3.3/4.5	1.3/4.6/5.9
High population scenario	1.2/3.3/4.5	1.4/5.4/6.8
Energy use per person, kw[b] (DCs/LDCs)	6.9/0.8	7/1.3
GNP per person, 1980$ (DCs/LDCs)	7,600/600	11,300/1,300
Total energy use, TW[c] (DCs/LDCs/total)		
Low population scenario	8.3/2.7/11.0	9.1/6.0/15.1
High population scenario	8.3/2.7/11.0	9.8/7.0/16.8

[a] Historical data, for comparison (World Bank, 1981)
[b] kW = kilowatts
[c] TW = terawatts

EPILOGUE

It is often pointed out that the "elements of a strategy" described above are going to be difficult to put into practice – in some cases perhaps impossible. Indeed, proponents of such approaches are regularly proclaimed "unrealistic". Of the difficulty of implementing the present proposals there can be little doubt, but it is necessary to ask what the "realists" offer as alternatives. If their answer is the usual one – an extension of or extrapolation from the status quo – one may be forgiven for wondering whether they have grasped the perils of the present human predicament at all.

DISCUSSION POINTS

1) Dr. Makhijani raised the difficulty of measuring the benefits and costs of Figure I, and discussed the meaning of GNP. In reply, Professor Holdren emphasised that the curve did not refer to GNP, which is not a good measure of wellbeing. The curve was intended to make one point only, that the costs of additioned energy supplies will eventually be greater than the benefits, and a sensible society would not go beyond that point.

It is however important to consider the meaning of GNP since this kind of parameter is basic to all models and economic variables are expressed in monetary terms. In market, capitalist economies GNP has a specific meaning, that is the amount and rate at which labour is monetised. Thus if you cook at home this is not part of the GNP whereas if you eat at a restaurant it is, and if you beat your wife it is not part of the GNP whereas to build a MX missile is.

A comparison of Brazil and Bangladesh illustrates the limited value of GNP as a measure of wellbeing. Although in Brazil the GNP has increased in monetary terms, the condition of 50% of the population of the Brazilian people has deteriorated and is comparable to that of the poor people of Bangladesh. The condition of 80% of the Brazilian people has not changed in economic terms, their environment is much worse and to achieve the economic growth whose profits are experienced by 20% of the internal population and by certain sections outside Brazil, most of the people of Brazil have to confront a level of repression which is much higher than it was before the economic growth took place.

2) Dr Hrabovszky pointed out that according to Figure III, the rate of primary energy use is an ever-increasing function of time, and Dr Gallopin thought that a phase IV of lower energy use should be added to the graph. It seems to be assumed that developing countries must move through the successive phases with time. This would ignore the impact of better technology and a more suitable social organisation, and it seems that what is stated is that the poor countries must follow the same predetermined path as that followed by the industrialised countries. The main point of confusion is that time is not a good choice for the horizontal axis. Either this is one possible historical path or a cross section across countries today. The graph should not be viewed as normative.

Professor Holdren agreed that developing countries do not have to follow the same path or make the same mistakes. Also some countries like the USA are quite likely beyond the optimum point. The total energy use has been going down in the USA for some years now, and if economic effects are normalised out, it is found that the energy use per unit out put has been declining quite drastically.

3) Dr Gallopin emphasised the need for caution in defining criteria for the choice of energy sources in relation to stages of development, so that a country may not in the future find itself locked into an inappropriate path. Thus in Latin America as a whole the inheritance of the past and present style of development is an adaptation to a pattern of development based on oil resources. Because of the forces of inertia it is much more difficult to change to cheaper, more national and sustainable energy sources than it is for the industrialised countries. One of the criteria that is particularly important for developing countries in choosing energy sources should be an inherent compatibility with food production.

Professor Holdren replied that he had emphasised criteria that have dominated in each phase, but had also noted that energy sources must be compatible with other social goals. For example, beyond economic criteria one should diversify as an insurance policy against mistakes.

4) Dr Hrabovszky disagreed on the link between nuclear power and nuclear weapons, since nuclear power is not the best route to get to nuclear weapons.

Professor Holdren emphasised that he was referring to a nuclear weapons capability. To develop a weapons programme from scratch without reference to a nuclear power programme requires a political decision and an open programme subject to internal sanctions and to external responses. It is almost impossible to conceal the facilities required to develop nuclear weapons with the dedicated technology for that purpose. If however a country has a nuclear power programme it has in place the technical infrastructure and almost all the facilities to start a weapons programme any time it choose, and it can postpone that decision while maintaining the option, and can have that option without any marginal economic or political cost.

References

Anderson, Kent (1981). *A Comparison of Various Energy Conservation, Energy Supply, and Passive Solar Design Alternatives for Space Conditioning in New Residences.* Report to the California Energy Commission. Berkeley, CA: University of California, Energy and Resources Group. July. 118 pp.

Barnaby, Frank (1981). World arsenals in 1981. *Bulletin of the Atomic Scientists* 37 (7): 16-20 (August/September).

Beijdorff, A.F. (1979). *Energy Efficiency.* London: Shell International Petroleum Co. Ltd, Group Planning. April.

Bogue, Donald J. (1969). *Principles of Demography.* New York: Wiley. 917 pp.

British Petroleum (1981). *BP Statistical Review of the World Oil Industry.* London: British Petroleum Co. Ltd.

Brown, Harrison (1954). *The Challenge of Man's Future.* New York: Viking.

Budnitz, R.J. and J.P. Holdren (1976). Social and environmental costs of energy systems. *Annual Review of Energy* 1: 553-580.

Clark, Wilson and Jake Page (1981). *Energy, Vulnerability, and War.* New York: Norton. 251 pp.

Cook, Earl (1976). *Man, Energy, Society.* San Francisco, Calif.: W.H. Freeman and Co. 478 pp.

Cottrell, Fred (1955). *Energy and Society.* New York: McGraw Hill.

Daly, H.E. (1977). *Steady-State Economics.* San Francisco, Calif.: W.H. Freeman and Co. 185 pp.

Darmstadter, J., J. Dunkerley and J. Alterman (1978). International variations in energy use: findings from a comparative study. *Annual Review of Energy 3:* 201-224.

Deese, David A. and Joseph S. Nye, eds (1981). *Energy and Security.* Cambridge, MA: Ballinger. 489 pp.

Deudney, Daniel (1981). *Rivers of Energy: The Hydropower Potential.* Worldwatch Paper 44. Washington, D.C.: The Worldwatch Institute. 55 pp.

Eckholm, Erik P. (1976). *Losing Ground.* New York: Norton. 223 pp.

Ehrlich, P.R. and A.H. Ehrlich (1981). *Extinction: On the Causes and Consequences of the Disappearance of Species.* New York: Random House. 305 pp.

Ehrlich, P.R., A.H. Ehrlich and J.P. Holdren (1977). *Ecoscience: Population, Resources, Environment.* San Francisco: W.H. Freeman and Co. 1051 pp.

El-Hinnawi, Essam and Asit K. Biswas, eds. (1981). *Renewable Sources of Energy and the Environment.* Dublin: Tycooly International Publishing, Ltd. 219 pp.

Ellis, A.J. (1981). Geothermal energy. In E. El-Hinnawi and A.K. Biswas, eds., *Renewable Sources of Energy and the Environment.* Dublin: Tycooly International Publishing Ltd. pp. 25-55.

Flavin, Christopher (1981). *Wind Power: A Turning Point.* Worldwatch Paper 45. Washington, D.C.: Worldwatch Institute. 56 pp.

Gibbons, John H., Chairman. Demand and Conservation Panel of the Committee on Nuclear and Alternative Energy Systems (1978). U.S. energy demand: some low energy futures. *Science 200:* 142-152 (14 April).

Gilinsky, Victor (1977). Plutonium, proliferation, and policy. *Technology Review* 79 (4): 58-65 (February).

Haefele, W. and W. Sassin (1977). The global energy system. *Annual Review of Energy* 2: 1-30.

Harris, DeVerle P. (1979). World uranium resources. *Annual Review of Energy* 4: 403-432.

Harte, J. and M. El-Gasseir (1978). Energy and water. *Science* 199: 623-634. 10 February.

Harte, J. and A. Jassby (1978). Energy technologies and natural environments: The search for compatibility. *Annual Review of Energy* 3: 101-46.

Hayes, Denis (1977). *Rays of Hope: The Transition to a Post-Petroleum World.* New York: Norton. 240 pp.

Holdren, J.P. (1978). Fusion energy in context: Its fitness for the long term. *Science 200:* 168-180.

Holdren, John P. (1980). *Integrated Assessment for Energy-Related Environmental Standards: A Summary of Issues and Findings.* Report LBL-12779. Lawrence Berkeley Laboratory. Springfield, VA: National Technical Information Service. October. 59 pp.

Holdren, John P. (1981a). Energy and the human environment. In G.T. Goodman, L.A. Kristoferson, and J.M. Hollander, eds., *The European Transition From Oil.* London: Academic Press. pp. 87-120.

Holdren, J.P. (1981b). Fusion-fission hybrids: environmental aspects and their role in hybrid rationale. *Journal of Fusion Energy* 1: 197-210.

Holdren, J.P. and P.R. Ehrlich (1974). Human population and the global environment. *American Scientist* 62: 282-292.

Holdren, J.P., Gregory Morris and Irving Mintzer (1981). Environmental aspects of renewable energy sources. *Annual Review of Energy* 5: 241-291.

Hoffman, Thomas and Brian Johnson (1981). *The World Energy Triangle: A Strategy for Cooperation.* Cambridge, MA: Ballinger. 219 pp.

Howe, James W. and James J. Tarrant (1980). *An Alternative Road to the Post-Petroleum Era: North-South Cooperation.* Monograph No. 14. Washington, D.C.: Overseas Development Council. 74 pp.

Hubbert, M. King (1969). Energy resources. In U.S. National Academy of Sciences, Committee on Resources and Man. *Resources and Man.* San Francisco: W.H. Freeman, pp. 157-242.

Hughart, David (1979). *Prospects for Traditional and Non-conventional Energy Sources in Developing Countries.* Washington, D.C.: World Bank. 132 pp. Staff Working Paper No. 346.

IIASA. (1981). *Energy in a Finite World.* Report of the Energy Systems Program of the International Institute for Applied Systems Analysis. Cambridge, MA: Ballinger. 225 pp.

Kemeny, John G., Chairman (1979). *The Need for Change: The Legacy of TMI.* Report of the President's Commission on the Accident at Three Mile Island. Washington, D.C.: Government Printing Office. October. 179 pp.

Kendall, Henry W. and Steven J. Nadis, eds. (1980). *Energy Strategies: Toward a Solar Future.* Cambridge, MA: Ballinger. 320 pp.

H.W. Lewis, R.J. Budnitz, H.J.C. Kouts, W.B. Lowenstein, W.D. Rowe, F. von Hippel and F. Zachariasen (1978). *Risk Assessment Review Group Report to the Nuclear Regulatory Commission*, NUREC/CR-0400, U.S. Nuclear Regulatory Commission, Washington, D.C., September.

Likens, G.E., R.F. Wright, J.N. Galloway and T.J. Butler (1979). Acid rain. *Scientific American* 241 (6): 43-51. October.

Linden, Henry R. (1980). *1980 Assessment of the U.S. and World Energy Situation and Outlook.* Chicago: Gas Research Institute. 75 pp.

Lovins, A.B. (1976). Energy strategy: The road not taken. *Foreign Affairs* 55: 65-96.

Lovins, A.B. (1977). *Soft Energy Paths: Towards a Durable Peace.* Cambridge, MA: Ballinger Publishers. 231 pp.

Lovins, A.B. (1980). Nuclear weapons and power-reactor plutonium. *Nature* 283: 817-823. February.

Lovins, A.B. and L.H. Lovins (1980). *Energy/War: Breaking the Nuclear Link.* San Francisco: Friends of the Earth.

Lovins, A.B. and L.H. Lovins (1981). *Energy Policies for Resilience and National Security.* Washington, D.C.: Federal Emergency Management Agency.

Makhijani, A. and A. Poole (1974). *Energy, Agriculture, and the Third World.* Cambridge, MA: Ballinger.

Meyers, R.F. and W.D. Dietzmann 91979). World geography of heavy crude oils. *Proc. 1st International Conference on the Future of Heavy Crude and Tar Sands* (Edmonton, Alberta, Canada, 4-12 June 1979), Ch. 2. 16-28.

Moglewer, Sidney (1981). IAEA safeguards and nonproliferation. *Bulletin of the Atomic Scientists* 37 (8): 24-29. October.

Morgan, M.G., S.C. Morris, A.K. Meier and D.L. Shenk (1978). A probabilistic methodology for estimating air pollution health effects from coal-fired power plants. *Energy Systems and Policy* 2: 287-310.

Morrison, Denton E. and Dora G. Lodwick (1981). The social impacts of soft and hard energy systems. *Annual Review of Energy* 6: 357-378.

Nader, L. and S. Beckerman (1978). Energy as it relates to the quality and style of life. *Annual Review of Energy* 3: 1-28.

Newcombe, Ken with Kaye Bowman, Marion Christie, and James Pokris (1980). *Energy for Development.* Canberra, Australia: Center for Resource and Environmental Studies, for UNESCO. 242 pp.

Nuclear Energy Policy Study Group (1977). *Nuclear Power Issues and Choices.* Cambridge, MA: Ballinger. 418 pp.

Office of Technology Assessment, U.S. Congress (1979). *The Direct Use of Coal.* Washington, D.C.: Government Printing Office. 411 pp.

Osborne, Fairfield (1948). *Our Plundered Planet.* Boston: Little, Brown.

Richter, Roger (1981). Testimony from a former safeguards inspector. *Bulletin of the Atomic Scientists* 37 (8): 29-31. October.

Ross, Marc H. and Robert H. Williams (1981). *Our Energy: Regaining Control.* New York: McGraw-Hill. 354 pp.

Sant, Roger, W. (1979). *The Least-Cost Energy Strategy: Minimizing Consumer Costs through Competition.* Arlington, VA: Energy Productivity Center of the Mellon Institute.

Sawhill, John C., ed. (1979). *Energy Conservation and Public Policy.* Englewood Cliffs, N.J.: Prentice-Hall. 259 pp.

Schipper, Lee (1979). Energy use and conservation in industrialized countries. In John C. Sawhill, ed., *Energy Conservation and Public Policy*. Englewood Cliffs, NJ: Prentice-Hall. pp. 46-78.

Schipper, Lee (1980). *Energy Demand and Conservation in Kenya: Initial Appraisal.* Report LBL-10538. Berkeley, CA: Lawrence Berkeley Laboratory of the University of California, Energy and Environment Division. 27 pp. March.

Schipper, Lee and A.J. Lichtenberg (1975). Efficient energy use and well-being: The Swedish example. *Science* 194: 1001-1013. December.

Simon, Julian L. (1981). *The Ultimate Resource.* Princeton, N.J.: Princeton University Press.

Solar Energy Research Institute (1981). *A New Prosperity: Building a Sustainable Energy Future.* The SERI Conservation/Solar Study. Andover, MA: Brick House Publishing Co.

Sorensen, Bent (1979). *Renewable Energy.* London: Academic Press. 683 pp.

Stobaugh, R., D. Yergin, I.C. Bupp, M. Horwitch, S. Koreisha, M.A. Maidique and F. Schuller (1979). *Energy Future.* Report of the Energy Project at the Harvard Business School. New York: Random House. 353 pp.

Taylor, T.B. (1978). Low cost solar power. In *Nuclear Energy and Alternatives,* ed., O. Kadigaroglu, A. Perlmutter and L. Scott. Cambridge, MA: Ballinger, Ch. 44.

U.S. Arms Control and Disarmament Agency (1980). *World Military Expenditures and Arms Transfers 1969-1978.* Washington, D.C.: Government Printing Office. 166 pp.

U.S. Council on Environmental Quality (1981). *Global Energy Futures and the Carbon Dioxide Problem.* Washington, D.C.: Government Printing Office. 92 pp. January.

U.S. Department of Energy, Energy Information Administration (1981). *Monthly Energy Review, September.* Washington, D.C.: Government Printing Office. 101 pp.

U.S. National Academy of Sciences, Committee on Sulfur Oxides (1978). *Sulfur Oxides.* Washington, D.C.: National Academy of Sciences. 209 pp.

U.S. National Academy of Sciences, Committee on Nuclear and Alternative Energy Systems (1979a). *Energy in Transition 1985-2010.* San Francisco: W.H. Freeman. 783 pp.

U.S. National Academy of Sciences, Committee on Literature Survey of Risks Associated With Nuclear Power (1979b). *Risks Associated With Nuclear Power:* Summary and Synthesis. Washington, D.C.: National Academy Press. April. 156 pp.

U.S. National Academy of Sciences, Panel on Demand and Conservation of the Committee on Nuclear and Alternative Energy (1979c). *Alternative Energy Demand Futures to 2010.* Washington, D.C.: National Academy Press.

U.S. Nuclear Regulatory Commission (1975). *Reactor Safety Study.* Rep. No. WASH-1400 (NUREC-75/014). Springfield, VA: National Technical Information Service. October.

Willrich, Mason and T.B. Taylor (1974). *Nuclear Theft: Risks and Safeguards.* Cambridge, MA: Ballinger.

World Bank (1980). *Energy in the Developing Countries.* Washington, D.C. 92 pp.

World Bank (1981). *World Development Report 1981.* Washington, D.C.: World Bank. August. 192 pp.

3.3 Energy, Agriculture and Rural Development

J.P. HRABOVSZKY

INTRODUCTION

In the era when humanity relied predominantly or exclusively on renewable sources of energy for its activities, rural areas were not only self-sufficient in energy, but in most situations represented a "net exporting situation." The rapidly growing use of fossil fuels in the last 100 years has changed this situation radically in most countries of the world, and the recent revaluation of fossil sources of energy demand a serious effort for analysis of future directions for energy use in agriculture and production of energy from agriculture. With the exception of hydropower, agriculture (including forestry) is the main provider of renewable energy for mankind, and, in many developing countries, especially those at the very early stage of development, it provides the bulk of total energy used in society.

In many drier parts of the developing world, the shortage of fuelwood and other fuel sources for cooking and heating have reached a severity which is as dramatic as the shortage of food. Thus the pressure is mounting not only from rising oil prices, but also from dwindling forest resources.

In this paper three of FAO's major analytical efforts are described in the area which links energy, agriculture and rural development. These are: the energy related analysis of the "Agriculture: Toward 2000" Study (AT 2000); the analyses completed for the FAO Expert Consultation on Energy Cropping Versus Food Production; and FAO's analysis and policy development with respect to wood for energy.

Before providing a description of these models, their results and the policy implications of their findings, first a short section is provided on the role of the concept of Rural Energy Systems in analysing energy issues in rural areas and in formulating development programmes in relation to them.

RURAL ENERGY SYSTEMS

Whether one analyses an individual village or the whole rural sector of a continent, the multiplicity of participants, often in the roles of both producer and user of various sources of energy, the flows of energy within the system and the alternative sources of energy and technologies for their use all point to the need to utilize rural energy systems as an organizing concept and as an analytical tool.

Three main components are usually recognized within such a rural energy system. The first is the sub-system of agricultural production with both its demand and output of energy. On the demand side this uses both non-renewable and renewable energy sources. It is fashionable these days to speak of energy input-output ratios, but in our view this has only limited application. The explicit role of agriculture is to convert "cheap and abundant" energy from the sun into products with energy contents which are of value to mankind. In this process it uses energy from many sources, including those it produces itself. In the final analysis the efficiency of agriculture in energy use needs to be measured in value terms because of the great variability in the unit value of the energy it consumes and produces. On the output side, in terms of value the main energy output of agriculture is food energy, but if one measures it in physical energy units then it is forest products.

The changing relative price of commercial energy sources is now influencing agriculture to adjust the use of energy in its production processes and, if one looks at the wider food system, then also in the transport, storage and processing stages. The potential adjustments could be seen to move in the directions of: making more efficient use of commercial energy, by various means, probably each involving some measure of substitution of other inputs or sources of energy for them; an explicit substitution of commercial energy by energy from renewable sources; general reduction of energy use in the production/processing system. A good example of the first type of adjustment could be efforts to increase the efficiency of inorganic fertilizer use by more careful placement, timing, and dosage of fertilizers, together with a finer tuned assessment of plant nutrient needs. These would require more labour and more skills and greater knowledge to enable the savings of fertilizers. A more direct substitution example is the use of farmyard manure or green manure for inorganic fertilizers. The two main driving forces behind these substitution efforts would be changes in relative prices and changes in technology which makes one input more attractive versus the other. It is useful to recall here that the great fertilizer "explosion" of the post-World War II years came about partly because of the low price of oil, but also because technology of nitrogenous fertilizers made major strides towards reducing costs. Fertilizer responsive varieties added to these forces calling for greatly increased use of fertilizers. This was reflected in the 12-14 percent annual growth of fertilizer use in developing countries during the 1960s.

The other main energy use in agriculture is for power inputs. Basically these can come from three major sources: human labour, animal labour, and machine power. The historical trend has been for larger volumes of power to be applied per unit of land, and to replace hand labour, at first by draught animals, and later by machine power. These historical substitution processes have also demonstrated that straightforward "energy unit substitutions (such as HP, KW or joule) do not measure well the actual substitution ratios, or, alternatively, if measured in these units, these ratios are highly varied and change rapidly over time and over ranges of substitution. Given these relations, the measurement of input/output relations in energy use for agricultural production becomes a difficult undertaking and many past efforts seem to have yielded highly unrealistic results.

Many fears are expressed these days concerning the potentially great reduction in agricultural production as a result of increases in commercial energy prices, and thus reduced use of energy intensive cash inputs into agriculture. It is often pointed out that because energy prices are rising fast, agriculture will not be able to use the needed inputs, production growth will not keep pace with food demand and prices of food will explode. Eight years have passed since the major price explosion for energy, and, after the shock of the 1974 food crisis, prices of food have returned to normal or below normal levels, as measured in relative prices to other goods and services. One could argue that this is a necessary corollary of the much lower commercial energy share in inputs used for agricultural production as compared with most other industries and services. Furthermore, as is pointed out above, there exist many opportunities for making much more effective use of energy-rich cash inputs, and their rising prices create the necessary incentives for this to happen. It is also important to remember that, for example in the cost of production of fertilizers only a relatively low share (about 25-30 percent– is from energy costs and the rest is much less responsive to energy prices, but often reflect pressures for higher rewards for labour and to capital. What often seems to be a more vexing problem, for developing countries at least, is to find the necessary foreign exchange with which to finance imports of energy inputs into agriculture. Recent years have seen a slow growth or even decline in the prices of tropical export products and thus for a large share of export earnings for the majority of developing countries. Under these circumstances there is a general tendency to allocate the limited imported commercial energy sources to non-agricultural sectors, and thus try to force more non-commercial replacements for them in agricultural production, or to reduce agricultural production. Given the rapidly rising demand for food in the developing world, these do not prove to be tenable solutions. Therefore a priority treatment of agriculture's limited energy requirements seems to be an essential component of balanced development strategies, as is discussed later under policy implications.

Agriculture, and especially forestry production, generates a large volume of biomass, part of which is available for use as energy sources. In addition, the provision of draught animal power is a direct energy output from agriculture. Food output also represents energy, but we prefer to account for it as food.

It is useful to distinguish between energy outputs from agriculture according to the criteria of whether they are used on the farm as "intermediate" products, e.g. feed, straw, manure, biogas, and draught power, or produced explicitly, or partly, for sale off the farm without processing or after processing, such as wood, charcoal, liquid fuels from agricultural products, or others. Many of the by-products of agriculture are used very widely in developing countries in rural households as fuel for cooking and heating, and with changed price relationships their use is coming back again in developed countries.

This brings us to the second component of rural energy systems, namely the rural household. Similarly to agriculture, the household is also a unit which uses and produces energy in various forms. The main energy forms it uses are, as mentioned above, for cooking and heating in developing countries, but the wide use of electricity and transport represents a heavy share in the energy use of developed country rural households. Income level is the main determinant of the switch from simpler forms of renewable energy use to fossil fuel based energy in rural households, influenced in addition by the development level of energy infrastructures. There is also fairly clear evidence that overall use levels of energy in rural households are tied closely to income levels and those are strongly influenced by the magnitude of productive assets owned.

The third component of rural energy systems is represented by non-agricultural commercial users and producers of energy. These include all non-agricultural services and production facilities in rural areas as users of energy. Among the energy producers, often better described as "providers", are commercial sources of energy coming from the outside, mainly liquid fossil fuels and electricity, though in some locations coal and charcoal can also be important.

Tying together these components into a real system are the flows of energy between them and the host of factors which influence these flows themselves. For the purposes of this paper, two specific aspects are singled out from among these many relationships: (1) those which are involved in the distributive aspects of rural energy flows, and (2) those which are related to existing or potential food and fuel conflicts.

Access to energy is open to most members of rural societies either through ownership of assets, such as land to utilize free sun-energy to grow crops, or through purchasing power in monetized societies, or historically established exchange relations in traditional ones. In developed countries the higher level of incomes permit all individuals to enjoy a reasonable use of energy sources. In most developing countries income and asset distribution in rural areas is unequal and thus those at the bottom part of the distribution suffer shortages of access to energy. Commercialization and monetization often further exaggerate these inequalities, especially in situations where the shortage is most pronounced.

Technological change can also cause major changes in energy flows in rural areas as shown by the income distribution effects of mecahnization of agricultural production in situations where un- or under-employment exists and where landless labourers have no alternative employment opportunities outside agriculture. It has been calculated that in densely settled rural areas of the developing world the access to firewood requires nearly a third as much area as that needed for food production, even under conditions of high productivitity and effective utilization. This shows that where population pressure on the land is high, the poor who have no access to land are also in a very disadvantageous position with respect to fuel sources.

The abovementioned pressure on land for food and for traditional forms of fuel is already causing major problems of soil degradation through deforestation on many fragile lands of the developing world, and, through its self-reinforcing dynamism, it is building up to tragic proportions in selected areas.

A different form of competition between food production

and fuel production for land and other agricultural productive resources is represented in rural energy systems by the growing programme of converting agricultural products, i.e. biomass into liquid commercial fuels, in most cases through the production of ethyl alcohol. While at the present price relationships are such that for a dynamic programme some form of subsidization is needed to make alcohol competitive with petroleum products, and in many situations in addition the rate of facility construction is limiting, in the longer run under changed oil-price conditions rising demand for raw materials to produce alcohol may have a major effect on world food production and consumption. Given the many complexities involved, caution is necessary at the world level before such programmes are given incentives to attain major dimensions.

RELEVANT FAO MODELS AND RESULTS FROM THEIR APPLICATIONS

In this section a description is given of two FAO modelling efforts which are directly relevant to the issues around energy and agriculture. The first of these is the model system underlying FAO's *Agriculture: Toward 2000 Study,*[1] and the second one is the analytical effort made for the Expert[2] Consultation on *Energy Cropping Versus Food Production,* which includes a discussion on a price equilibrium model for estimating the impact of energy based on food commodities.

The "Agriculture: Toward 2000" (AT 2000) Model

Between 1977 and 1981 FAO completed its study on "Agriculture: Toward 2000". When embarking on the AT 2000 Study FAO's aim was to provide an analysis of agricultural development needs and opportunities for the world in the period up to 2000. This knowledge can help agricultural planners

[1] *Agriculture: Toward 2000* – preliminary report to the 20th FAO Conference, FAO, Rome, November 1979. The final report was made available at the 21st FAO Conference, FAO, Rome, November 1981.

[2] *Energy Cropping Versus Agricultural Production,* report of the FAO Expert Consultation on Energy Cropping Versus Food Production held in Rome, 2-6 June 1980. FAO Agricultural Services Bulletin no. 46, Rome, 1981.

in individual countries to use it as a frame against which to plan the development of their own country's agriculture, it helps international aid efforts in identifying areas of need and it assists FAO in formulating its own programme of activities in support of developing country agricultures. The results of the analysis also serve as the main input by FAO into the development of the International Development Strategy of the UN for the Third Development Decade, covering the agricultural sector.

AT 2000 is a normative, demand driven analysis. Its starting point is the identification of objectives for agricultural development within overall economic and social development. Among the objectives which are relevant here, quantification has been possible for: food and industrial demand, export earnings, employment and income generation. However, formally it is only food and other demand (including exports) for agricultural commodities which drives production, the other objectives are treated as constraints or as non-pre-specified, but ex-post quantified objectives, such as employment or income generation. The unit of analysis is the individual country, and 90 developing countries were analyzed in detail. This excludes China and also a number of small countries, the latter together make up only 3 percent of the population of the developing regions. For the developed countries less detailed analyses were completed, basically relying on existing studies. The disaggregative detailed production analysis including input and investment requirement estimates was developed and used only for the 90 developing countries.

World commodity balances and their international policy implications are derived by aggregating developing countries into one group and developed countries into another and analysing the trade flows, both net and gross, between the groups. To be able to arrive at world commodity balances, in most cases the production adjustment is made in developed countries, reflecting their assumed equilibrating role. In some cases, such as sugar and vegetable oils, an explicit surplus was left at the world level to provide starting points for trade policy and adjustment discussions.

The Study assumes constant relative prices for the future. On first sight, this may appear to be a weak point of the methodology. We believe it is not so. It is our belief that over the long run, it is not relative prices which drive agricultural production forward, but shifts in the demand function, as influenced by changes in population, taste and income, which in turn is parallelled by shifts in the production function under the impact of new technology and investment. The assumptions underlying the shift in the production function appear much less "heroic" once one is willing to look back into the past, and not only into the future. The world has doubled its agricultural output in about the last 25 years, and, if anything, has changed on the price front, the terms of trade for agriculture have worsened. Recent rises in food prices may be a sign that long term price relationships have been out of line.

The Study developed three overall Scenarios. Of these, one reflects more or less a continuation of past trends, and thus arrives at an indeed untenable world situation calling for changes. Scenario A represents an optimistic outlook on overall development in developing countries and a successful response of agriculture to the demands made on it. In that sense it is more like a planned normative, target oriented scenario. Scenario B, which is also normative, assumes lower overall economic growth rates and describes a relevant agricultural response.

In the operational methodology, called General Agricultural Production System (GAPS), Supply Utilization Accounts are established for the base period, then demand projections are carried out, with population and income estimates coming from the UN, but using standard FAO projection methodology. The derived total domestic consumption plus preliminary targets for self-sufficiency ratios (SSRs) combined provide the preliminary targets to be aimed at in the production analysis. The latter starts with an assessment of land and water resources and their development potential both in terms of totals and rate of growth. Multiplying the arable areas by cropping intensity estimates provides the outside boundary of harvested area for future cropping plans. In the next step the 1975 production information is decomposed into area yield and production of 28 crops or commodity groups on six types of land and water situations. For future periods the maximum levels of these available types of areas are allocated between competing crops according to demands made from the use side of commodities, allowing for differential yield increases, depending on crop, type of land and country situation. With respect to new technology the assumption is made that the technologies which will be the backbone of agricultural production in 2000 are already known and tested, but they need to be applied on a wider scale and often with higher input levels.

After the crop-by-crop analysis, chesks are carried out as to whether land and irrigation constraints have been respected, and whether on a judgement basis a better combination of proposals could bring one closer to the desired output level. The bulk of this analysis is similar to a budgetary analysis made in farm management.

For livestock a similar procedure is used at this first part of the GAPS methodology. It takes into consideration numbers of animals, offtake rates and carcass weights yielding meat output, or share of milking cows in the herd, their numbers and their average production per animal. Efforts are under way to further develop the livestock analysis component, to bring it into line with the sophistication of the crop analysis.

In the crop analysis the next step is to evaluate the input requirements of the specified cropping programme. This is done by using a set of input/output coefficients, called the Global Technology Set, which specifies yields and the related input bundles for 28 crops, 6 land types and 4 production (conceptually management) levels. The GAPS routine then interpolates between the specified points in this set the input requirements of the yield estimated in a given country, for a specific crop in a specific land type. Then all input requirements are aggregated across land types and crops to arrive at national figures. Requirements for 15 inputs are estimated.

These identified inputs then provide an opportunity to estimate the commercial energy input requirements which are implied. Table I shows the main items of commercial energy use in developing countries for 1980 and 2000. Before embarking on discussing the magnitude of the changes implies, it is useful to recall the main features of the methodology which underlie these estimates.

For machinery the estimated energy use comes from estimates of total tractor numbers and their annual use. Tractor numbers are estimated as part of the "power model" of the AT 2000. This starts with a parameter for total power inputs into agriculture expressed in terms of Man day Equivalents (MDE), which comes from the input estimation procedure mentioned above. This unit is the common accounting unit which brings human labour, draught animal power and machine power to a common denominator. It represents the work done in a day by a male worker. Substitution relations between it and the other sources of power have been estimated on the basis of empirical data. The total power requirement emerging from the input analysis is then broken down into "resources" from which it could be covered. At first the volume of work done by draught animals is deducted, based on exogenous estimates of draught animal numbers and their annual work performance. The agricultural labour force is estimated on the basis of past trends. Tractor numbers are estimated on the basis of a statistically derived historical relationship between per caput income levels and the ratio of tractors to the number of people engaged in agriculture. Future income levels, together with the growing labour force, thus define the number of tractors to be used in the future. This reflects the rising substitution of hand labour by tractors as their price ratio changes with rising wages reflecting rises in incomes, and it also reflects the availability of capital to invest in tractors. Then the number of tractors is multiplies by their estimated annual use and by applying a variable substitution ratio between tractor hours and man-days, the man-day equivalent of work to be done by tractors is derived. This is deducted from the total man-days demand for power and the residual reperesents work to be done by hand labour.

On the basis of numbers of tractors, the energy used in their production and repair and the fuel used for their operation, energy use in mechanization is estimated. (This estimate, by the way, excludes energy required for the manufacture of hand tools and draught animal equipment, which represent a relatively smaller volume.) The next item, energy for irrigation, is related to the share of irrigation which is served by pumped water and includes both manufacturing costs as well as operational needs for energy. Energy requirements for fertilizers (inorganic) are derived by multiplying the volume of NPK by their unit requirements for energy in manufacture, packaging and transport. The estimates for energy requirements for pesticides are only very rough, based on crude indicators of energy requirements related to estimated volume of pesticides expressed in value terms.

These items exclude some other energy uses in agriculture, such as drying, energy used in livestock production, and in forestry.

TABLE I
Agricultural energy use in developing countries, 1980-2000.

	Machinery		Irrigation		Fertilizer		Pesticides		Total	
Aggregate grouping and year	Amount (10^3 mt oil eq.)	%	Amount (10^3 mt oil eq.)	%	Amount (10^3 mt oil eq.)	%	Amount (10^3 mt oil eq.)	%	Amount (10^3 mt oil eq.)	Amt./ha (kg. oil eq.)
90 developing countries										
1980	11,119	31	4,259	12	18,781	53	1,142	3	35,301	47
2000	54,947	35	7,200	5	92,988	59	2,795	2	157,930	169
African developing countries										
1980	1,011	41	122	5	1,128	46	188	8	2,449	12
2000	5,111	44	232	2	5,811	50	488	4	11,642	46
Latin American dev. countries										
1980	5,107	45	451	4	5,317	47	410	4	11,285	59
2000	27,018	57	761	2	18,745	40	929	2	47,452	163
Near East developing countries										
1980	2,803	43	1,096	17	2,535	39	145	2	6,580	78
2000	7,041	36	1,734	9	10,351	53	333	2	19,458	210
Far East developing countries										
1980	2,199	14	3,011	20	9,802	64	397	3	15,409	58
2000	15,777	20	5,135	6	58,081	73	1,046	1	80,039	269

Source: *Agriculture: Toward 2000*, FAO, Rome.

As Table I shows, commercial energy use is to rise from a total of 35 million tons of oil equivalent to 158 million tons of oil equivalent 1980 and 2000, if the desired production targets are to be met. This represents an annual growth rate of 7.5 percent. Relating it to a 3.7 percent growth in agricultural output, the implication is that each percentage growth in output requires a nearly 2 percent growth in commercial inputs. Experience, however, shows that as commercial energy use becomes wider spread, this ratio comes down much closer to 1:1.

Given these growth rates in agricultural use of commercial energy, over time it is likely that the share of agriculture in commercial energy use will stay fairly stable at around 4.5 percent, or would rise only slightly.

It is useful to see these estimates in comparison with projected commercial energy use in the agriculture of developed countries. Whilst in these countries the rate of growth can be expected to be much lower at 2.4 percent per annum, the volumes involved are much larger. The 1980 consumption could be put at about 134 million tons oil equivalent which could rise to 216 million tons by 2000 if recent trends continue. Being related to an output growth of 2.2 percent per annum, the implied ratio is 1.09 which is low in comparison with the figures for the developing countries, which is estimated at 1.92 for 1980-2000. There is also a strong possibility that under the impact of changed ratios between the prices of commercial energy sources and agricultural output there will be major efforts made in the agricultural sector of developed countries to increase the efficiency of use of commercial energy in agriculture, and this would probably narrow the above ratio.

Many of these estimates depend on the relative price of energy to agricultural output. In turn this ratio will be strongly influenced by the fact that the commercial energy input share in agricultural production is lower than in most other branches of the economy, thus higher energy prices are likely to exert higher pressure on the price of non-agricultural commodities. Often there will be a need for governments to ensure that agriculture is not "priced out" of the commercial energy markets, if the world's future food supply is to be secure.

The indications from the estimates are that the composition of commercial energy use in the agriculture of developing countries is not going to change drammatically, but mechanization and fertilizers will gain some more weight and irrigation will lose some as its rate of growth is relatively slow. There are notable regional differences in the energy use patterns. Irrigation and fertilizer use are more important in the two regions with more intensive agricultural development as a result of shortages of land, while the two regions with the highest incomes (Latin America and the Near East) clearly show the importance of mechanization, with Africa also being influenced by the inclusion of the Maghreb countries in the total.

The analysis has also clearly shown the changes over time in the structure of sources which provide power for agriculture. Table II shows the changes in the estimated mix of human labour, draught animals and machines in the provision of power to crop production over the period 1980-2000.

This information indicates that while tractor (machine) power will increase at a rapid rate of 6.7 percent per annum, its share in the provision of total power inputs will rise only from 8 to 19 percent, and in the lower income regions only from 3 to 8 percent. The share of draught animals in the provision of power is expected to decline from 29 to 23 percent under the impact of rising incomes in some countries and growing population pressure on the land in others. The rate of growth estimated for the use of human labour indicates that it could grow somewhat faster than the agricultural labour force, and thus average annual utilization per caput could rise from 114 days to 134 days per year between 1980 and 2000.[1]

In deriving these estimates, a maximum annual number of days in crop production per member of the agricultural labour force was postulated at 250 days. This has meant that in a number of countries (especially in South America and some in the Near East) the number of tractors had to be increased from the levels indicated by the income/tractor relationship so as to be able to meet MDE requirements indicated by the large expansion in production, under-pinned by a large increase in cultivated area.

With respect to fertilizer (inorganic) use, the question could be raised of whether no alternative energy sources, preferably

[1] It needs to be recognized that on the average labour use in crop production represents only about 40-60 percent of total use of the agricultural labour force.

renewable ones, could be used to meet the desired level of NPK inputs to crop production. In the analysis, the implicit assumption was made that organic sources of fertilizer are going to increase, but their share would remain approximately the same as in 1980. This assumption had to be made because of the lack of data on which to base a more detailed assessment of present NPK availability from these sources and the rate at which they could be developed. A large share of the potential is already used and there are many local, difficult to change, factors which constrain their greatly increased utilization.

At the same time the Study makes explicit proposals for increases in the leguminous plants' component of cropping patterns. This is to be achieved through enlarged areas for production of pulses for human consumption, increases in the area of soybeans and groundnuts, and through a very large increase in the area of cultivated fodders, of which a good share is to be leguminous crops. The proposed rise in cropping intensities in low rainfall areas of winter rain patterns and those in tropical areas of high rainfall both reflect proposals for reduction of fallows in the rotations and their replacement by nitrogen fixing leguminous crops.

The price equilibrium model of the expert consultation on energy cropping versus food production

The objectives of the Expert Consultation were:

- To evaluate the land resources and production capability available for potential energy crops, including an appraisal of their extent and distribution on a worldwide basis.
- To review potential energy crops and establish their efficiency, applicability and net energy balances in various agricultural systems and agro-ecological conditions.
- To evaluate the competition of food production with energy production and its economic, social and political implications at different price levels for agricultural products and fossil fuel.
- Based upon A, B, and C, establish policy guidelines for energy cropping consistent with food production requirements.

The discussions in the Consultation on the issue of price responses of agricultural production to increased demand for

TABLE II

Power requirements for crop production

(a) Levels of requirements[a]

Region	1980 MDE[b] Requirements				2000 MDE[b] Requirements			
	Total	Labour	Draught animals	Mechanical	Total	Labour	Draught animals	Mechanical
	.... Billion MDE							
90 Developing Countries	82.6	55.1	20.8	6.7	128.4	80.5	23.8	24.1
Africa	12.0	10.1	1.5	0.4	20.7	17.0	2.0	1.7
Far East	51.9	34.1	16.1	1.7	76.0	50.1	18.9	7.0
Latin America	13.0	7.3	2.2	3.5	22.8	8.4	2.1	12.2
Near East	5.7	3.7	1.0	1.0	8.9	4.9	0.8	3.2
Low income countries	56.0	37.9	16.6	1.5	84.3	58.4	19.3	6.6

a Power requirements are set equal to power output in the calculations

b Man Day Equivalent

(b) Percentage share of different sources in total power output

Region	1980 Labour	1980 Draught animals	1980 Machines	2000 Labour	2000 Draught animals	2000 Machines
90 Developing Countries	67	25	8	63	18	19
Africa	84	13	3	82	10	8
Far East	66	31	3	66	25	9
Latin America	56	17	27	37	9	54
Near East	64	18	18	55	9	36
Low income countries	68	29	3	69	23	8

(c) Growth rates of power input from different sources in power provision (percent per annum)

Region	Model MDE	1980 – 2000 Labour	1980 – 2000 Draught animals	1980 – 2000 Machines
90 Developing Countries	2.3	1.9	0.7	6.7
Africa	2.8	2.7	1.3	7.2
Far East	1.9	2.0	0.8	7.6
Latin America	2.8	0.8	–0.2	6.5
Near East	2.3	1.5	–0.7	5.7
Low income countries	2.1	2.2	0.8	7.6

agricultural commodities for their conversion to liquid fuels were based on the preliminary analysis prepared by FAO for this purpose, the results of which are available in the "Background Paper"[1] which formed the basis for the Consultation's deliberations.

The aim of the price response analysis reflected the view that although the majority of the world's population lives in countries where domestic production covers a very large share of total food consumption, a substantial volume of world agricultural production is traded on the international markets. Thus resource allocations in one part of the world do have repercussions on consumers in another part of the world. In addition to global effects, it is these potential implications on various groups of countries and producers and consumers within them which was analysed.

The value of agricultural trade in the period 1975-76 (valued at the same price as non-traded volumes) was 10 percent of the value of world agricultural production. However, for the two most important potential or actually used feedstocks for alcohol production, i.e. sugar and coarse grains, the traded share of total production was much higher at 24 and 11 percent respectively. Thus one could argue that any additional demands made on a given country's sugar and grain production will rapidly be translated into price and supply effects on the world market. These price and supply effects in turn would have important influence on consumption of commodities, especially hurting badly those groups whose reliance on imported grain is high and whose income is low.

Under the discussion of major food/energy issues by the Expert Consultation a number of implications were hypothesized which could emerge from using increasing volumes of agricultural commodities for energy production. There the emphasis was on the policy choices which are to be faced by individual countries given their food production and energy endowment and the worldwide changes in prices and in supply and demand were touched upon only in general. In this part the preliminary results of an analysis estimating the quantitative effects of a hypothesized large scale programme for energy production (liquid fuels) from agriculture are reported.

---

[1] *Energy Cropping Versus Agricultural Production*, FAO. op. cit.

The basic analytical framework is rather simple: it is assumed that energy use represents a new added demand to previously existing uses and this can be interpreted as a shift in the whole demand curve to a higher level, meaning that at any given price,

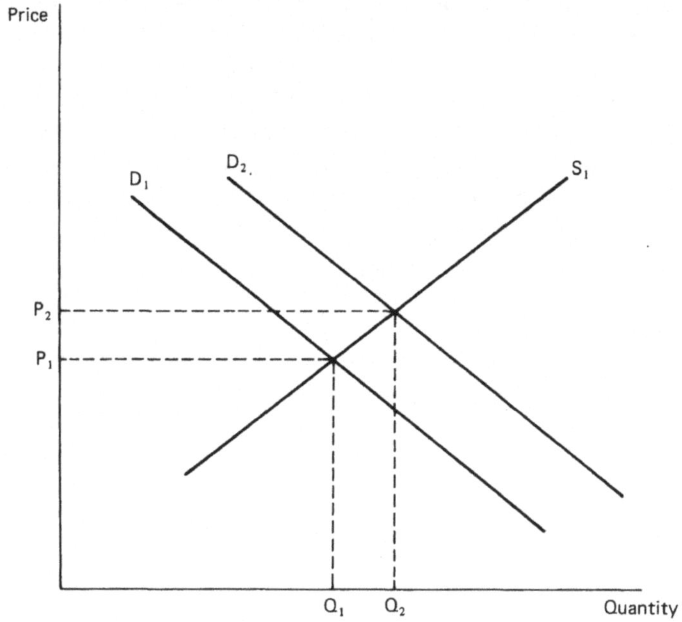

D_1 demand for commodity without alcohol use
D_2 demand for commodity with added alcohol demand
S supply schedule
Q_1 quantity demanded and supplied without alcohol demand
Q_2 quantity demanded and supplied with alcohol demand added
P_1 price of commodity without alcohol demand
P_2 price of commodity with added alcohol demand

FIGURE I Adjustments to added demand for energy use.

the quantity demanded would be larger by the added volume of the energy-derived demand (see Figure 1). Given the new volume demanded, supply would respond to the resulting new higher price, thus lowering demand and price will settle at a new equilibrium level. The price and volume change will depend on the responsiveness of supply to higher prices and on the decline in quantity demanded as a response to higher prices.

To test combined simultaneous medium term effects (up to 1985) on prices and quantities produced and consumed, two levels of energy added demand were postulated – a low and a high level. At the low level it was assumed that a relatively small "shock" will only cause very small changes as its volume compared with total production and consumption is very small. This level of energy use of agricultural commodities is likely to come about if the most important potential participants in an energy (fuel) from agriculture programme do not "force" the pace of their programmes but keep other considerations also in view.

The higher level of"energy shock" is assumed at a level where substantial price and volume changes could be expected and economic and social implications could become visible and possibly problematic. For this level to be reached widespread large scale programmes are required with major government support. The two "levels of assumed volumes of added demand from energy" in 1985 are shown below in Table III together with the total volumes produced and traded in 1985 excluding alcohol demand of the commodities involved.

While both the low and the high added energy demands represent only a small share of total coarse grains volume (2 and 4 percent respectively for low and high) they do come to a substantial portion, 16 percent of production and 64 percent of the traded volume for sugar in 1985. Sugar prices, especially non-contract sugar prices on the world market are notoriously sensitive to short-run demand/supply imbalances but even in the medium run such a large addition to demand can be expected to have a major impact on prices and production.

The results of the experimental runs for testing the world implications of added demand for fuel production from agriculture are summarized below in Table IV. Given the large differences in the situation of the three commodities involved, they do show very different responses. The most sensitive commodity is cassava, but a large part of this sensitivity can be

TABLE III
Feedstock volume demanded for alcohol in 1985 (mill tons)

Commodity	Sugarcane		Cassava		Coarse grains	
Level of alcohol demand	Low	High	Low	High	Low	High
Countries						
U.S.A.					12	25
Brazil	150	225	1.5	3.0		
South Africa	2.0	3.0	4.6	4.6		
New Zealand					2.0	3.0
Austria					0.2	0.3
Australia					0.6	1.2
Sudan	0.1	0.2				
Sudan (molasses)	(0.11)	(0.11)				
Thailand	1.1	2.5	2.2	6.6	0.2	0.4
India (molasses)	(3.8)	(3.8)				
Philippines	2.0	3.8				
Kenya (molasses)	(0.3)	(0.3)				
Central America	4.9	10.0				
World Total[a] for alcohol 1985	160.1	244.5	8.3	14.2	15.0	29.9
World Production[b] 1985	1031.8	1101.8	126.2	137.2	838.3	870.8
World Trade[c]	249.2	253.5	13.2	15.5	95.2	77.9

[a] excluding molasses
[b] FAO commodity projections 1979
[c] imports

explained by the relatively small share of the total volume which is traded, and the very small number of countries which are involved in exports. In the analysis it was assumed that those countries which grow cassava only for domestic use will not enter into world trade in the future. If this assumption is relaxed and if it can be thought that if the international cassava market is attractive then more countries would enter the trade, then it

TABLE IV

Quantitative changes in demand under the impact of alcohol programmes

	LOW					HIGH				
	Base Demand	New demand incl. alcohol adjust.	Reduction in other uses	Reduction in other uses as % of total original demand	Price change under alcohol impact %	Base Demand	New demand incl. alcohol adjust.	Reduction in other uses	Reduction in other uses as % of total original demand	Price change under alcohol impact %
	Mill. tons					Mill. tons				
SUGAR										
Total World	103379	115345	4034	3.9	12.5	103379	123647	4152	4.0	12.1
Developed c.	52876	51622	1054	2.0		52876	50026	3150	6.0	
Developing c.	50503	63723	2980	6.0		50503	73621	1002	2.0	
CASSAVA										
Total World	126214	134181	333	.26	5.6	126214	139539	875	.7	14.6
Developed c.	12842	17113	329	2.6		12842	16573	869	7.0	
Developing c.	113372	117068	4	—		113748	123342	6	.01	
COARSE GRAINS										
Total World	834657	846119	3858	.46	1.6	834657	855972	8685	1.04	4.0
Developed c.	531841	543890	2751	.52	1.6	531841	554660	6981	1.3	
Developing c.	302816	302229	1107	.37		302816	301312	1704	.6	

is likely that under the assumption of the high alcohol programme the response in production would be even higher than what the present analysis shows as a 10 percent increase in production, as compared to a 11 percent increase in original demand through alcohol use, calling for a 15 percent increase in prices. This in turn depresses demand by 1.1 percent or .9 million tons. However, it is noteworthy that nearly all of this depression in demand would take place in developed countries, where cassava is hardly used for food but mainly for feed and for industrial starches. At the low alcohol programme level, the price change is only about one-third of the high and correspondingly the reduction in demand for non-alcohol use in developing countries is an even lower percent as compared to the high alternative, and the price rise is only 6 percent.

For sugar the relative magnitude of the "alcohol shock given" is much higher when expressed as percent of total production, but because the supply response is spread over a large number of countries, there is a relatively bigger increase in production and thus a smaller price adjustment required for equilibrium. Under the high alternative sugar production response will cover 82 percent of the alcohol caused additional demand and prices would rise only 12 percent above the base situation level. Both for the low and the high alternative, the outcome of the analysis is highly dependent on what assumptions are made for Brazil. If one postulates that Brazil will be able to produce the sugarcane volume needed for its alcohol programme without reducing its export availabilities, then a large share of the alcohol induced demand volume will be met and everything hinges on Brazil's ability to produce this volume without running into much higher production costs. On the other hand, if the past supply behaviour of Brazil is taken as a guide, then Brazil would have difficulties in meeting its alcohol demand plus its export volumes of the past and the rest of the world would have to fill this gap. The postulated volume of sugarcane output for Brazil, of 294 million tons of cane could be grown at a yield of 55 t/ha, on 5.3 million ha, about 8 percent of the likely total arable land of Brazil, or expressed as share of the total harvested area it would come to close to 13% of the total. This needs to be compared with the present share of sugarcane in total harvested areas in the country, which comes to 5 percent (1978). Together with the assumed cassava programme for alcohol, sugarcane for alcohol would represent 9 percent of the total harvested area, and thus it is unlikely that it could be brought about without influencing the area and production of other crops in Brazil.

Looking at the picture worldwide, the earlier mentioned sugar surplus from the AT 2000 Study is worth considering, together with recent sugar surplus stocks on the world market. The AT 2000 surplus sugar in 1985 could be around 3.5 million tons, and the recent (1975-79) sugar stocks which were surplus above normal carry over stocks and which caused a major collapse of the sugar market were at 12 million tons. Compared to that, the postulated additional demand for sugar from the higher alcohol programme is very large at 24.4 million tons, and thus even stronger price effects than the 12 percent increase emerging from the analysis of the higher alternative may be possible.

The volume of coarsegrains assumed to be used for alcohol are the smallest percentage additions to existing demand: they reach 2 and 3.5 percent respectively for the low and high alternatives, and as a result the percentage changes in prices (1.6 and 4.0 percent) and in the volume of other uses are also the smallest of the three commodities analyzed. Nevertheless, they are the most important from the point of view of nutrition and welfare consideration for the poor of the world. The starting point of this analysis has been the production and consumption levels estimated in the FAO commodity projections. Those are "positive" estimates, i.e. aim at identifying the most likely outcome in 5 years time. In the AT 2000 the consumption and production estimates are "normative" as they assume that governments will strive for them as targets which are desirable and feasible. The difference in coarsegrain production in 1985 consumption estimates between the results of these two studies come to 17 million tons, i.e. under the postulated conditions of world agricultures as seen to be feasible by the AT 2000 Study, a good share of the additional coarsegrain demand for alcohol could be accommodated in terms of producing it. Looking at the price change which is appropriate for this level of production, at +4.0 percent this seems to be small and well within the price variations of the past.

The magnitude of the indicated price changes and the sensitivity of prices to changes volumes of demand and production (supply and demand elasticities in the medium run) are of the magnitudes as found by other researchers. At the same time the violent fluctuations of grain and sugar prices in the period around the world food crisis in the first half of the 1970s provide a warning of the added variability once psychological expectations influence short-run markets. It appears, however, that the fact that any major alcohol programme needs major investments in facilities and they take time, provides a "cushion" to the shocks and major problems which are likely to occur only in years when world production is set back by natural or other forces, and when the added demand from alcohol exaggerates an already tight market situation.

The above discussed analysis to some degree also may overstate some of the price responses and they also do not differentiate between food uses and other uses which are effected by the higher prices. In general, both for cassava and for coarse-grains the most likely use that will first decline will be feed use, as was demonstrated during the 1973-74 period, and thus the effect on the diet of the population in the developing countries could be small. This, however, is dependent on the willingness of major food aid donors to provide an appropriate level of cereal food aid, even when the prices for them are attractive on the world market.

The above analysis has shown that if energy use of agricultural commodities reaches the higher level of potential use, it will have limited but substantive impacts on agricultural commodity prices and on consumption. On the other hand it would appear that the volume of commercial fuel produced by these assumed programmes will have very little impact on commercial energy prices, as it would represent only a fraction of one percent of worldwide commercial energy use, though obviously in individual countries, e.g. Brazil, its effects will be large.

Since that analysis was carried out, the world sugar market has seen a major boom and its collapse, and the changes in the import policies of the EEC have exerted a more important influence on world cassava trade than its use for alcohol would represent. The U.S. alcohol programme is showing signs of a slowdown under the dampened support which the new administration is apparently willing to make available.

A seemingly forgotten, but nevertheless important area where agriculture can substitute for fossil energy sources is in the production of agricultural raw materials, such as fibres, rubber, and raw materials for chemical processes. It is rarely recognized that the fossil fuel energy equivalent which would be required to replace the world's use of natural fibres, rubber, oils, soap, leather, and other commodities which compete with synthetics based on fossil fuels, amounted in 1979 to 41 million tons[1] of oil equivalent. A recent study by FAO provides an example and demonstrates the important effects of changes in petrochemical prices on the longer term competition between sisal and polypropylene. The revitalization of the natural rubber industry has benefitted from the same effects.

POLICY IMPLICATIONS OF THE FAO ENERGY–AGRICULTURE RELATED STUDIES

These are discussed below under four major headings: (1) energy use in agricultural production and food systems, including substitution of non-renewable by renewable energy sources; (2) fuelwood and charcoal production for rural households and other uses; (3) the food and fuel conflict; and (4) general considerations on energy policy and planning.

Energy use in agricultural production and the food system

In a recent meeting of FAO's Committee on Agriculture and in the 21st FAO Conference held in November 1981, national and international policy issues related to energy use in agriculture were discussed and agreed upon, and these provide a useful summary of the policies that are central for future rural development.[2] They also reflect the major conclusions with respect to energy and agriculture of the UN Conference on New and Renewable Energy Sources, held in Nairobi, August 1981.

---

[1] This calculation assumes that on the average 1 ton of natural product from agriculture replaces 1 ton of mineral oil equivalent. The figures include: all rubber, cotton lint, other fibres, hides and skins, and 25 percent of total vegetable oil production.

[2] "Energy in Agriculture and Rural Development", COAG/81/9, December 1980. Document discussed at the Sixth Session of the Committee on Agriculture, FAO, Rome, 25 March-8April 1981; and "Energy in Agriculture and Rural Development" C 81/25, August 1981, FAO Conference paper.

Under the heading of national and international issues the above analysis demonstrates the extremely wide ramifications of energy problems in relation to the agricultural and rural sectors. The formulation of agricultural and rural energy policy, within the overall context of national and international energy policies, is obviously a complex undertaking, but it is one that should be carried out without delay in the many countries that have not yet done so.

An overriding consideration is that the agricultural sector must be assured the small but crucial supplies of commercial energy that it requires for the production and use of energy-intensive inputs. Until new technologies are developed, these are essential to maintain the expansion of agricultural production, and in particular to accelerate the production increase in the developing countries sufficiently to keep up with the rapidly rising demands of their growing populations without further burdensome increases in their imports. It appears that in a number of countries agriculture itself will be increasingly called on, through energy cropping, to provide (at least in the short run) part of the renewable energy resources to substitute for the diminishing resources of fossil fuel. It is therefore appropriate that agriculture should be given priority in the use of the energy for such essential purposes as the production of chemical fertilizers and pesticides, and power for farm machinery and irrigation equipment.

In many countries price policies will be a major factor in ensuring that farmers can both obtain and afford the energy-intensive inputs they require. In general such inputs contribute so greatly to increased yields that their use should continue to be profitable even at higher prices, but it will be necessary to keep a close watch on farm price relations in the light of the volatile energy situation. In some cases selective subsidies will have to be used to raise food prices, while also taking compensatory measures in favour of the poorer consumers. Such policies as transport equalization schemes for fertilizers and other inputs may require reconsideration in the light of higher transport costs.

At the same time, agriculture, like all other sectors, must play its part in the more economical use of all forms of energy, and in the gradual transition from non-renewable to renewable sources of energy. In general the measures that are required have the advantage of not only reducing the depletion of irreplaceable resources, but also being better for the environment. They should also contribute to the greater self-reliancc of developing countries that is a basic element of a New International Economic Order.

There is a major task, particularly for research and extension services, in developing and promoting farming systems that save fossil fuel. These would include the more efficient use of chemical fertilizers, greater use of organic fertilizers, crop rotations, catch cropping (especially with legumes capable of fixing atmospheric nitrogen), improved tillage and other practices to save tractor fuel, more efficient use of irrigation water, weed control by better tillage and mechanical or hand weeding, integrated methods of pest control relying as far as possible on natural elements, greater use of draught animals, and improved hand implements and animal draught equipment. Basic research should be stepped up in such fields as enhancing the efficiency of photosynthesis in crop plants, increasing their responsiveness to and reducing their requirements for energy-intensive inputs like fertilizers, breeding varieties resistant to pests and diseases, and transferring the capacity to fix atmospheric nitrogen to such basic food crops as cereals and roots.

The fisheries sector offers considerable scope for energy-saving, in view of the high incidence of fuel in operating costs. Measures to control the number and size of fishing vessels would not only save fuel but also contribute to the conservation of an important natural resource by limiting access to heavily exploited stocks.

An obvious, but so far neglected, area for saving energy is in the off-farm parts of the food systems of the developed countries. In some cases it may be necessary to plan the relocation of the production of certain agricultural commodities in the light of higher transport costs. The developing countries should be on their guard against the growth of energy-lavish food systems of the kind found in developed countries.

Fuelwood and charcoal production for rural households and other uses

Similarly to the policy issues identified in the COAG meeting, FAO's Committee on Forestry endorsed the following policy recommendations at its last session.[1]

The evolving energy situation is placing new emphasis on every possible energy source but especially on those renewable sources among which wood is traditionally one of the utmost importance. In addition to being a renewable and versatile form of stored energy, wood is socially and environmentally acceptable; when available and locally accessible its sustainable supply can contribute substantially to energy self-reliance, at least in rural systems. The dependence of large number of people in developing countries on wood for energy is likely therefore to be reinforced by the increasing difficulty to shift to alternative fuels in the foreseeable future. A major effort is indispensable to ensure the maximum contribution of the forest biomass to the basic energy requirements of these populations through its close integration in rural energy systems and in a way which is compatible with the environment.

Producing and using wood for energy on an industrial scale also deserves particular attention as the cost of energy from wood is continuously improving along with the steady increase of the price of fossil fuels. Either through direct conversion or through the production of alternative fuels, the contribution of wood to new energy systems could be substantial either in decentralised energy systems or when important land and capital resources are available. There are no inherent barriers to the use of wood for energy as limitations are in fact essentially related to comparative cost and properties of fossil fuels of limited supplies.

The prospective development of the production and use of wood for energy are related to the following major issues which the Committee has agreed:

1) Wood related energy must be seen in the context of national energy policies and considering energy balances within forestry, the advantages of wood as a low energy requirement material as well as a renewable source of energy. The new energy dimension may require some adjustment in forest policies.

2) Most fuelwood and charcoal presently do meet basic energy requirements in rural energy systems and they should therefore

[1] *Wood for Energy* (COFO/80/2). Document discussed at the Fifth Session of Committee on Forestry, FAO, Rome, 26-30 May 1980.;

be closely integrated into these systems and into other rural development activities. Growing trees for fuel outside the forests by farmers requires institutional arrangements for forest services to provide the necessary extension and support services.

3) Particular concern should focus on those areas of heavy dependence on wood as a primary source of energy mainly for domestic requirements and where shortages are severely affecting the poorest as well as deteriorating the environment. Research and extension are necessary to promote a more efficient use of available wood fuel resources.

4) Available information on wood for energy needs to be substantially complemented and improved regarding available resources and potential supply including wood residues as well as regarding consumption and requirement levels and people's present and future needs; it should form the basis for planning future developments and for identifying and formulating programmes and projects.

5) Organizing the supply of wood for energy deserves particular attention as it should be made available on a physical and economic basis which meets the requirements of the users; distribution systems should usually be substantially improved.

6) Making energy a primary production objective would require developing further the multiple use management of forest resources with due considerations to other production and environmental functions and with increased attention to those parts of the forest biomass not presently included in active management and suitable for energy use.

7) Energy plantations' potential requires a careful assessment based on the possible biomass productivity, the availability of suitable land and the comparative costs as it could involve competition with other land uses: one solution could be integrating tree growing for fuel with agricultural practices.

8) The potential use of wood for the industrial production of energy is mostly based on the comparative economics in relation to other fuels: a continuous analysis of the relative positions of energy from wood could therefore be required as well as of the implications of new technological developments.

The renewed importance of wood as a source of energy is placing a new emphasis on the role of forestry in meeting some basic requirements of human life especially for billions of people in rural areas. The new energy dimension requires closer integrated management of resources in conjunction with other rural development activities in the framework of comprehensive energy policies. It may require further adaptation of the professional capabilities of foresters and of the forestry institutions in order to meet the new importance of the energy demand placed on wood.

The food and fuel conflict

The abovementioned Committee on Agriculture session and the 21st FAO Conference also agreed that the most controversial and uncertain policy area concerns the use of biomass, and in particular of agricultural crops produced specifically for the purpose, to provide liquid fuels to replace those derived from fossil fuels. In the longer run, the production of alcohol from wood and from crop residues, which would not compete with food crops, could become an important source of liquid fuel and high priority should be given to perfecting the necessary conversion technologies as quickly as possible. Energy cropping based on food crops might then prove to be no more than a temporary expedient. In the meantime policies encouraging energy cropping should be pursued with great caution and flexibility, bearing in mind that they may quite soon become obsolete, as well as their possibly serious repercussions on food supplies and prices. The FAO Expert Consultation recommended that countries planning the production of energy from biomass should carefully examine their situation and policy options, and that they should be assisted by FAO, together with other organizations such as UNIDO, in the monitoring of relevant technology and making their national assessments.[1]

The specific recommendations of the Expert Consultation on Energy Cropping Versus Food Production are also relevant here. They state:

1) Individual countries planning biomass energy programmes as part of their energy systems related to agriculture should

[1] FAO Expert Consultation on Energy Cropping Versus Food Production, Rome, 2-6 June 1980. op. cit.

carefully examine the range of factors which are relevant in their situation and examine policy options with respect to timing and flexibility. They should be helped in this endeavour by international technical assistance agencies including FAO. More specifically, the support from FAO working together with other agencies such as UNIDO could encompass the following components:

– monitoring and assessment of relevant and appropriate technology;

– provision of a clearer understanding of the complex interrelationships influencing food/energy decisions;

– assistance in individual country level assessments.

2) Trade and aid mechanisms transmit the influence of food/energy conversion programmes throughout the world; thus it is recommended that countries take into account the potentital effects outside their boundaries, which their own action could create. To be able to do so, there is a need to provide the necessary information which permits meaningful assessment of worldwide implications. International bodies interested in keeping track of world supply/demand balances – such as FAO, the World Bank, WFC – should provide the necessary overall assessment and monitoring function.

Within FAO, ongoing programmes of monitoring the short-, medium- and long-term developments in the food situation should make a special effort to integrate potential effects from food/energy conversion into their analyses.

3) Underlying the two above recommendations is the dominant role of the nature and magnitude of response in production to the changed circumstances brought about by demand for food/feed commodities for energy conversion. While shorter-term responses of production are reasonably well known, large uncertainties surround long-term responses. Thus the Consultation recommended that more long-term analysis be conducted to improve our understanding of longer-term impacts of policies related to food/energy conversion programmes. Evaluation of the impacts of existing biomass energy programmes through time in both developed and developing countries would be important for this purpose.

General considerations on energy policy and planning

It is essential that national energy policy in developing countries should pay proper attention to the needs of their poor rural communities. It will be necessary to ensure that their traditional sources of fuel are not made unavailable to them by their wholesale conversion to more costly commercial energy that is out of their reach. It is also necessary to allocate resources for the development of such renewable sources of energy as windmills and small hydroelectric units, which (although perhaps insignificant at the national accounting level) can make a substantial impact on the levels of living of the rural poor.

There is a danger that national energy policy in developing countries may tend to view the agricultural and rural sectors mainly as a ready source of renewable energy, to provide substitute fuels for the use of the predominantly urban, affluent minority, and ease the balance of payments. This danger springs particularly from the usually weak position of the agricultural ministry in the government hierarchy and the dispersed departmental responsibility for rural development. One way of avoiding the danger would be to establish institutional machinery for the formulation of energy policy that involves all the many interests concerned, including agriculture, fisheries, forestry, rural development and nutrition, as well as consumer organizations.

Finally, there are a number of international policy aspects. The international research effort should clearly pay greater attention to the various national policy concerns outlined above. The increased pressure on fragile soil resources that could result from energy cropping adds to the urgency of establishing machinery for the international monitoring of the state of the world's soil resources. The potentially important role of natural fibres and rubber in replacing the synthetic products based on petro-chemicals that have recently tended to oust them could be assisted by the establishment of further formal and informal international commodity agreements, including price stabilization measures.

Perhaps the most important international policy issues are those concerning world food security. The world food situation is already highly insecure, without the new competition of energy cropping for scarce land and other agricultural resources. While it is probable that some of the present fears may be exaggerated, it is essential that a close watch should be kept on the situation, and especially on the longer-term implications, once large-scale programmes have had time to get under way. International agreement is needed that, especially in the event of the tight food supply situation that must always be expected in a year of widespread bad harvests, priority will be given to food exports and the honouring of food aid commitments, and that the maintenance of food security reserves will not be jeopardized. Suitable international machinery already exists in the FAO Committee on World Food Security (CFS) and the Committee on Food Aid Problems and Policies (CFA), as well as the FAO Council and Conference, and the World Food Council.

While these policy recommendations emphasize specific areas, it is quite clear that effective impacts on the energy problems of rural areas can be made only if the proposed action takes into full consideration the implications of that action on all parts of the rural energy systems, including technical, economic and social aspects.

DISCUSSION POINTS

1) Dr Leach pointed out that relatively small quantities of commercial fuel are involved both now in developing country agriculture and even in the high growth scenario in 2000, which implies a doubling of food production in the 90 countries studied. The 37 million tons of oil being used at present is about a quarter of the oil savings made in OECD countreiss in 1979-80, and the 175 tons projection to 2000 is one half of the forecast reduction in OECD oil use in 1978-90. Not all of this energy is oil, one half is fertilisers and feedstock. Thus there is no fuel constraint in the next 20 years, with the FAO style of agricultural development. It is then a question of prices, and of whether there is an effective demand for the technologies that use the energies. It is important to consider the size of the fuel component of the input package, including fertiliser, tractor fuel and tractor, for the small, poor farmers in the developing world.

Dr Hrabovszky asked what would be the real effective constraint, since the input quantities of fuel are so small. In those areas where energy use is direct, such as fuel use for pumping, tractors, or heating hot-houses, the energy pinch has already been felt, and it will be a severe constraint. The end price pinch on some other factors, especially fertilizers, will not be so great. There is still a lot of fat in the fertilizer business, and fertilizer use will be the least affected among the energy users, while it is the most important in terms both of volume and its effect on production. For small farmers the importance of these factors, especially of draught power, is extremely mixed. Effects differ across the world, and possibilities and constraints are also very divergent. The servicing of small farmers has improved tremendously in some countries, but in others not at all.

2) Dr. Norse agreed with the potential production aspect of the work, but was uneasy about the environmental dimension, both in the Third World and the developed countries. In a range of developed countries the nitrate levels of both surface water and ground water supplies are now close to or above the WHO recommended level. At the moment those levels are increasing at about the same rate as the increased use of nitrogen fertilizer, so if this is extrapolated forward the problem is obvious. Also there is the problem of soil loss in the grain growing areas, particularly in the USA but also in Canada and Australia. Soil losses of about 30 tons per hectare went up to 90-100 tons per hectare when production was greatly expanded in the 1970's. The sustainable soil loss is up to about 10 tons per hectare, so here is a major constraint which feeds back into the energy system because of the increased fertilizer requirement.

Dr. Hrabovszky admitted that the modelling was not good on these aspects. However the FAO is helping major improvements in existing irrigation schemes, part of which is putting in drainage, are proposing that of the naturally flooded areas about 50% should be drained and flood protected by the year 2000, and putting about 30-40% of non-irrigated lands under soil-conservation measures.

Dr. Holdren asked to what extent the agriculture 2000 projections included not only potential gains in arable area but potential losses due to salination, water logging, erosion, etc. . . Dr. Hrabovszky replied that potential losses from land degradation are not in the study, but that the FAO and others are hopefully putting in enough investment to stop a good part of it. Another FAO study, "Land resources for future populations", has develpped a special methodology for trying to estimate strategies to account for soil losses.

3) Dr. Makhijani pointed to an imbalance of trade between the Third World and the capitalist countries which can be measured in calories, but is most clearly seen in terms of trade oriented acreages. Generally about 10 million hectares in the capitalist world is devoted to growing cereals for export to the Third World, while 50 million hectares in the Third World is devoted to growing exports to the capitalist world. The energy balance is similar. Moreover in the Third World the energy is largely provided by bullocks which use valuable grazing land. In a comparison of rice exports from Thailand with automobile imports it is found, for a given value of products, that the exports of energy and labour are about ten times that of the imports of energy and labour. Such facts illustrate a very systematic structure in the world which must be taken into account in any projections.

Dr. Hrabovszky agreed that developing countries were net food exporters but felt that the trade acreage comparison was unfair, since the production figures vary considerably, possibly 7 tons of maize per hectare in a developed country, 1½ tons of maize per hectare in a developing country. One problem is that some countries rely heavily on agriculture exports, while the importing countries may not have to rely on agriculture at all.

There are still substantial comparative advantages to some developing countries in producing certain tropical products. For example in Malaysia about 80% of rubber land could not grow any food crops, and a large part of Ceylon tea land is also not suitable for food production. It is necessary to consider the local situation, which means that the problems of aggregating trade balances for 90 developing countries are formidable, and under certain circumstances may be misleading.

4) Dr. Frohberg wondered if enough account is being taken of comparative advantages. After all, from a economic point of view, self-sufficiency may be very costly. While realising the political factors involved it is important to clarify the economic costs.

Dr. Hrabovszky noted that country intentions indicate a desire for improved self-sufficiency. The modelling must accept this, and not develop plans which member countries refuse and are not interested in. After all for a number of non-economic, security reasons it is sensible to have a reasonable degree of self-sufficiency, and it is often sensible economically. Free trade in agricultural commodities needs to be very carefully examined. If a country can be self-reliant in major staples and if it is not a very costly enterprise, it should do that.

5) Dr. Norse raised the question of competition for labour between agriculture and energy at seasonal peaks. Dr. Hrabovszky indicated that fundamentally the employment model is a residual one. This is a demand driven study, not employment oriented. Employment estimates exogenous to the study are introduced into the analysis.

6) In connection with the fuel wood problem, Dr Hrabovszky stated that the study emphasis is on agroforestry and on what is called forestry for the people, being mainly fuel and construction wood oriented. They consider village wood lots, roadside plantations, stove improvements, generally trying to make the best use of available resources for growing food and for the best use of wood. There is also a major effort under way to improve charcoal production, both to increase efficiency and to determine if it can substitute for other fuels.

7) Dr. Frohberg enquired whether estimates have been made of changed energy needs in the substitution of tractors for draught animals. Dr. Hrabovszky first noted that they do not believe that it is possible to sensibly express human labor in energy terms. Substitution ratios in work performed in energy units are totally different in many different operations. They have not looked specifically at the relationships between draught animals and tractors, but do note that in general animal populations are either constant or decreasing slowly, being now used for meat and milk rather than as draught animals previously.

8) Dr. Glaeser pointed out that China is excluded from the study, but that poor harvests and subsequent grain purchases can tremendously influence the world market. Dr. Hrabovszky noted that for China, as for developed countries, the FAO has not carried out any detailed supply analysis, but have taken the results of existing country studies. Imports and exports of China are in fact linked to the world market, and by the year 2000 18-24 million tons of cereal imports into China are assumed.

9) Professor Richardson asked which of the related studies have been used, which have been found most useful, and what was learned from the comparison process. Dr. Hrabovszky said they made a number of comparisons and found their results to be in general agreement with those of other studies. The closest cooperation was with IIASA.

10) Dr. Lévy noted that most figures are given for 1985, and wondered if further trends had been identified. There is a danger that, for example in Brazil, the situation might become a run-away situation, almost uncontrollable. The figures are relatively reassuring, but what are the prospects beyond 1985?

Dr. Hrabovszky replied that the projections have not been carried past 1985, but that the situation may not change substantially. The fundamental issue is the question of the world economic long-term agriculture supply function. The best assumption to make in the long-term is a constant relative price of food. Food has about the lowest commercial energy component of any consumer good, so that if energy prices go up it is likely that the relative prices of other goods will increase before that of food. An estimate for the past 60 years showed a fluctuation about a relatively constant mean price.

3.4 Energy for Agriculture in the Long Term: Towards more Self-Sufficiency[1]

M. JAMEEL and A. PAPIN

INTRODUCTION

The next few decades are crucial for mankind. The world population is currently around 4 billion people and is expected to double in some 50 years. Food situation is already very serious. With total arable area being in short supply, to feed 8 billions of people at better standards is possible only through further intensification of agricultural production in the industrialized countries and major shift to the energy intensive agricultural

[1] Presented to the Conference by Dr. G. Parker.

system in the developing world.

The energy required by agriculture is a small fraction of the over-all demand, more so in industrialized countries but also in the developing parts of the world. Rational decision-making should therefore ensure that the rather small demands of the agricultural sector would continue to be met. However, since the finiteness and growing cost of fossil resources is an inescapable reality, practically every nation in the world will need to implement some program of energy conservation. In agriculture, efforts to substitute commercial energy with renewable forms represent a first-hand conservation option. At first sight, it seems that similar attempts elsewhere could probably have greater national impact. There are several reasons why, in spite of this plausible argument, efforts to reduce commercial energy input to agriculture (without, of course, adversely effecting productivity) are worthwhile. The first belongs to the domain of social philosophy and arises from the simple observation that any reduction of affluence, facility or even convenience is more equitable – and easier to implement – if it is shared by all segments of society. There are more practical reasons as well. Given the unique role of agriculture as the "food-provider" of society, it is important that it strives to achieve as much self-sufficiency as possible in the systems sense, i.e. some measure of independence from the other national systems, such as power system, oil distribution system, etc. Another incentive for introducing renewable resources in the agricultural system is that some agricultural operations seem almost custom-made to benefit from soft decentralized energy technologies.

Having briefly seen why it makes sense to exploit renewable energy resources in agriculture, we now turn to consider the practical possibilities of realization in long term. In so doing, we will refer first to the question of what global long-term agricultural development is going to look like. Then we will show a likely global agricultural energy demand futures. Finally, we will concentrate on the prospects of agriculture for being more self-sufficient in energy. Here we will refer to the prospects for application of renewables in some categories of farm operation, as well as to potential contribution of agriculture to society in terms of energy.

Much of the material below either directly refers to or is supported by the results of the global energy study carried out at the Energy Systems Program of the International Institute for Applied Systems Analysis (Haefele, 1981). Therefore we feel necessity to forego the main sections of the paper with a brief outline of the over-all IIASA approach to energy.

IIASA APPROACH TO LONG-TERM GLOBAL ENERGY DEVELOPMENT

Geographically, the world was separated into seven regions on the basis of economic structure and energy resources availability. The regions can be characterized as follows:

Region I(NA)	North America
Region II(SU/EE)	The Soviet Union and Eastern Europe
Region III(WE/JANZ)	Western Europe, Australia, Israel, Japan, New Zealand, and South Africa
Region IV(LA)	Latin America
Region V(Af/SEA)	South and Southeast Asia, and sub-Sahara Africa excluding South Africa
Region VI(ME/NAf)	The Middle East and North Africa
Region VII(C/CPA)	China and other Asian centrally planned economies

The period studied was the next half century, from 1980 to 2030, as one carrying the main burden of transitional changes in the world energy system, and being characterized with the steepest increase in global population.

The development of the two global scenarios (labelled "High" and "Low") was a principal tool of exploring future uncertainties. These scenarios are hypotheses which are believed to cover a "middle-of-the-road" range of future economic and energy development.

Within each scenario and for each region an iterative procedure of quantitative analysis was used supported with a number of mathematical models. Starting with the initial estimates of population and economic growth, it calculated consequent energy consumption patterns, alternative energy supply mixes, and their impact on regional economies, gave feedbacks for making necessary corrections in the input estimates, until a balanced state of the world was reached.

It should be noted that the IIASA scenario estimates of the

development of the agricultural sector of global economy were necessarily aggregate. In the next section these aggregate estimations are illustrated in terms of useful food energy supplies.

FUTURE FOOD SUPPLY IN THE TWO BENCHMARK SCENARIOS

There is a direct relationship between population and demand for food. If we set a goal of an energy diet of some 3200 kcal/day for each inhabitant of this earth as a way of solving the food problem, then by the year 2030 when the global population is expected to make up around 8 billion people (see Table I for IIASA population projections), this ideal would require production of some 9.3×10^{15} kcal of food per year world-wide. In

TABLE I
Population projections by region, high and low scenarios (mill. people).

Region	Population 1975	2000	2030
I	237	284	315
II	363	436	480
III	560	680	767
IV	319	575	797
V	1422	2528	3550
VI	133	247	353
VII	912	1330	1714
World	3946	6080	7976

the period of 1975-77 according to FAO estimates (FAO, 1980), the global food production averaged 3.7×10^{15} kcal yearly and provided for the world-wide average diet of 2590 kcal/cap/day. So, a 2.5-time, or a 1.6-percent yearly, increase in global agricultural production over the next 50 years expressed in calories would be sufficient to catch the target.

Under the assumptions of the High IIASA scenario as can be seen from Table II, a 3 fold increase in global agricultural value added in 2030 against the 1975 level can be expected. And the

Low scenario supposes a 260-percent increase. Is it much enough to feed the people at the best today's standards?

TABLE II
Agricultural value added by region, high and low scenarios.

Region	1975 $ bill	1975 %%	2000 $ bill	2000 %%	2030 $ bill	2030 %%
High scenario						
I	46.8	2.8	86.6	2.1	118.9	1.5
II	137.6	10.7	222	8.1	306.3	4.0
III	138.3	5.8	270	4.5	292.3	2.5
IV	41.5	12.2	96.7	7.6	164.2	4.6
V	122.7	36.1	307.8	25.5	565.1	16.2
VI	13.3	7.0	36.9	4.1	67.1	2.3
VII[a]	92.8	29.0	197.2	21.0	367.5	15.0
Low scenario						
I	46.8	2.8	70.1	2.3	83.4	2.0
II	137.6	10.7	228.2	9.4	329.9	7.0
III	138.3	5.8	195.9	4.4	199.7	3.0
IV	41.5	12.2	87.2	9.5	144.9	6.5
V	122.7	36.1	273.5	29.6	462.8	23.2
VI	13.3	7.0	32.1	5.0	52.4	4.0
VII	92.8	29.0	165.6	24.0	255.5	19.0

[a] All estimations given in the paper on Region VII should be considered in the light of general lack of reliable data on that region.

To make this sort of estimation, we referred to the history of global agricultural development. Namely, we sought for correlations between (1) agricultural value added per capita of economically active population in rural areas (as a measure of labour productivity), (b) agricultural contribution to GDP as a reflection of general level of economic development, and (c) calorific content of agricultural value added. A cross-country analysis covering some 90 nations of the world was performed on the basis of FAO data (1980) and the data released by the System Research Center (SRC) of Case Western Reserve University (1975 and 1976).

The results of the analysis show that in primitive agricultural economies characterizing with labour productivity of some 250 $VA per person employed per year or below, the calorific value of each dollar of output is very high – up to 60 kcal/$VA/day. Pure food orientation of agriculture, low rates of food wastes, small share of livestocks, lack of food stocks, low wage rates, etc., are the reasons for that. With steps toward the regular agricultural system along the extensive path the number of calories per unit of output decreases steeply. For instance, to the level of 1000 $VA/cap it falls down to 15-15 kcal/$VA/day. And it becomes flat after a certain (high) level of agricultural advancement is achieved. Extrapolating this dependency we made a rough estimation of long-term trends in the energy diet consistent with those assumptions. Table III reports on that.

TABLE III
Energy diet assumptions, high and low scenarios, in kcal/cap/day.

Region	1975	2030 High	2030 Low
I+II+III	3377	3450	3420
IV+V+VI+VII	2263	2940	2800
World	2590	3070	2960

So, both IIASA scenarios imply that the 2030 world-wide average energy diet would be around 3000 kcal/cap/day. Though being still lower than the "ideal" diet above, it meaningly overlaps the level of daily food intake per person necessary for survival. One should not forget regional and national differences, however, which are only partly displayed in Table III. For Region V, for example, the analysis envisages the energy diet being in the range of 2580 to 2740 kcal/cap/day by 2030, upper bound of which is only slightly above the survival level. This tells that actual food supplies available for the developing world would remain dependent on the willingness of industrialized countries to produce more food than could be consumed domestically and the ability of developing countries to pay.

Now let us turn to the question as what the prospects of agricultural development assumed in the two benchmark scenarios mean in terms of energy requirements.

AGRICULTURAL DEMAND FOR ENERGY IN THE TWO BENCHMARK SCENARIOS

In 1975, the world was consuming primary energy at an average rate of 8.2 TWyr/yr[1] or 5.7 TWyr/yr in terms of final energy (that is converted and delivered to end-users). From the latter some 0.1 TWyr/yr (or less than 2%) was direct input to agriculture (including mechanization and irrigation) and roughly 0.12 TWyr/yr more went to the fertilizer production as feedstocks and process energy. Thus, total contribution of agriculture in global energy consumption made up around 3.5% at that time.

From the viewpoint of energy intensiveness of agriculture (energy input per unit of value added) there are distinct differences between regions. For instance, in 1975 in Region 1 energy intensiveness of agriculture (without fertilizers) was some 25 times higher than that in Region V. In average, the developed regions consumed 10 times more energy per dollar value added than the developing regions (excluding region VII).

The future scales of agricultural needs for energy largely depend on general conditions of long-term agricultural development. The key point here is availability of arable land. It is generally agreed that the land resources of the world are potentially capable of supporting large increases in agricultural output. However, because of a number of reasons not much from these resources can be assumed to be developed in a long time to come.

1) an estensive low-yielding agricultural system is not capable of improving global food situation for very long.

2) great share of potential arable land falls on the countries which already enjoy sufficient land availability.

3) development cost for a large portion of land resources is too high, and

4) in many cases new arable land is available in remote areas unattractive for people.

These points imply that essential agricultural productivity improvements must come from increases in the use of fertilizers, irrigation, and farm mechanization. The IIASA scenarios took

[1] TWyr/yr equals roughly 1 bill. tce or 5.15 bill. boe.

these factors into account. Below we will outline the scenario projections on energy requirements by agriculture separately for on-farm operations (including irrigation and mechanization) and fertilizer production.

Direct energy inputs to agriculture

While projecting almost no changes in energy intensiveness of agriculture in the developed regions, the scenarios assumed a 10-time increase in this factor in the developing world over the next 50 years. Thus, by 2030 the average energy intensiveness in regions IV, V, and VI would be about the same (2.8 kWh per dollar value added) as the present average value for the developed regions. The final energy used in agriculture in 2030 would be about 45 and 37 times the 1975 level in the developing regions and just 2.4 and 2 times in the developed regions (High and Low scenarios correspondingly – see Table IV.) If in 1975 total agricultural consumption of commercial energy in the developing regions was 10 times lower than in the industrialized regions then already by the year 2000 these two are expected to become of the same order of magnitude, and by 2030 the developing regions would overrun the developed regions in this respect by nearly a factor of 2.

TABLE IV
Direct energy input to agriculture in high and low IIASA scenarios (GWyr/yr).

Region	1975	High 2000	High 2030	Low 2000	Low 2030
I+II+III	89.8	174.4	217.3	150.5	183.0
IV+V+VI+VII	8.8	139.0	415.5	197.4	329.5
World	98.6	313.4	632.8	343.6	512.5

Here is a rough idea of the contribution of main on-farm operations in total consumption of final energy in the year 2030 as implied in the scenarios. In regions I, II, and III some 2% of total energy input would be used for irrigation (compared with 1.3% in 1972/73 – FAO, 1977), another 8 to 10% for drying crops, and the remainder would feed farm machines (including transportation). In the developing regions, 4 to 5% of total commercial energy input would be used for irrigation purposes and the remainder would be consumed by machines. As far as drying energy is concerned, it is expected to be a negligible share in consumption of commercial energy – in developing countries drying is believed to continue to be invariably accomplished with sunshine.

Energy to fertilizers

In the IIASA analysis energy needed for fertilizer production was originally counted in the basic materials manufacturing sector and in feedstocks production. In this subsection we try to estimate the order of magnitude of this component of agricultural energy use in pure form.

In 1972/73, total production of fertilizers in the USA was 18.8 mill. ton and around half of this amount were nitrogenous fertilizers (FAO, 1974). At the same time, almost 90% of all process energy consumed in fertilizer production was associated with nitrogenous fertilizers. As much as 79% of process energy were consumed by ammonia production plants alone (Sittig, 1979). Keeping these in mind we based our approach to the assessment of future energy requirements by fertilizer production on more detailed considerations of nitrogenous fertilizers with rough accounting – by generating adjusting coefficients – for other types of fertilizers.

In 1975 global production of nitrogenous fertilizers made up 42.4 mill. tons (N). More than 80% of this amount were consumed in industrialized countries. It resulted in an average "nitrogen intensiveness" of agriculture in these countries equalled to 1.2 kg(N) per dollar value added. In the developing world the corresponding average value was 0.3 kg(N) per dollar. According to the UNIDO estimation (UNIDO, 1978), in the period from 1978 to 2000 average growth rates of demand for nitrogenous fertilizers will be 4% annually in the developed regions and 6% per year in the developing regions. This demand being met would mean growth rates of nitrogen intensiveness during the same period of time of around 1.8% yearly in the developed regions and some 5.6% yearly in the developing regions. Taking these figures for granted and assuming (quite arbitrary) that half these growth rates would be the case after the year 2000, we estimated consumption rates of nitrogenous fertilizers by IIASA world regions over the 50 years to come. Assuming specific process energy requirements as 1.9 kWya per ton (N) and feedstock requirements as 0.82 kWyr per ton (N) (Sittig, 1979) and taking adjusting coefficients in the range of 1.05 to 1.1 (depending on the fertilizer production structure of a region considered) we came to the following rough figures of total energy needed for fertilizer production in 2000 and 2030 (Table V). As this table shows, in the developed regions total energy input to fertilizer production might increase by a factor of 3 by 2030 when compared with 1975. In the developing regions corresponding figure would be 20.

TABLE V
Energy needs in fertilizer production in high and low IIASA scenarios (GWyr/yr).

		High		Low	
Region	1975	2000	2030	2000	2030
I+II+III	103.5	237.0	350.0	202.5	337.0
IV+V+VI+VII	22.3	146.0	448.0	127.0	354.0
World	125.8	383.0	798.0	329.5	691.0

Share of agriculture in global demand for energy

As was already said above, in 1975 around 3.5% of total commercial energy consumed in the world came to agriculture (including fertilizer production). In the long term the share of agriculture in total energy consumption is going to increase, largely due to rapid growth of energy intensiveness of agriculture in the developing world. With all above reservations, under the assumptions of High scenario the share of agriculture in global demand for final energy may reach 6% by 2030. In Low scenario it may make up some 8% by that time. In Low scenario the figure is greater which is consistent with the assumption that under any economic circumstances accelerated development of the world agriculture should be provided for.

From all forms of commercial energy, liquid fuels and electricity are the two which direct agricultural energy needs are overwhelmingly satisfied from. As far as oil is believed to be in short supply in the long term, it is important to check against agricultural demand for liquid fuels on the background of overall demand for liquids of the globe as a whole. It is done by the means of Table VI.

TABLE VI
Agricultural demand for motor fuel as share of global demand for liquids in high and low IIASA scenarios (TWyr/yr).

		High		Low	
	1975	2000	2030	2000	2030
Global demand for liquids	2.85	5.46	10.36	4.37	6.69
of which agricultural demand	0.08	0.22	0.43	0.19	0.34
agricultural demand in %% to total demand	2.8	4.0	4.1	4.3	5.1

As the Table shows, no dramatic increase in agricultural share in liquid fuel demand of the world is going to occur in long term under the assumption of both scenarios. The values of 4.1 to 5.1% which might be reached by 2030 are very modest.

Thus, the IIASA estimations of agricultural energy demand indicate that, in spite of continuation of the energy intensive agricultural system in the developed regions and impetuous growth of energy intensiveness of agriculture in the developing regions which are projected to be the case over the next 50 years, agricultural development is not going to make any special requirements of the world energy development, above and beyond those experienced with today or those by other sectors of the economy. Nevertheless, in view of growing oil price and possible oil shortages any measures toward increasing self-sufficiency of agriculture in energy through wider use of renewable energy resources are of high importance, especially in the developing regions, as these measures can help to secure interruption-free energy supply necessary for stable food production. Some prospects on utilization of renewable energy resources for agricultural purposes are examined in the next section.

CONTRIBUTION OF RENEWABLE ENERGY SOURCES TO AGRICULTURE

Not all agricultural operations can switch to the use of renewable, instead of fossil, energy. There are a certain irreducible demand for fossil fuels, e.g. when used as feedstock in manufacturing fertilizers. Also, for operating high-power machinery like tractors and threshers, oil or grid electricity would appear to be irreplaceable for a long time to come. However, there are at least two categories of farm operation, pumped irrigation and crop drying, which seem to hold the greatest promise for application of renewables and the prospects for these will be discussed in this section.

The indirect use of sunshine, through photovoltaic conversion to electricity, for running low-lift water pumps is considered to have a high potential for the medium-term future in countries without extensive power grids (Fick and Sprague, 1980). The characteristics of irrigation pumping needs make the well-known disadvantages of solar energy (low energy density, interruption, and supply-demand matching difficulties) largely irrelevant: 1) since irrigation pumps usually employ motors in the 0.3-0.5 kW range, the already achieved 10% conversion efficiency would imply only a 3 meter x 3 meter collection array in middle latitudes and correspondingly smaller in the sunnier arid areas, 2) water can be stored without much difficulty or expense, so that the inability to run solar devices at night is not a constraint here (the actual usage of even diesel pump sets is only 2000-3000 hours per year), and 3) since one does not need to irrigate when and where it rains, there is an automatic matching between supply and demand.

Thus, the convenience advantages of oil or grid electricity over decentralized renewables are unimportant in this case and there remains only the competition in cost. In case studies done for Tanzania, Chad, Bangladesh, India, and Pakistan (Fick and Sprague, 1980), solar pumping has been shown to be competitive with diesel power at 1977-78 oil prices provided solar-cell module costs decrease to a level around $1 per peak-watt. The cost of solar cells has dropped in less than three years from $20/Wp to $5/Wp and there is every likelihood that the USERDA goal of $0.5/Wp by 1985 would be closely approached if not actually met. With climbing oil prices, it is thus reasonable to expect that photovoltaic water pumping would become preferable to diesel pumping by the mid-eighties even taking the longer duty cycle of the latter into account.

The next question is: how much penetration of the photovoltaic technology could one expect in the irrigation power market in the next 50 years? We estimate very roughly that the photovoltaic technology could capture about 7% of the irrigation operation energy market by the year 2000 and about 20% by the year 2030. The total magnitudes are, world-wide, about 0.66 GWyr/yr by 2000 and 4.2 GWyr/yr by 2030. These appear insignificant in the context of the total energy market but do alleviate the situation in regions where the crisis matters most. Also, considering the fact that the present sales of photovoltaic systems are in the range of 1 MW (peak) annually (Fick and Sprague, 1980), this provides a sufficient expansion potential to stimulate development and production for penetrating other power markets, particularly the rural areas of developing countries.

Wind energy can also be harnessed to lift water and it has in fact been so used since antiquity. It is more variable than solar energy and is in addition locally dependent. These factors make it more difficult to estimate the likely contribution of this resource. The IIASA study provides a rough indication of the user-oriented potential of wind as 50 GWyr/yr. How much of this can one expect to be realized by, say, 2030? There are several points to be considered. First, although wind-mills appear to be "low-technology" compared to solar cells and thus more appropriate for rural areas of developing countries, in fact their maintenance and operation demands more skill from the end-user than that of solar cells. Second, industry's interest in solar cell technology development is largely a consequence of the success story of semi-conductors; one does not discern a similar enthusiasm for windmill design improvement and so cost reductions are uncertain. However, under Kenian conditions of labor and energy costs, locally manufactured wind-pumps systems can be economically viable. With these uncertainties and the strong dependence on local factors, we anticipate that wind power's maximum contribution to irrigation would be around 5% of IIASA potential, i.e. around 2 GWyr/yr. This is still about 10% of the global irrigation energy demand estimated for 2030. To summarize, we feel it is feasible that as much as 30% of the irrigation energy requirement in the year 2030 could be supplied by solar photovoltaic systems and wind-mills, with higher penetration in the developing than industrialized countries.

Let us now turn to another farm operation in which solar

energy can and does even now play a direct and vital role, vis. crop drying. For the developed regions, this is a very small amount of energy and it is unlikely that farmers would look toward some other drying methods (e.g. exposing to sun under polyethylene tents) at least until the turn of the century. After that, with oil price rising, solar energy could be utilized where available. A penetration of 15-20% would, we believe, be the upper bound and a figure of 10% more likely. Thus 2-4 GWyr/yr of fossil fuel could be displaced by solar energy by the year 2030. This is of the same order of magnitude as in water pumping for irrigation and, locally, the substitution could have significant benefits. In developing countries, drying is almost invariably accomplished with sunshine – as standing crop or after harvest – so there is no question of substitution of fossil energy by renewables.

ENERGY FROM BIOMASS

For the sake of completeness and comparison, we include in this section some activities such as forestry which are not directly agriculture-related but do nonetheless contribute to biomass energy resources. Specifically, we will consider crop wastes, animal manure, fuelwood, forestry residues, energy plantations, and municipal solid wastes.

The biomass resources base

Before discussing possibilities of exploitation, it is necessary to form a rough idea of the magnitudes one is talking about. The aim therefore is not four-figure accuracy but to present estimates and to see how poor they really are. In Tables VII and VIII the estimates on the biomass resources base of the world in the late 1970's are presented, as well as extrapolated to the years 2000 and 2030 on the assumption that the residues would increase in proportion to the total or urban population as applicable. Population projections are taken as in Table I. Suggested calorific values are: crop/animal/human wastes: 3500 kcal per kilogramm of dry organic matter; wood residues: 4200 kcal per kilogramm of air dry wood; green wood: 4 m of green volume yields approximately 1 kWyr thermal;

TABLE VII
Estimated biomass resources base of the world in the late 1970's

Resource	Region I	II	III	IV	V	VI	VII	World
Crop residues	250	340	250	170	380	70	280	1740
incl. cereals	248	340	234	63	287	65	277	1510
sugarcane	7.4	0	13	103	92	3.6	10	230
Animal waste	190	205	235	300	385	65	155	1530
incl. cattle & buffaloes	175	150	170	270	330	36	83	1210
sheeps & goats	3	24	37	15	41	29	24	175
pigs	10	24	17	11	4	0	38	105
poultry	3.4	8.8	9	5.6	7.8	2	10.7	45
Human refuse	150	160	245	45	135	15	95	850
incl. night -soil	19	29	45	26	100	10	65	295
municipal solid waste	130	130	200	20	35	5	30	550
Nat. forests[a]								
gross increment[b]	800-1000	700-1050	430-730	1200-3000	530-1700	60-210	70-260	3800-8000
logging residue	65-125	50-100	35-70	20-60	35-100	5-15	20-60	230-530
mill residue	125	100	70	17.5	30	3.5	17.5	365

a in mill. tons of air dry wood per year; remaining figures expressed in mill. tons of dry organic matter per year if not mentioned otherwise

b in mill. cubic meters of green volume

Exploitation of bio-resources

Several exhaustive studies have recently appeared dealing at length with various aspects of biomass utilization. Our purpose here is to explore the possibilities of obtaining energy from biomass in a regional context using the resource estimates developed in preceding subsection. We will look one by one at the various ways in which energy is or can be extracted from organic matter.

TABLE VIII
Estimated energy content of biomass residues (GWyr/yr)

Resource	Region I	II	III	IV	V	VI	VII	World
Crop residues								
late 1970s	120	160	115	80	180	30	135	820
2000	145	190	140	145	320	55	195	1190
2030	160	210	160	200	450	80	250	1500
Animal waste								
late 1970s	90	95	110	140	175	30	75	710
2000	110	115	130	250	310	55	110	1080
2030	120	125	150	350	435	80	140	1400
Human refuse								
late 1970s	70	75	110	20	60	7	45	380
2000	80	85	135	40	120	13	45	550
2030	90	100	145	65	200	22	125	750
Wood residues								
late 1970s	105-140	85-110	60-80	20-45	35-70	5-10	20-45	330-500

Direct combustion

This is the traditional technique for releasing the energy of the chemical bonds and is still practised all over the world for different purposes and with different efficiences. In the United States, an estimated 11.6 million dry tons of wood residues were used to provide heat and cogenerated electricity to the timber industries in 1970, while in Region V, the IIASA estimates put "non-commercial" fuel use in 1975 at 370 million tons of coal equivalent mostly in the household sector. These two categories of use represent the extremes of the spectrum of industrial development as well as of end-use efficiency.

Against this background, we do not envisage domestic burning of wood or waste in the developed region. However, considering the size of the resource, industrial and commercial use ought to forge ahead, the most promising applications being (a) the on-site burning of mill residues for raising steam and co-generating electricity (logging residues are not suitable for this and would be considered later), and (b) incineration of municipal waste to provide district heating.

In the developing regions, fuelwood and user-collected waste would doubtless continue to serve the needs of some fraction of the low-income population. In 1975, according to the IIASA study, non-commercial fuel supplied 44 GWyrs (all from wood) of primary energy to Region II, 109 GWyrs (80% wood) to Region IV, 344 GWyrs (67% wood) to Region V and 10 GWyrs (all wood) to Region VI. Using the postulated end-use efficiencies, we find that – as a fraction of cooking energy requirement in useful terms – non-commercial sources provided 25% to Region II, 60% to Region IV, 70% to Region V and 15% to Region VI. Obviously, all these people would be greatly beneffited if more efficient end-use devices were developed and popularized among them. In the spirit of our paper, we assume that this would be done so that over the next fifty years the efficiency could be raised from the present value of 7.5% to 30% in the developing regions. In this case, all the heat requirements of rural households in the year 2030 could be met by 55 GWyr/yr primary energy of wood and wastes in Region I, 235 GWyr/yr in Region V and 35 GWyr/yr in Region VI, i.e. less than the non-commercial fuel usage at present. This would not only ease the pressure on natural forests but would also leave a comfortable surplus of organic matter for other uses. With anaerobic digestion supplementing direct combustion, the effect would be even more pronounced.

Biogas

Biogas is the name given to the combustive gas (approximately 50-60% methane, with average calorific value of 5500 kcalories per cubic meter) obtained by the fermentation of animal dung and other organic matter in the absence of air. The technology is now firmly established in China where an estimated 2 million biogas plants are already in operation. Different designs are being evolved in other parts of Asia and Africa. The technique is particularly attractive because of its three-in-one functions: providing a convenient energy carrier reclaiming the fertilizing value of manure and improving village hygiene. Depending upon the circumstances, one or the other of its functions may take on the primary role, the others being treated as bonuses. There are however problems associated with ambient temperature, proper maintenance, social acceptance etc. which would need to be resolved in the local context.

In the industrialized countries like the USA, which have large beeflots, anaerobic digesters could be installed primarily for waste disposal and, secondarily, to get the fertilizing slurry for the associated corn fields. The biogas, being of variable supply, may not be preferred for domestic use and in winter most of it may be needed to heat the digester itself. In developing countries, largely blessed with a moderate climate, the gas production is likely to be uniform and more welcome as a domestic fuel supply. It is estimated that 1 kg of dry waste produces 0.4 to 0.5 cubic meters of biogas and yields 0.7 kg of dry sludge with 2% nitrogen content (Parikh, 1978). Thus, 1 ton of animal dung, processed through a biogas plant, can deliver 2.2 to 2.7 million kilocalories of direct energy and save 0.2 million kilocalories of fossil fuel used in making chemical fertilizer. The local potential of energy production in the developing regions, by optimal utilization of animal and human waste alone (not counting crop residues), works out to nearly 300 GWyr/yr in the late 1970s and would increase in the future. Of course, the fraction of the potential actually realized will differ from region to region but the figure serves to underscore the substantial possibilities of this technology. Combined with more efficient combustion of wood and crop wastes, it has the promise of delivering energy to areas which are traditionally neglected by commercial energy suppliers.

Liquid fuels

Ethanol can substitute for motor fuel up to about 20% by volume without changes in the engine. Although it has a lower specific energy content than gasoline, some tests indicate that volume for volume there is no reduction of engine performance by the substitution presumably because of the higher octane value of ethanol. In view of the global thirst for liquid fuel, the production of ethanol from biomass is a particularly attractive use of renewable resources. As feedstock, one may use sugar-bearing crops (conversion efficiency of around 77%), starchy crops (55%) or cellulosic material (30%); process energy has to be supplied in addition and the overall energy balance depends critically on the extent to which associated residues may be burnt to provide the required heat. At present, Brazil and the United States are front-runners in producing ethanol from sugarcane and maize respectively.

Quite apart from the possibility of growing crops (sugarcane, maize, sorghum, etc.) or trees (e.g. eucalyptus) specifically for energy, there is considerable potential of exploiting existing biomass resources more profitably. For instance, crop wastes and logging residues may be collected and used both as feedstock and process fuel to distill ethanol. In this case the overall efficiency is such that 1 ton of wood yields 200 litres of ethanol without any external heat supply. Since the energy content of crop residues is lower than wood, 1 ton of residue would yield about 150 litres of ethanol. On this basis the gasoline substitution potential is significant if 80% of logging residues and 40% of cereal residues are collected and processed for ethanol. For illustration, in Region II this potential would be 28-35 billion litres, in Region IV, 6.5-12.5 billion litres, and in Region V, 22.5-33 billion litres. For comparison, the 1975 consumption of gasoline by cars in the three regions was respectively 22, 17, and 14 billion litres. Thus, we see that the energy is there (even without growing specific energy crops), and the technology is there; the size and siting of the plants would have to be determined in the context of the country concerned.

Another route for obtaining liquid fuels from biomass is by gasification followed by conversion into methanol; the ethanol route is sufficient to illustrate the potential.

Gasification

It may happen that certain regions either have liquid fuel or have surplus biomass not convenient for conversion to ethanol; the technology of gasification may then be developed to supply heat as such. Gasification may, for instance, compete with incineration of municipal refuse; depending upon convenience or economics, the supply of district heat or gas may be considered in a city. If gas pipelines already exist, the production of "synthetic natural gas" may turn out to be more advantageous.

In summary one may say that varied possibilities exist for exploitation of biomass as an energy resource. The situation may be summarized in the Table IX where relative importance of different directions of biomass use is shown.

TABLE IX
Relative importance of different uses of biomass resources

Resources	Uses 1	2	3	4	5	6
Cereal residues	xx	x	x	xxx		xx
Bagasse	xx	xxx	x	xx		x
Animal and human faeces	x		xxx			x
Municipal solid waste		xxx	x		xx	
Fuelwood	xxx	xx		x		xx
Logging residues	xx			xxxx		
Mill residues		xxx			x	
Energy crops and plantations				xxxx	x	

1 non-commercial use; 2 industrial or commercial combustion and co-generation; 3 anaerobic digestion; 4 liquid fuels; 5 synthetic natural gas; 6 non-energy uses

DISCUSSION POINTS[1]

1) Professor Meadows praised the new technique by which participants give one another's papers at conferences which is being pioneered here; since Dr. Pauker presented this paper. This guarantees that at least one person reads each paper carefully, and if the present high standard could be repeated it would increase the clarity and reduce the length of presentations. Dr. Richardson continued with the suggestion that Dr. Pauker should attend all conferences, and present all papers.

2) Dr. Pauker felt that new biotechnologies may be of great importance in the future.

Dr. Norse pointed out that innovation often follows a logistic curve and that over a 20-year time horizon it is unlikely that biotechnology will have a significant impact. Dr. Hrabovszczky

[1] The authors were not present to answer questions on this paper.

agreed with this point, that with a time horizon of 20 years little significant impact can be expected.

Dr. Richardson felt that the time horizon to the year 2000 has assumed too great an importance. In the models with a longer time horizon, a number of sharp discontinuities appear in the period 2015-2020. There is a need in modelling not only to look to boundary conditions such as environment and population, but also to consider extended time horizons.

3) Dr. Hrabovszky stated that taking into account available technological improvements, the real impact may come from the production side, not from the technological conversion side. Thus the improvement of ecoconversion by 10-20% and reduction of heat requirements may be overshadowed by a change in production from 35 tons per hectare to 120 tons per hectare of sugar cane. Also with some of the tropical tree-crops, such as palm oil and coconuts, experiments with modern varieties have given as much as 4 tons of oil per hectare, oil that can be used to drive tractors.

4) Dr. Makhijani warned that the low proportion of energy used in agriculture (3% of energy today, increasing to 5% in 2000) must not lull us into a sense of false security, because such figures refer to the total agriculture sector. In some developing countries, such as India, a majority of the energy is used in a highly commercial surplus generating area, and it is this relatively energy intensive area that is feeding the cities. Most of the population is living in a subsistence agriculture system, which is low energy intensive. As energy costs go up, the costs of food production in the high energy intensive areas must go up, probably faster than the rest of the economy.

References

FAO (1980). *1979 FAo Production Yearbook*, 33.

FAO (1977). *The State of Food and Agriculture 1976.*

FAO (1974). *Annual Fertilizer Review 1973.*

Fick, G., and Sprague, R.M. eds. (1980). *Decision Support Systems: Issues and Challenges.* Proceedings of International Task Force Meeting. IIASA. June 1980. Pergamon Press, Oxford.

Haefele, W. Program Leader (1981). *Energy in a Finite World.* A Global Systems Analysis. Report by Energy Systems Group of IIASA. Ballinger Publishing Company, Cambridge Mass. USA.

Parikh, J. (1978). *Energy Systems and Development. Constraints.* Demand and Supply of Energy for Developing Regions IIASA, Laxenburg, Austria.

Sittig, M. (1979). Fertilizer Industry. Processes, Pollution control and Energy Conservation. Noyes Data Corporation Park Ridge, New Jersey, USA.

Strauch, P.G., *et al.* (1975). Food Data Base. Data Presentation: for Regionalized Multilevel World Food Model. Vol. I. Systems Research Center, Case Western University, USA.

Summers, K. (1976). *Aggregation and Estimation of Macroeconomic Data for 1960-1974. Systems Research Center.* Case Western University, USA.

UNIDO (1978). *Second World-Wide Study of Fertilizer Industry: 1975-2000.*

3.5 The Food and Energy Barriers as Determinants of Development in East-West Economic Relations

P. BOZYK and A. BUDNIKOWSKI

INTRODUCTORY COMMENTS

Whether we look at it from the viewpoint of particular countries or the whole world economy, one of the basic conditions of economic development is the complete fulfilment of the ever-growing food and energy needs. Even though the strategic importance of fuels and food, as well as some specific characteristics of the food and energy economy (i.e immobility of the main factor of production in agriculture – land; the need to overcome social barriers to increased farm output; the rational use of energy), call for solving food and energy problems on the national scale, in most cases these problems can only be solved through international cooperation. Among determinants of such cooperation one should have the differentiation in the natural conditions of fuel extraction and, to lesser extent, food production, as well as uneven distribution of the means of production.

A particular place in economic cooperation, including co-operation on food and energy, is attributed to relations between highly developed market economics and European countries with centrally planned economies (designated further as East-

West relations). These two groups of countries are the world's basic food and energy producers and consumers. They also contribute a major part of the world's industrial output and trade turnover.

The East-West relations have so far played an important role in meeting each other's demand for food and energy. During the last few years, however, these relations have been developing at a slower pace. Should the difficulties in the East-West relations be regarded as transitory, or should they become a barrier to further development of these relations?

An answer to this question becomes the foundation of our considerations.

TRADITIONAL FORMS OF EAST-WEST LINKS

Since a number of years East-West links reflect these countries' position in the traditional international division of labour.

The Western countries have traditionally belonged to the group of industrialized countries, while the East (with the exception of the GDR and Czechoslovakia) was a part of the group specialized in raw materials and agricultural goods. Thus, the first group exported manufactured goods and imported energy and food; the opposite was true of the second group. Such a pattern of economic links, dating back to the XIXth century, still exists in its unaltered shape.

In the past such division of labour has been justified by the relative abundance of the factors of production. The Eastern countries were abounding in natural resources, whereas the West disposed of much capital.

The process of industrialization in Eastern and Central European countries has changed these proportions. The industrial development in those countries has seriously increased their demand for raw materials and energy. At the same time, the resources have become less accessible and their extraction more capital-intensive. The fast rise in the living standards of the population has sharply stepped up the demand for food; the Eastern countries have switched from being exporters to being importers of food. Such a move had its impact on the East-West links.

In order to substantiate our thesis, we will quote the data

published by UNCTAD in 1978.[1] In that year the East has contributed 8.7% of the world exports and 8.5% of the world's imports. The trade with the developed market economies has accounted for 26% of the exports and 31% of the East's global imports.

The fuels have made up for 38% of the revenues from the East's exports to the West. The share of fuels in the Soviet exports to the West was as high as 61%, while in the exports of the other CMEA countries it comprised a mere 16%. The deliveries of fuels from the East covered about 6.7% of the Western import demand, while the average share of the East in the Western countries' imports was as low as 3.4% (1978); it was half as low as the share in the delivery of fuels.

Food and agricultural products have accounted for 18.3% of total Eastern exports to the West, which is much higher than is the case of exports to other groups of countries. This share is much lower in the USSR (14.1%) than in the other countries in question (22.4%). The West buys 56% of all the agricultural products sold by the East.

In 1978 the manufactures have constituted some 31% of the East's total exports to the West (14.2% of the Soviet exports, 48% of the other countries' exports). The total exports of manufactures from the East constituted only 2% of the Western demand for this type of products.

The situation was quite different when we look at these countries' imports. The West has been meeting nearly 31% of the East's import demand (37% in the case of the Soviet Union. 27% in the case of the other countries). This significant share was related mainly to the East's demand for imported food (27% of the USSR's demand for imported food, 38% for the other Eastern countries). For Western exporters of food, the Eastern countries bought 4.6% of their total food exports (and 9.1% of food exports from the USA).

The fuels had a nearly imperceptible role in Western exports to the East; in 1978 their value accounted for about 1% of total Eastern imports. This share has been quite significant, however, in imports from the developing countries, which covered 17.5% of import needs in the East (58% of USSR imports and 12% of the remaining countries).

[1] UNCTAD. Handbook of International Trade and Development.

A major share in the East's imports from the West was held by the processed manufactured goods. Their share in Soviet imports from the West amounted to 62%, it was as high as 75% in the remaining Eastern countries. Western deliveries covered 36% of total import demand for these goods; these deliveries, however, accounted for only 4% of total Western exports of manufactures.

To depict the scope of energy and food links between the East and the West our analysis will resort to the use of chosen structural coefficients i.e. delta coefficient and correlation coefficients.[1]

The delta coefficient expresses the relation between the real value of exports from region "i" to region "j" to its hypothetical value. This value bases on an assumption that the geographical distribution of international turnover is proportional to the share of total imports and exports of particular regions in the world trade (the so-called level of independence). The coefficient of correlation, adapted by J. Kotyński to evaluate the intensity of international trade links, is comprised within −1 to 1 interval. It is closely related to the delta coefficient since it is a product of delta coefficient (decreased by a unity) and the normative factor whose value depends on the share of the exporting and importing regions in the world trade.[2]

In the Tables at the end of this paper one may find delta and correlation coefficients, both magnified 100 times, for global network of trade exchange and world turnover in food, agricultural products and fuels in 1978.

---

1 delta coefficient: $\delta_{ij} = \dfrac{Z_{ij}}{Z_{i.}\, Z_{.j}}$

correlation coefficient:

$$r_{ij} = \frac{Z_{ij} - Z_{i.}\, Z_{.j}}{\sqrt{Z_{i}\,(1-Z_{i.})\, Z_{.j}\,(1-Z_{.j})}} = (\delta_{ij} - 1)\sqrt{\frac{Z_{i.}\, Z_{.j}}{(1-Z_{i.})\,(1-Z_{.j})}}$$

where: Z_{ij} – share of the export flow from region i to j in the total world trade

$Z_{i..}\ Z_{.j}$ – share of region "i" ("j") exports (imports) in the total world trade

2 J. Kotyński: Structure of International Trade. Methods of Appraisal (Struktura handlu miedzynarodowego. Metody oceny, PWE, Warszawa 1979).

As the analysis of delta coefficients shows, in 1978 the total exports of the East to the Western countries accounted for 39% and the imports from the West – 46% of the value attributed to the "independence level". In the case of fuels, however, the value of Eastern exports to the West corresponded on the average to two-thirds of the "independence level". The exports of fuels to Western Europe were some 20% higher than the value corresponding to the distribution of the Eastern exports to particular regions proportional to their share of global imports in the world turnover in fuels. In particular, the exports of fuel from the East to the EFTA countries was three times higher than the matching level of independence. The exports to the EEC constituted nearly 90% of such a level. The above-average coefficients showed in the exports of the remaining Eastern countries (USSR excluded), which was mainly related to the exports of Polish coal. Seen both from the side of the imports and the exports, the East-West trade in food corresponded to about a half of the "independence level". The imports from Northern America were close to the hypothetical value, while exports of food from the East (USSR excluded) to EFTA were some 70% higher from the value deducted from the null hypothesis. Relatively high delta coefficients were also observed in exports of Eastern agricultural and forest products, especially to the EFTA countries, as well as in imports of these particular goods to the East, especially Soviet imports from Australia and New Zealand.

The same general conclusions can be reached through the use of normative measures such as the coefficients of correlation to evaluate the intensity of trade links between the East and the West. Calculations show that the intensity of East-West links in fuels and food in 1978 was higher than the mean intensity of links in East-West trade. It confirms the fact that the traditional structure of the international division of labour between the East and the West has remained unchanged.

BARRIERS TO FURTHER GROWTH IN EAST-WEST RELATIONS

The preferential treatment of industrial development in the European centrally-planned economies was reflected in the growing share of the industry in these countries' national income. The growth rate of the industry in the discussed period was higher than of the remaining sectors of the economy, especially higher than the growth rates of the agriculture and the extraction industry. In all these countries the share of the industry in generating national income has reached (or surpassed) 50%.

The highest dynamics of growth were noted in the processing industry, especially in the heavy, electromechanical and chemical industries. Their development stimulated the demand for raw materials and energy. This demand was satisfied through own resource base or, if necessary, imports from the USSR and Poland.

In the long term this solution became unfeasible. The development of the extraction industry is increasingly capital-using; the deposits of minerals are less accessible and their extraction more difficult. This called for more mechanization and automation of work. The investment outlays were sharply rising. Given the existing potential of the processing industry, this type of expansion creates a barrier to development. The arising capital-intensive methods used in the extraction of raw materials (and, therefore, its falling efficiency) have a negative impact on the dynamics of development in the East.

This phenomenon is further aggrevated by the highly resource-using and energy-using types of industrial production developed in the European centrally-planned economies. These branches of production were developed as the result of extensive methods of industrial development; the technological modernization performed in the Seventies has not really altered the general development trends.

For these general reasons the Eastern countries are becoming net importers of raw materials. Such a change affects the prospects of the future East-West relations. These relations can no longer base on the exports of energy-bearing raw materials and food, since these commodities are in short supply in the East.

This type of reasoning has been confirmed by a prognosis elaborated in the Foreign Trade Institute in Warsaw.[1] Using

[1] This prognosis has been prepared by Dr. A. Czepurko. He has used a number of mathematical functions to describe the dependence of East-West trade on general evolution of trade in the West. The basic equations used were these of a trend. The author has analysed four different equations, estimated by the method of minimal squares, and finally chose two of them in which the empirical sequence was closest to the equation of the trend.

the development tendencies characterizing the East-West trade in the 1967-79 period, a projection of its commodity structure in 1985 was established. This projection points to further falling dynamics in the exports of fuels, raw materials, materials for production and food. As the result, the general development rate of East-West economic relations can be slowed down if the socialist countries fail to step up their exports of industrial goods to the developed Western economies. In other words, the commodity structure of East-West trade in the Eighties has to include a growing share of manufactured goods, the exports of which have to be stepped up by the Eastern countries.

To what extent will the changed commodity structure affect the energy and food problems of the two groups of countries? In our opinion, the solution of these problems will be possible when the East-West trade links experience profound changes, not only quantitative but also qualitative.

THE DIRECTIONS OF CHANGE IN THE EAST-WEST TRADE LINKS

These changes should be complex and should embrace different fields of economic activity.

The most important issue is the coordination of economic policies pursued in the East and in the West. So far, this policy has been conducted autonomously, as expressed in each group's desire to achieve self-sufficiency in food and energy. There's a lack of long-term projects aiming to solve the food and energy problem. Our suggestion can be narrowed down to wider use of the simplest form of coordinating economic policies – the exchange of information on capacities and needs of the two groups of countries. In some chosen fields it is possible to consult each other on the directions of the conducted economic policy. There's not much opportunity, so far at least, for common economic policy. The existing forms of coordination in economic policy can be regarded as an introduction to further stages of coordination,

To practice, there exists a number of institutionalized channels for exchange of information such as the United Nations, other international organizations, specialized agendas etc. An important role in developing East-West relations is played by the European Economic Commission. On the whole, all these channels fail to provide a unique, coherent system of exchanging information between particular countries, groups of countries and economic regions. Hence, we see the need to elaborate such a system and the United Nations University should play a key role here.

Special importance should be given to the exchange of information on energy and food policy. In the conditions of growing food and energy barriers this type of information is very essential to general economic policy. The opposite is also true: the informations about the country's economic policy, its natalistic policy and other factors determining the demand for food and energy are essential for the food and energy policy.

It is obvious that the national economic policy cannot be subordinated to any transnational strategy. The coordination of economic policies (through the exchange of information) can take form of promotion for certain energy-saving patterns of consumption, model of small family etc. These informations are quite important as far as their impact on supply and demand is concerned.

Of much importance is also the energy-saving direction of industrial development. This field concerns especially the Eastern countries. Their industry is extremely energy-using. Should this situation change, these countries could find reserves of energy, which would affect the energy situation of the East-West group. Our reasoning bases on the assumption that the rising energy needs of the world will to a lesser degree be satisfied by new access to natural resources but they will be increasingly satisfied through improvements in the use of energy such as energy-saving technical progress, rationalization in the structure of production etc.

This rationalization should be promoted by the development of East-West cooperation in production and science.

The cooperation in production should be founded on developing specialization. In the field of the power industry it can include both the search and processing of traditional and new sources of energy, as well as more rational use of all available sources of energy. The development of cooperation in this field should also include the deepening division of labour in the production of power machines and equipment. The Eastern countries, especially the USSR and Poland, have reached a high level of production in some mining equipment, furnishings for mines, drilling equipment, great power units, power equipment etc. This production offers a real basis for specialization. The development of specialization in other industrial branches might also improve the energy situation for instance through lengthening the scale of production and thus a decrease in the input of energy per unit of output.

The deepening specialization of production should be accompanied by the development in the production cooperation. This process is favoured by the similar level of economic development and the complementary economic structure of the two groups of countries. This complementarity is reflected in the lead of the West in production of high-technology goods, whereas the East has an advantage in the production of energy-using goods. The development of cooperation between the two groups can lead to less energy-using production in the West (transferring this production to the East) and to less energy-using production in the East as well (due to the transfer of Western technology). Taking into consideration how energy-using is their industry, the Eastern countries are still more lavishly provided with fuels and energy. This advantage means that they can still continue the exports of fuels and energy not only through the increase in their output, but also through economies in their use. The same criterium can serve as a basis for cooperation in production.

In order to counteract the growing energy barrier, the East and the West should cooperate in investment projects. Of particular interest here is the more effective use of the resources in the East. So far, these countries have resorted to extensive use of their resources. The opportunities for cooperation lie in the development and transfer of new technologies and capital engaged in extraction and processing of resources. The experience of the Seventies shows that the existing opportunities are enormous and the effects drawn by the West (increased deliveries of fuels and energy) quite perceptible.

Far behind is the development of joint ventures. As previous experience has shown, the interest of Western firms in joint ventures with Eastern partners is limited. Also the interest of Eastern enterprises seems to be slight. Among the numerous barriers to such undertakings one should name the differences in functioning of the economy, seller's market in the East and buyer's market in the West etc.

Countering the food and energy barriers calls for development of East-West scientific and technical cooperation. This cooperation conditions the acceleration of technical progress in the East and it opens the way to less energy-and material-using production in these countries. Such an opportunity might be created through the shift from relatively simple forms of cooperation (the transfer of scientific and technical information, licensing, know-how) to more complex cooperation links.

In the Eighties, we should expect certain changes in the geographical structure of East-West economic links. Their direction will be determined by the opportunities to develop trade, cooperation in production, science and technology. These opportunities are determined by the political situation, business outlook, credit facilities etc.

The potential conditions for further development of cooperation in food and energy exist especially in Europe. It's the Eastern European countries which are most concerned with this type of cooperation. Both the official statements and economic practice point to the fact that economic integration of these countries within CMEA is of open type and it doesn't limit the development of cooperation with countries from outside the CMEA. At the present stage of development, the wider exchange with the capitalist countries becomes one of the conditions of further development. This exchange is stimulated by the situation in the balance of payments, which forces the socialist countries to step up their exports rather than reduce imports. Through cooperation with the West, the East can take advantage of the access to modern technology, modern construction design, know-how etc.

The Western Europe has also a number of reasons to be interested in developing cooperation with Eastern Europe. It is especially concerned about the possibility of enlarging the scale of production due to the expansion of markets. In question is the reduction of barriers in fuels and raw materials which can be overcome through the changed structure of production etc.

The processes of regional economic integration are an objective phenomenon. They result from the tendency to internationalize economic life in the East and the West. Even though these processes take place in different political, social and economic conditions, they don't have to "close" each of the blocks and narrow the opportunities for development of their economic relations. Both sides, however, have to engage in a number of undertakings, especially EEC-CMEA agreements, in order to preserve the permanent character of development in their mutual trade links.

It's hard to define the development of economic relations between the Eastern Europe and the United States; the state of these relations is always subordinated to political factors. Potentially, the United States' market interests all countries in the world, the East Europe included, due to its most modern technology and science. At the same time, this market disposes of the world's largest agricultural output and plays the role of the most important exporter of grain and fodder.

Despite the enormous potential, one shouldn't expect that in the Eighties the hitherto modest trade links between the East and the USA will be transformed into quantities significant to both sides. It applies especially to the relations between the Soviet Union and the United States. Due to political reasons, the Soviet Union has no "Most Favoured Nation" clause in the United States; it also has no access to government crediting institutions and American experts of technology, machines and equipment to the USSR are subject to strict strategic control. While this control has been slightly eased in the beginning of the Seventies, it has now become very strict again.

Our guess is that in the Eighties the United States will treat small and medium-sized socialist countries selectively. Some of the countries, enjoying the MFN clause, will be treated more liberally than the rest of the socialist countries. The access to government credits, technology, sources of energy, grain and fodder will probably remain differentiated.

Some influence on the state of relations with the North American region will be exerted by the economic situation in the socialist countries, their level of output and the competitivity of their products. The North American market has always been very demanding on the exporters due to high requirements of American customers in the field of technical level, quality and delivery time. This market will continue to have a different system of legal institutions protecting the customer, different security, sanitary and technical standards etc. For the socialist exporters who are used to less strict quality standards and the seller's market, the American market will remain difficult. This market is a place of tough competition between domestic producers, West European exporters and lately – from exporters coming from developing countries.

In the Eighties, there should be a boost in the economic relations between the East, West and South. While at present these relations enter their initial phase, their development depends on the ability to bring together the capital, technology, technical supervision and deliveries of machines and equipment.

The potential increase in the socialist countries' share in such undertakings is related to the growing interest of all partners in question. Such cooperation offers many advantages to the developing countries mainly because it prevents the creation of capital, technical and organizational monopolies controlled by the industrialized Western economies. At the same time there appears an opportunity to integrate the technologies, qualifications and experiences coming from all participants of the undertaking. The socialist countries have a lot of experience in the construction of large industrial plants. They also have the capacity to erect them and manage them with qualified technical staff. The capitalist countries can offer modern technology and crediting facilities. Finally, the developing countries may contribute their raw materials, materials for production and relatively unskilled labour.

The use of the existing potential in East-West-South relations calls for elaboration of institutional links, especially crediting institutions. We are thinking here of institutions created directly to develop this type of triangular cooperation, as well as general institutions corresponding to shifts in the international economic order. One doesn't expect the concept of the so-called New International Economic Order to be fully operational in the Eighties. Some success might be reached in the elaboration of some partial concepts related to this "new order". Support should be given to the approach which calls for such "order" which will favour the global optimalization of development processes, including optimalization for all groups of countries, regardless of their social and political systems. Such a global solution favours the promotion of both the East-West and the East-West-South relations.

TABLE I

Intensity of total world trade /SITC 0-9/ in 1978 by selected regions, as measured by structural coefficients.

Destination	Planned economies of Eastern Europe			Developed market economies				Developing countries
Origin	Total	USSR	Other	Total	EEC	USA	Japan	
0	1	2	3	4	5	6	7	8
A. Delta coefficients /x100/								
E. Europe	653	527	742	39	42	10	22	65
USSR	558	.	954	41	44	6	39	92
Other	735	985	558	37	41	13	8	42
Developed	46	55	40	106	116	86	59	102
EEC	44	44	44	115	154	50	19	79
USA	31	45	20	90	65	.	169	154
Japan	39	73	14	71	34	199	.	197
Developing	44	66	29	106	78	179	240	104
B. Correlation coefficients /x100/								
E. Europe	52.0	25.2	45.3	–26.4	–12.7	–10.7	–5.7	–6.0
USSR	28.6	.	40.0	–17.0	–8.2	–7.5	–3.0	–0.9
Other	42.7	37.4	23.1	–19.5	–9.3	–7.3	–4.9	–7.1
Developed	–23.4	–12.3	–19.6	13.0	16.2	–8.0	–14.0	1.3
EEC	–12.7	–7.9	–9.6	16.0	28.5	–14.3	–8.7	–8.6
USA	–7.3	–3.6	–6.3	–4.7	–8.6	.	5.7	10.4
Japan	–5.3	–1.5	–5.6	–11.7	–13.3	10.9	.	15.2
Developing	–9.3	–3.6	–8.9	4.8	–8.5	16.7	18.3	1.3

Source /to Tables 1-4/: Calculated by J. Kotyński on the basis of data published by UNCTAD: Handbook of International Trade and Development Statistics. Supplement 1980.

TABLE II

Intensity of world trace in food /SITC 0+1+22+4/ in 1978 by selected regions, as measured by structural coefficients.

Destination / Origin	Planned economies of Eastern Europe: Total	USSR	Other	Developed market economies: Total	EEC	USA	Japan	Developing countries
0	1	2	3	4	5	6	7	8
A. Delta coefficients /x 100/								
E. Europe	478	423	568	50	49	31	20	82
USSR	287	.	751	28	20	7	71	190
Other	515	504	533	55	54	36	10	61
Developed	50	43	61	107	118	62	102	98
EEC	18	7	35	124	169	47	15	67
USA	98	105	87	82	62	.	216	148
Japan	4	3	5	69	38	216	.	247
Developing	137	163	96	97	77	196	103	93
B. Correlation coefficients /x 100/								
E. Europe	27.3	17.9	20.2	–16.1	–9.6	–5.0	–5.0	–2.2
USSR	5.3	.	11.1	–9.2	–5.9	–2.7	–0.7	4.2
Other	27.3	20.5	17.1	–13.4	–7.8	–4.3	–5.2	–4.2
Developed	–20.8	–18.2	–9.7	13.4	19.1	–16.1	0.7	–1.1
EEC	–17.6	–15.4	–8.4	23.2	38.6	–11.5	–15.8	–11.6
USA	–0.3	0.5	–1.2	–11.9	–14.5	.	14.7	11.6
Japan	–2.6	–2.0	–1.6	–3.8	–4.4	3.2	.	6.6
Developing	7.9	10.3	–0.5	–2.5	–12.5	20.6	0.5	–2.3

TABLE III

Intensity of world trade in agricultural raw materials /SITC 2–22–27–28/ in 1978 by selected regions, as measured by structural coefficients.

Destination	Planned economies of Eastern Europe			Developed market economies				Developing countries
Origin	Total	USSR	Other	Total	EEC	USA	Japan	
0	1	2	3	4	5	6	7	8
A. Delta coefficients /x 100/								
E. Europe	393	83	535	77	77	6	91	59
USSR	466	.	680	75	73	0	133	47
Other	240	256	233	81	86	17	3	84
Developed	52	62	47	111	118	121	81	80
EEC	47	36	53	123	179	28	9	36
USA	38	19	32	89	54	.	227	177
Japan	36	20	43	35	34	43	.	366
Developing	97	166	65	85	69	90	138	156
B. Correlation coefficients /x 100/								
E. Europe	28.2	–0.9	34.2	–11.9	–5.6	–10.5	–1.1	–5.9
USSR	28.5	.	36.9	–10.4	–5.4	–8.9	3.3	–6.1
Other	7.5	4.5	5.8	–5.4	–1.9	–5.1	–6.5	–1.3
Developed	–18.1	–7.8	–16.3	22.6	17.6	9.2	–8.9	–11.0
EEC	–7.2	–4.8	–5.3	17.3	28.0	–11.4	–15.6	–13.2
USA	–8.4	–5.2	–6.4	–6.8	–13.8	.	18.5	13.4
Japan	–2.6	–1.8	–1.9	–14.2	–6.9	–2.7	.	16.3
Developing	–0.6	6.7	–5.3	–14.8	–14.6	–2.1	8.8	15.7

TABLE IV

Intensity of world trade in fuels /SITC 3/ in 1978 by selected regions, as measured by structural coefficients.

Destination	Planned economies of Eastern Europe			Developed market economies				Developing countries
Origin	Total	USSR	Other	Total	EEC	USA	Japan	
0	1	2	3	4	5	6	7	8
A. Delta coefficients /x 100/								
E. Europe	781	355	841	66	89	7	8	35
USSR	810	.	923	63	87	4	9	38
Other	651	1983	465	78	97	17	4	21
Developed	16	33	13	114	161	71	52	46
EEC	11	17	10	105	206	25	1	42
USA	31	119	19	107	63	.	161	102
Japan	125	910	16	23	2	57	.	402
Developing	25	81	17	101	87	121	122	123
B. Correlation coefficients /x 100/								
E. Europe	55.4	7.1	56.3	–19.5	–2.5	–15.5	–12.0	–10.3
USSR	51.8	.	56.1	–18.8	–2.6	–14.2	–10.7	–8.8
Other	18.1	21.2	11.2	–5.1	–0.2	–5.6	–5.1	–5.1
Developed	–9.5	–2.6	–9.1	11.4	19.2	–6.6	–8.8	–11.9
EEC	–7.6	–2.4	–7.2	2.9	25.0	–13.0	–13.5	–9.7
USA	–2.2	0.2	–2.4	1.6	–3.3	.	3.2	0.1
Japan	0.2	2.3	–0.7	–4.5	–2.2	–0.7	.	4.9
Developing	–28.8	–2.5	–29.7	3.3	–13.9	16.5	13.5	16.9

DISCUSSION POINTS

1) Professor Cole noted two main elements in the initial theoretical discussion, the need to improve technology within Eastern Europe, and the need to rationalise the international production structure. Both could contribute to an overall improvement in the efficiency of the world production system. In East Europe there has been a tradition of subsidies of food, for the domestic consumer. Because oil has been relatively cheap, it is not surprising that industry is rather inefficient in the use of energy relative to the high priced oil economies. Now both are beginning to change.

In Hungary food subsidies have turned out to be relatively ineffective, and the USSR is adjusting its oil prices to other East European countries in line with world prices. So there is a historical reason for the present situation.

Since that situation still prevails to some extent, it is not surprising that when Western companies put part of their production into East Europe, they want to take advantage of the cheap energy. The question is, if you are trying simultaneously to improve the technology of the production sectors within the economy and at the same time to open up the economy and at the same time to open up the economy and use the international market as a way of improving the efficiency of the domestic economy, to what extent do you have to regulate this process? If the technology is too inefficient, there is a danger that some sector of the economy could be practically wiped out, which is something we often observe. In this process of cooperation how does the model show the controls and regulatory mechanisms that should be exerted on this process of opening up the economy?

Dr. Bożyk agreed that it is impossible to suddenly open up the economy, and simultaneously improve techniques of production and increase the level of technology in a country. As the case of Poland showed, the deficit in the balance of payments may increase sharply, and it is impossible to have a short term equilibrium. It is therefore necessary to open up the economy step by step. This has been shown by the experience of Hungary, which started to open the economy in 1958. For 12 years they slowly opened different sectors of industry, and only in the 1970's did they start to open the economy more radically.

2) In answer to a query from Professor Meadows, Dr Božyk felt that the East European countries and the USSR will prefer energy saving technologies and production in the future. There are few possibilities for an increase in the production of energy at the same rate of the increase in industrial production.

3.6 The Global 2000 Report and its Implications for the United States

G.O. BARNEY[1]

This paper describes the basic findings of the U.S. *Global 2000 Report to the President*, the limitations and implications of these findings generally, and the Reagan Administration's plans for follow-up through the Council on Environmental Quality in the Executive Office of the President.

BACKGROUND

The Global 2000 study was requested by President Carter in his 1977 environmental message to the Congress. In the message, President Carter directed the Council on Environmental Quality and the Department of State to undertake a study of foreseeable trends in population, resources and environment for the entire world out to the year 2000. Thirteen government agencies participated in the work. These agencies ranged from the Department of Agriculture and the Bureau of the Census to the Central Intelligence Agency. The study required three years for its completion and cost approximately $1 million.

The Global 2000 Report is the first global study of population, resources, and environment by any national government, and as such it has become quite well known throughout the world. It is available now in six languages and has sold well over half a million copies. The German translation alone has sold more than 400,000 copies, and the Japanese translations have sold approximately one hundred thousand copies. Government, industrial, and academic leaders have discussed the report in

[1] Presented by A. Matthews.

professional meetings in People's Republic of China, Italy, Germany, Netherlands, Norway, Sweden, Panama, Mexico, Cuba, Egypt, Jordan, and India, among others.

ASSUMPTIONS

Many assumptions are necessary in conducting analysis as broad as the Global 2000 study, and three of the assumptions used in this study are particularly important. The first assumption concerns the fact that it was possible to examine only one policy scenario. As a result, the Global 2000 study implictly assumes a continuation of present policies throughout the world. Consequently, the Global 2000 study's findings must be understood to be projections, not predictions. There will be many policy changes before the turn of the century, and it is hoped that the timely warning provided by the Global 2000 study will lead to changes of policies that produce a much brighter future than projected by U.S. Government agencies assuming a continuation of current world policies.

A second major assumption is that a rapid rate of technological advancement will continue and that market forces will effectively limit demand whenever supply constraints are encountered. Without the technological advances assumed in the study and without the incentives provided by market forces, many problems would be even more difficult than projected.

A third major assumption was that there would be no wars or major economic disruptions. This assumption has been violated already by the revolution in Iran, the invasion of Afghanistan and the assassination of Egyptian President Anwar Sadat. If such events had been included in the analysis, many problems would have appeared even more complex and difficult than projected.

THE CONCLUSIONS IN BRIEF

The overall conclusion of the Global 2000 study is that, if present policies continue, the world in 2000 will be more crowded, more polluted, less stable ecologically, and more vulnerable to disruption than the world we live in now. Serious stresses involving population, resources, and environment are clearly visible ahead. Despite greater material output, the world's people will be poorer in many ways than they are today.

For hundreds of millions of the desperately poor, the outlook for food and other necessities of life will be no better. For many it will be worse. Barring revolutionary advances in technology, life for most people on earth will be more precarious in 2000 than it is now – unless the nations of the world act decisively to alter current policies and trends.

THE FINDINGS

The population projections developed by the U.S. Bureau of the Census point to an increase in the world's population growth from about 4 billion in 1975 to more than 6 billion in the year 2000, an increase of 55 percent. While the percentage rate of growth is projected to decline slightly from about 1.8 percent per year down to about 1.7 percent per year, the number of people added annually to the world's population will actually be increasing about 30 percent *faster* by 2000 than today. Where now there are about 77 million persons added to the world's population each year, the annual increase by 2000 will be up to 100 million, a number almost as large as the total population of Japan today.

Ninety percent of the population growth will occur in the developing countries, and much of that growth will be concentrated in urban areas. Many cities in the developing countries will become almost unimaginably large. Mexico City, for example, is projected to become the largest city in the world with more than 30 million people, three times the entire metropolitan New York population today.

The gross national product (GNP) projections point to a 145 percent increase in worldwide economic activity over the 1975 to 2000 period. On a per capita basis, however, this is a substantially smaller increase – about 57 percent.

Rapid economic growth is expected in the newly industrializing countries, especially in Latin America, but per capita GNP of the populous nations in South Asia are expected to remain below $200 per year.[1] Furthermore, the gap between the

[1] All dollar figures are in 1975 dollars.

world's richest and poorest people is expected to widen. For every \$1 per capita increase in the GNP projected for the poorest countries, a \$20 per capita increase is projected for the richest countries.

The food projections prepared by the U.S. Department of Agriculture show that the world's food production will increase substantially by the year 2000, but only at the expense of increased production costs and a doubling (relative to 1970) of real food prices. Food production is expected to increase 91 percent over the 1970-2000 period. On a per capita basis, this amounts to a 15 percent increase. Unfortunately, this increase will not be uniformly distributed. Countries that are already relatively well fed – Japan, Eastern Europe, and USSR – are expected to receive the highest per capita increases. The populous countries of South Asia and the low income countries of North Africa and the Middle East are not projected to improve much at all. The per capita food consumption in sub-Saharan Africa is actually projected to decline to about 20 percent below the minimum nutritional standards established by the U.N. Food and Agriculture Organization. At this level of food consumption, adults would not be receiving enough food energy to permit them to carry on typical work functions, and children could not be expected to mature to their normal mental and physical capabilities.

Several factors are involved in the projected doubling of world food prices. First, the U.S. Department of Agriculture expects the world's net stock of arable land to increase only 4 percent over the 1970 to 2000 period. Therefore, to achieve a 91 percent increase in production, it will be necessary to increase production yields per hectare throughout the world. The Department of Agriculture feels that production yields will be increased primarily through use of more energy-intensive inputs – fertilizers, pesticides, and fuel for irrigation pumps and tractors. Unfortunately, yield-enhancing inputs are already being applied in sufficiently large quantities that the point of diminishing returns has been reached in many areas. This fact means that with each additional kilogram of fertilizer applied to the world's fields, a smaller increment in production yield is achieved than with the previous kilogram. As a result of both diminishing returns on energy-intensive inputs and increasing energy prices, the U.S. Department of Agriculture expects that throughout the world farmers' costs will increase significantly in the years ahead. To provide incentives to farmers to produce more, food prices will have to increase. The Department of Agriculture has projected that to achieve a 91 percent increase in production, farmers will have to be paid twice as much for each kilogram of grain as farmers were paid in 1970. Thus, the world food issue of the future is not simply that of growing more food, but rather that of growing more food at prices sufficiently low that those people most in need of food can afford to buy it.

The fisheries projections developed by the U.S. National Oceanic and Atmospheric Administration (NOAA) suggest that fish will not provide an answer to the world's food problems. NOAA points out that the catch of traditional marine species leveled off in 1970 at about 70 million metric tons, and experts at NOAA do not expect the world's catch to increase significantly above 70 million metric tons on a sustained basis. In fact, they feel that improved management and increased protection of coastal habitat will be necessary to maintain even this catch.

NOAA scientists do feel that there is potential for increasing the catch and use of non-traditional species (such as lantern fish and krill) and that there is potential for increasing the use of aquaculture. With aquaculture, however, it is important to emphasize again the importance of energy intensiveness. Most aquacultural techniques now in use are quite energy-intensive, and before it is assumed that aquaculture will become a major part of the world's fish economy, the effects of yet further energy price increases should be examined carefully.

According to analysis by the Central Intelligence Agency, the world's forests are declining now at about 18 to 20 million hectares each year. Forests in the industrialized countries are nearly stable, and most of the deforestation is occurring in the tropical forests of less developed countries. The rate of deforestation could easily increase as a result of further needs for agricultural land, for fuelwood, for lumber and for pulp, but even if one assumes no increase, the current rate of deforestation would mean a decline of about 40 percent in the forests of the developing countries by the year 2000. On a per capita basis this would amount to 63 percent loss of growing stock of wood in the developing countries, a disastrous loss.

The water projections prepared jointly by the Department

of the Interior and the Central Intelligence Agency, point out that there are already regional water shortages scattered around the world and that by the year 2000 these regional shortages can be expected to become more extensive and more severe. Over the 1970 to 2000 period, population growth alone will cause water requirements to double in nearly half of the countries of the world. In many of the developing countries river flows will become more erratic as a result of deforestation and the loss of hydraulic stabilization now provided by forests. Virtually everywhere the development of new water supplies will become more costly, and increased silt-loading of river-waters will accelerate the accumulation of silt in existing dams and irrigation facilities.

The nonfuel minerals projections suggest that there will be adequate supplies of nonfuel minerals to meet demands through the year 2000. While further investments and discoveries will be needed to maintain the world's reserves of nonfuel minerals, the projections do not point to any major problems in the nonfuel minerals sector of the world's economy.

The nonfuel minerals projections, however, are based on two assumptions which need further examination. One of these assumptions is that energy prices will have no effect on the production of nonfuel minerals. Since the production of nonfuel minerals requires large amount of energy, this assumption seems somewhat unrealistic. Second, the nonfuel minerals projections assume that the quarter of the world's population in the industrialized countries will continue to absorb three-quarters of the world's mineral production. This assumption may take inadequate account of political developments that are taking place around the world today.

The Department of Energy began its energy projections by noting that on geological reasoning alone one must expect the world's production of petroleum to peak sometime in the 1990s and that for political or conservation reasons the world's petroleum production could peak during the 1980s. Since the world now derives about half of its commercial energy from petroleum, the anticipated peaking of world petroleum production implies the need for a transition to other forms of commercial energy.

The beginning of such a transition can be seen in the projections prepared by the Department of Energy. The transition projected is primarily to nuclear energy: a tripling of nuclear power production is projected for the 15-year period, 1975 to 1990. Coal use is also projected to grow somewhat, but not as rapidly as nuclear power.

Assuming this type of transition, the Department of Energy was able to make world energy projections out to the year 1990 – but not beyond. The reason is that for world petroleum prices the Department considered likely, the demand for OPEC oil exceeded OPEC's oil-production capacity by 1990, or sooner. As a result, the energy projections in the Global 2000 study stop at 1990 rather than going on to 2000 as all of the other projections in this study do.

Since the original energy projections were made for the Global 2000 study, the United States had an unfortunate experience with a nuclear power plant at Three Mile Island, Pennsylvania, and in 1979 the world price of petroleum doubled, achieving a level that the Department of Energy (in 1978) did not expect until near the end of the century. The Department of Energy has now published revised projections in which the growth of nuclear power is less rapid and growth in the use of coal is more rapid. With the higher prices the Department is now able to project balanced supply and demand in the world's petroleum markets at least as far as 1995, but the Department has still not published world energy projections extending over the 18 years to the end of this century.

The Global 2000 study also discusses the developing countries' fuelwood crisis. In many developing countries fuelwood prices have increased as rapidly as or even more rapidly than world oil prices. While data on fuelwood are limited, the best projections available point to a 25 percent short-fall in fuelwood supplies before the end of the century. As a result, it is likely that animal dung and crop residues will be burned in increasing amounts, reducing the nutrients that are available from these sources to maintain the fertility of soils in the developing countries.

The environmental projections for the Global 2000 study analyzed the environmental consequences of all the other projections. The projections outlined previously, of course, have many environmental implications. It is not possible to describe all of the environmental implications here but three environmental problems that I personally think are among the most serious will be described briefly: deterioration of soils, global atmospheric problems, and the worldwide loss of genetic resources.

Virtually every projection in the entire Global 2000 study points to increased pressures on the world's soils. Further overgrazing, for example, will expand deserts. Further deforestation and even more intensive cultivation of existing farm lands will also accelerate erosion. Use of heavy equipment and declining humus content will lead to compaction of soils and reduced capabilities to absorb and retain water. Cultivation of new lands in arid areas and on steep slopes will accelerate erosion. More widespread irrigation will reduce fertility in many arid areas as traces of salt in the irrigation water accumulate in the soil, causing salinization. Fertility will also be reduced in areas of poor drainage as these areas become waterlogged following irrigation. Even increased combustion of hydrocarbons, especially coal, will adversely affect soils as more oxides of nitrogen and sulfur are exhausted into the air where they react with water vapour and precipitate as acid rain over wide areas. The acid rain gradually makes poorly buffered soils more acidic, reducing soil fertility. All of these prospects lead me to think that protection of soils will be the world's number one environmental concern in the decades ahead.

The second environmental problem that I would emphasize is the alteration of the world's atmosphere. There are two kinds of alteration in prospect. The first is the growing accumulation of carbon dioxide in the atmosphere. The second is depletion of the ozone layer in the stratosphere.

Atmospheric studies show quite clearly that the world's climate will change significantly as a result of continued increases in the carbon dioxide concentration in the atmosphere. The carbon dioxide acts much like the window in a greenhouse, allowing the solar energy to enter and preventing the heat from escaping. In this way further concentrations of carbon dioxide could increase the temperature of the entire planet by several degrees. The concentration of carbon dioxide in the atmosphere is increasing steadily and carbon dioxide concentration of the planet will continue to increase as more fossil fuels are burned and forests cut. If current energy and forestry policies continue, it is only a matter of time before significant climate changes must be expected. The best estimates now available point to disruptive changes in the world's climate by about the middle of the twenty-first century – within the lifetime of today's children – if present energy and forestry policies continue.

Although disruptive climate changes are about 70 years ahead, the carbon dioxide problem requires urgent attention now. The reason is simply that if an agreement among nations limiting the production of carbon dioxide is not reached before the turn of the century, the momentum of capital investment and technological decisions will have foreclosed largely the possibility of avoiding disruptive changes of the world's climate during the 21st century. If the world is to avoid human-caused accumulation of carbon dioxide and associated changes in the world climate, major changes of policies will have to be made in the next two decades. Since global agreement on the needed policy changes will be very difficult, this problem requires prompt and determined attention.

Like the carbon dioxide problem, the depletion of the atmospheric ozone layer deserves prompt attention. The ozone layer in the stratosphere shields the entire planet from large amounts of ultraviolet light, which causes skin cancer and damages crop plants. The factors causing depletion of the ozone layer are known to include chloroflurocarbon chemicals used as propellants in aerosol cans and as refrigerants, high-altitude flights, and nitrogen fertilizers. The prospects for controlling these factors varies widely. The nations of the world might someday agree to reduce the use of chloroflurocarbons in aerosol cans and to restrict supersonic high-altitude aircraft flights, but restricting the use of refrigerants and nitrogen fertilizers would be much more difficult.

The tragedy is that the application of both organic and inorganic nitrogen-fertilizers leads to the release of nitrous oxide, which is now known to contribute to the depletion of the ozone layer. Thus we face a vicious circle: Efforts to increase agricultural yields by the application of nitrogen fertilizers contribute to the depletion of the ozone layer, increasing the ultraviolet exposure and the associated damage to crop plants.

The U.S. National Academy of Sciences has studied this vicious circle and concluded that nitrogen fertilizers provide such important benefits to agriculture that the world need not consider limiting fertilizer use for at least a decade. Since it is very doubtful that the human species can continue to feed itself without the use of still more nitrogen fertilizers, I find only very limited comfort in the Academy's conclusion.

The third environmental problem that I will mention briefly is the loss of genetic resources. In the course of the Global 2000 study an effort was made to understand where major modifications of habitat would occur and also to assess the numbers of species localized in particular geographic areas. On the basis of this information, the number of species likely to be extinct by 2000 was estimated. The conclusion was that the loss of genetic resources will be enormous, perhaps as much as 0.5-1.5 million species – approximately one-fifth of all the species on the planet. Losses of this magnitude are unprecedented in human history.

Some of the most important genetic losses will involve the extinction not of species but of subspecies and varieties of cereal grains. Four-fifths of the world's food supplies are derived from less than two dozen plant species. Wild strains are needed for breeding resistance to pests and pathogens into the high-yield varieties now widely used. These wild strains are diminishing rapidly as wild lands are brought into cultivation. Local domesticated varieties often uniquely suited to local conditions, are also being lost as higher-yield varieties displace them. And the increasing practice of monoculture of a few strains – which makes crops more vulnerable to disease epidemics or plagues of pests – is occurring at the same time that the genetic resources needed to resist such disasters are being lost. The ultimate result could be a series of agricultural disasters similar to the 1970 epidemic of corn blight that reduced the U.S. corn crop by 710 million bushels.

THE GLOBAL 2000 STUDY IN PERSPECTIVE

The Global 2000 study is not a prediction of what the world will be like in the year 2000. It is projection by the most highly skilled professionals available within the United States Government of what the world would be like if public policies in effect in 1977 were to continue unchanged through the year 2000. Only this one scenario could be examined during the course of the study.

The future described by the Global 2000 projections is not

a desirable future. The gap between the richest and the poorest in the world becomes substantially wider. By every measure of material welfare the study provides – per capita GNP and consumption of food, energy, and minerals – the gap will widen. For example, the gap between the GNP per capita in the LDCs and the industrialized countries is projected to grow from about \$4,000 in 1975 to about \$7,900 in 2000. Great disparities within countries are also expected to continue.

The Global 2000 study makes clear that increasingly the world's problems of poverty are not independent of population growth, resource availability and environmental conditions and that these problems are inter-related. The study also makes clear that the time has passed when any nation can plan the future of its population, its economy, its resources, and its environment independent of what is taking place in the rest of the world. In essence, the Global 2000 study implies that the most important problems facing nations today are gradually becoming part of one highly-interconnected macro-problem.

There is temptation among segments of the populations of some industrialized countries to assume that the problems of the poorest of the world's poor are remote and of no immediate consequence to industrialized countries. While perhaps not of *immediate* consequence to industrialized countries, the problems of the world's poor are certainly of consequence ultimately to everyone because a vicious circle of cause and effect is at work.

An example illustrates this vicious circle. In parts of sub-Saharan Africa and in parts of South Asia, population pressure is already putting sufficient pressure on the available land that the carrying capacity of that land is being reduced by the expansion of deserts and the massive erosion of topsoil. Tragically, the environmental, economic and social conditions that follow are precisely the conditions that make it most difficult to slow population growth. When fuel, shelter and food supplies are low, infant mortality rates remain high and in some cases actually increase. When infant mortality rates are high, many parents wish to have large families simply to increase the chances that some of their children will survive. Thus a vicious circle is at work in which population pressure is reducing and destroying the carrying capacity of the land, and at the same time creating the very conditions under which it would be most difficult to bring population growth under control.

While such problems may not effect industrialized countries immediately, they certainly will in the long run. As deserts expand, populations are forced to move ahead of the deserts. When lifestyles must be abandoned, need for assistance from industrialized countries increase. Refugees and migrants find their way indirectly to industrialized countries, as they have in quite large numbers to the United States over the past decade.

It seems likely that no nation will be able to isolate itself from the inter-related macro-problem. Difficult as the macro-problem is, leaders and nations must find ways of dealing with it.

The Global 2000 Report concluded that there are no quick fixes. The only solutions to the problems of population, resources and environment are complex and long-term. These problems are inextricably linked to some of the most perplexing and persistent problems in the world – poverty, injustice, and social conflict. These problems together are the world's macro-problem.

NEW APPROACHES NEEDED

The Global 2000 Report offers no solutions to the world's macro-problem. Having investigated only one scenario – continuation of present policies – the study was not in a position to recommend one set of policies over another. It could conclude (1) that continuation of present policies leads to an undesirable future for everyone, (2) that an era of unprecedented cooperation and commitment is essential, (3) that the United States must cooperate generously and justly, and (4) that new and imaginative ideas – and a willingness to act on them – are essential. It is my conviction that global models will be essential in finding the new and imaginative ideas the world so desperately needs and in building the willingness to act.

Global models, for example, can greatly facilitate negotiations and the development of agreements. Disagreements occur for three reasons: (1) the parties disagree on the facts; (2) the parties disagree on the implications of the facts; (3) the parties disagree on which implications and consequences are desirable. Global models can and will over time largely eliminate disagreements over the first two of these three issues.

In a global model, the "Facts" as perceived by both parties are stated clearly in the model, and if either party disagrees, one can focus immediately on which facts are in dispute and take steps to reduce the uncertainty on the disputed facts.

The global model used by each party also indicates clearly how each party estimates the consequences of the facts. If the parties have different perceptions of how the world works, various tests are possible with the models, the results of which can be compared with observations in the world as a basis for deciding which notion is more nearly correct.

While disagreements over valuing alternative consequences differently cannot be resolved as easily as disagreements over the facts and implications of the facts, global models can provide a basis for helping all parties to understand better the overall consequences of alternative courses of action. Furthermore, if agreement can be reached on the facts and their implications, consideration of the remaining value questions might proceed in a much more orderly manner. For these reasons and others, I believe decision-makers will be using global models much more extensively in the years ahead.

Even now every decision-maker uses a global model. Most of the global models used by decision-makers are mental models, a mental conception carried around in the mind of the decision-maker of how the world works. Such models are very versatile, but generally narrow in scope and limited in both precision and detail.

Computerized global models are nothing more than an explicit statement of, and representation of, one or more mental modes. Computerized global models have the advantage that the assumptions can be examined and verified, and that the logical consequences of the assumptions can be worked out by the computer in vastly more precision and detail than the human mind can do.

Computerized models also have numerous shortcomings, one of the most important of which is the apparent accuracy of the results. The Global 2000 study, for example, reports the median gross world product for the year 2000 as being \$14,676.724 thousand million. It would be far more realistic and less misleading to report a projection of \$15±5 trillion, indicating clearly that the uncertainty in this projection is on the order of one-third.

It has come as a surprise to many people that the Government has had a global model for quite some period of time. The Government's global model is different from many other global models in that the sectors of the Government's global model do not reside in a single computer, but are located around Washington (and even in other cities) in computers of different departments and agencies and non-Governmental organizations. The food sector, for example, is at the Department of Agriculture. The population sector is at the Bureau of the Census. The GNP sector is at the World Bank. The location of the sectors, however, is only a matter of convenience. All of the sectors – including those that are now computational procedures performed with a calculator or adding machine – could be placed in a single computer without altering their functioning in any significant way.

One might argue that the Government's global model was not a single model until the Global 2000 study brough the various elements together and used them as a collective whole. However, the sectors of the Government's global model were interacting long before the Global 2000 study was initiated. The Bureau of the Census, for example, developed population projections and sent them off to be printed at the Government Printing Office (GPO). The Department of Agriculture then purchased the Bureau of the Census report from the GPO and inserted the Census Bureau's projections from the population sector of the Government's global model into the food sector to provide a basis for world food projections. The Global 2000 Study did not initiate the linkage between these sectors; it only accelerated previously-existing, very slow interactions in what already was and must continue to be a single, integrated set of interacting sectoral models.

The Government's global model has been used extensively to support decisions and policies of the United States Government in the Nixon, Ford, and Carter Administrations, and its sectoral models continue to be used by the Reagan Administration. The Bureau of Census population projections are used extensively throughout the government in policy formation. The projections from the food model were used as a basis for the United States position at the World Food Conference in Rome. The energy sector of the Government's global model provides the only detailed world energy projections developed within the United States Government.

While the Global 2000 projections are based on the very best analysis now possible with the United States Government's global model, the Global 2000 projections are riddled with inconsistencies and unrealistic assumptions, most of which bias the projections in a way that minimizes the apparent seriousness of the emerging world problems. Virtually every sectoral model assumes that its sector of the world economy (food, for example) will have access to all of the land, energy, water and investment capital that the model calculates that particular sector will need. At present the Government's global model does not even have the capability to add up all of the resources and capital assumed to be available to see that the total is within the limits of what is available on the planet. Furthermore, in comparing the Global 2000 study's environmental projections with the other projections in the study, it becomes clear that most of the other projections have assumed that the environment will continue to provide its goods and services in substantially increased amount, without interruption, without maintenance, and at no cost. The environmental projections cast grave doubt upon these assumptions.

In spite of its limitations, the U.S. Government's global model is capable of assisting in the search for and evaluation of the new and imaginative ideas that are so urgently needed.

The single most important piece of policy advice that I can abstract from the Global 2000 study stems from the following conclusion: ". . . the executive agencies of the U.S. Government are not now capable of presenting the President with internally consistent projections of world trends in population, resources, and environment for the next two decades." The policy advice would be: It is time for the United States Government (and every other government that does not have a better global model than the U.S. Government's global model) to put its global model in order. It is dangerous and unacceptable for one of the most powerful nations in the world to be establishing long-term policies on the basis of such inconsistent, contradictory and inadequate models and analyses.

In spite of its limitations, the U.S. Government's global model is, even in its present form, capable of assisting in the search for and evaluation of the new and imaginative ideas that the world so urgently needs. As a next step, it would be very beneficial if the United States Government would undertake a serious analysis of alternative scenarios. For example, how would the projections from the government's Global model be different if all of the recommendations of the Brandt Commission were followed? What kind of future is implied by following the recommendations presented in the World Conservation Strategy? How would the projected future of the world change if the recommendations of the Carter Administration's *Global Future: Time to Act* were followed?

Short of analyzing alternative scenarios, there is no analytical basis to recommend any particular policy action that might be considered.

THE REAGAN ADMINISTRATION FOLLOW-UP TO THE GLOBAL 2000 REPORT

Mr. A. Alan Hill, Chairman of the Council on Environmental Quality, in the Executive Office of the President, stated at his confirmation hearings that, "A high priority for the Council will be the coordination of White House efforts to address global environmental issues." Mr. Hill reaffirmed this priority in further testimony during December 1981, noting that the Administration is convinced that many of the difficulties projected in *The Global 2000 Report to the President* and other contemporary reports on world problems are avoidable. Mr. Hill reports that President Reagan wants the policies of the United States to alleviate, not compound, such difficulties. The President believes in the ability of intelligent human beings to take the actions necessary to alter the course of projected events.

Mr. Hill feels that the first steps in confronting global issues are to recognize that a number of serious global issues exist and to make a realistic evaluation of their effects on the environment and the human condition. *The Global 2000 Report* makes clear that the U.S. Government's ability to evaluate global problems analytically is quite limited. To correct this deficiency Mr. Hill and the Council on Environmental Quality (CEQ) are planning to improve the data and the models that were the basis for the Global 2000 projections and are the continuing analytical basis for the Government's international policies and programs.

The Reagan Administration's approach to global issues is based on pragmatism. In the next several decades, the world will face increasing stresses with regard to resources and environment on a global scale. President Reagan noted at the Cancun Summit Meeting on Cooperation and Development that the United States – as a technological leader, a major food producer, and a leading energy consumer – has an important stake in these issues. The U.S. economy, the welfare of the American people, and U.S. national security are involved.

Mr. Hill feels that the United States should be paying more attention to forecasting, and doing more to anticipate the kinds of problems discussed in *The Global 2000 Report* so that nascent problems can be dealt with before they become difficult. Too many times in the past, Mr. Hill feels the United States – and other nations of the world – have been caught in international crises, (the 1929 economic crash, World War II, the 1973-74 energy situation, the Iranian revolution, the Iraqi-Iranian War, etc.), because of insufficient information on important, changing global conditions. The United States has not had the capability to foresee a problem until its options for action were highly constrained. In other cases, decisions on global policy have been made without considering their total impact. These shortcomings, stemming from inadequate analysis of long-term global issues, are serious.

The Reagan Administration's objective is to correct these shortcomings, which have caused needless loss in U.S. and world productivity in the past. The Administration wants to develop and put into action improved methods which will allow Government officials to consider more carefully policy alternatives affecting future global conditions. The Administration wants to use these improved methods to analyze U.S. global policies and to help guide decisions on government actions that affect global conditions.

The Reagan Administration has taken steps to establish, under the leadership of the CEQ, an Interagency Working Group on Global Issues. The global issues working group is expected to have policy representatives from 20 agencies. The members will be policy-level representatives, persons with sufficient authority to commit their respective agencies to action. Approximately twelve of the members had been appointed as of December 1981. There will be a meeting as soon as members from the various agencies have all been named.

While each of the representatives is expected to provide his or her own supporting staff person, the Interagency Working Group on Global Issues will not have a separate staff beyond that provided by CEQ. CEQ expects to devote the equivalent of four staff-person-years to the work annually. Mr. Hill expects the Working Group to meet at least quarterly, forming small groups to address specific issues. This Working Group, which is part of the Cabinet Council system, will be analyzing and making policy recommendations on global issues of population, resources, and the environment. The Working Group will review the findings of recent global analyses, such as those in *The Global 2000 Report*, and will conduct its own review of global issues. As an early step, the Working Group will identify those global issues of the most immediate and greatest concern to the U.S. Government, and will determine how these issues may affect or be affected by U.S. domestic and foreign policy. Finally, the Working Group will recommend to the Cabinet Council on Environment and Natural Resources and to the President what actions should be taken.

The Interagency Working Group will also develop a program to improve the capability of the Government to gather, analyze, and assess information about the global issues of populations, resources and the environment. To coordinate this work, the Interagency Working Group on Global Issues will have a subgroup – the Global Forecasting Team – working specifically on improving the Government's abilities to assemble and analyze data on global issues. While the Working Group on Global Issues is composed exclusively of policy-level people, the Global Forecasting Team will be composed principally of experienced technical professionals.

The primary purpose of the Global Forecasting Team is to provide a coordinating mechanism for the Government's data and models on global issues. It will begin immediately to improve the Government's global data and analytical models. Depending upon the issue addressed, all 20 agencies might not want to participate, but there will be a policy person accountable from each of these agencies to participate in the work of the Global Forecasting Team. If there is hesitancy on the part of a particular agency, Mr. Hill sees it as his job to work with the agency to get them involved. A high priority will be improving the linkages between the various Government agencies' sectoral global models. The agencies' sectoral global models deal only with individual sectors (for example, food or population or energy) and need to be interlinked for improved consistency of assumptions.

Something that concerns Mr. Hill is that at present each U.S. Government agency looks at the world out of its own little window, seeing its own little segment of the work and making its own judgments within its own limited area of responsibility. But no agency in Government has responsibility for putting the whole picture together. Mr. Hill noted that the Bureau of Labor Statistics looks at its particular interests; the Nuclear Regulatory Commission deals with its limited business; the Energy Information Agency does its separate analysis; and the EPA does its own thing. Each agency is looking out his own little window, and nobody is coordinating these efforts so that the top policy makers in the United States get an integrated, consistent, analytical picture of what lies ahead.

Mr. Hill believes that one of the therapeutic developments that followed from the Global 2000 effort is a clear sense of the tunnel vision now affecting each of the U.S. Government's agencies and their models. Currently, the models cannot adequately relate to each other. A typical example is the linkages – or more accurately, the lack of linkages – between our food and energy models. At present these two models could easily be assuming that the same investment capital, water, land, labor, and energy will be used in both sectors. Such models must be able to interrelate so that priorities for limited resources can be established. Mr. Hill has indicated that he feels strongly about these matters and that he will be working on these issues through the Interagency Working Group and the Global Forecasting Team.

The Global Forecasting Team will work closely with and report to Mr. Hill. The Forecasting Team will have an early opportunity to demonstrate its usefulness and capabilities by doing technical analysis of issues that are undergoing policy consideration. Mr. Hill is already working on a list of issues for which he would like to update and extend earlier analysis, issues such as those presented by *The Global 2000 Report.*

Mr. Hill notes that the United States amasses an enormous quantity of data each year but too little effort seems to be expended in analyzing this data to ascertain what is happening, what can be done better, and where the problems are that lie just over the horizon. The U.S. Government, for example, has had difficulty anticipating what will occur concerning groundwater pollution, air pollution, resource exhaustion, loss of topsoil, loss of different minerals, timber problems, desertification, and climate changes, among others. Mr. Hill notes that the United States has a few individuals in colleges and universities looking at these issues, but they are generally regarded as oddballs, concerned that "the sky is falling." Mr. Hill makes clear that he doesn't know whether "the sky is falling," or whether it will fall tomorrow or in a hundred years. What he would like, however, is to have Government agencies looking at these issues more professionally to inform Government officials more precisely what lies ahead under alternative policy scenarios.

Mr. Hill is seeking an analytical capability which would permit decision-makers to have a more educated estimate of the consequences of various alternatives they are considering. This capability would be useful to policy officials in both the executive and legislative branches of the federal government, to officials in the state and local governments, and to decision-makers in the private sector.

Mr. Hill believes that the Interagency Working Group on Global Issues can accomplish much even in the short run. He believes that steps can be taken immediately to improve coordination of data and forecasting models. He feels that he has already established the institutional mechanisms for reevaluating the issues raised by *The Global 2000 Report* and for identifying new areas of concern. He is less certain as to what the longer term solutions may be, but is certain that the Reagan Administration will want to work with international organizations, other nations, and the U.S. Congress and public to evaluate future alternatives.

CEQ intends to work closely with three other U.S. Government agencies, the Environmental Protection Agency (EPA), The Department of the Interior (DOI) and the Department of State. These agencies will provide supporting resources to CEQ, which has had its staff and budget reduced by about 70 percent recently. CEQ already has an interagency agreement with EPA. EPA will assist CEQ in securing needed consultant specialists and other support. EPA's Office of Research and Development, for example, has a technical staff expert in environmental modeling and forecasting. In the natural resources area, Mr. Hill has had assurances of cooperation from the Director of the Geological Survey within the U.S. Department of the Interior, an agency with which CEQ has worked closely and successfully in the past. CEQ will continue a close working relationship with the Department of State, which in addition to its prime responsibility in setting U.S. foreign policy, also plays a vital role in U.S. participation in the international organizations.

The Reagan Administration also intends to seek cooperation, advice, and technical assistance from many other appropriate sources outside of Government. The Global 2000 study was designed to see what information the U.S. Government had within its agencies itself. Mr. Hill thinks *The Global 2000 Report* achieved this objective very well, and feels that there is a need now to have this information generally available to others and to have the Government benefit from what is available from sources outside the Government.

When *The Global 2000 Report* was issued, the Council received hundreds of comments from environmental organizations, commercial institutions, and individuals. These letters, and lengthy comments, and analyses indicate a broad public interest in the issues for which CEQ will continue to be responsible. In the Council's future work, Mr. Hill will consider a broad range of public views and involve the domestic and international talent necessary to improve Government's own analyses and actions. To this end, Mr. Hill hopes to achieve better technical coordination among the domestic and international organizations that collect, store, and analyze information on population, resources, and environment. Mr. Hill wants to determine early on the number of cooperative agreements the United States has with international organizations, the problems being studied under these agreements, the relative productiveness of these agreements, and the ways these agreements can be made more effective for everyone concerned.

DISCUSSION POINTS

1) Professor Levy asked for a clarification of the treatment of linkages. It was concluded in the Global 2000 model that while

there are some constraints in the individual sectors, the effects of constraints become more severe when the sectors are linked.

Mr. Matthews replied that in the 3 years available the Report was unable to study linkages in any great detail. However in order to derive some estimate of the effect, or direction, of such linkages a special study was made of several models produced by other groups outside of the USA government. In all cases it was found that the linkages increased the problems rather than assisting in any way. The outcome was more severe, more unfavorable when the linkages were taken into account, demonstrating the very great importance of doing so.

Professor Meadows added that both Professor Mesarovic with the World Integrated Model and themselves with the Limits to Growth model had carried out such tests, and that the results were the same for the different models. Where there was no accountability and all the capital could be put into each sector simultaneously, the futures were far more optimistic than in the runs with linkages where capital, food, etc., had to be allocated among the sectors.

Professor Lesourne however thought it was curious to have based conclusions on what had not been done in the Global 2000 model. The conclusion must depend on how the constraints are built into the particular model. It is possible to imagine ways of introducing the constraints which make the partial model results worse than the global model results. It is unfortunate that with these partial models which are amongst the best in the world it was not possible for the control team to fully discuss the assumptions made by individual departments.

Mr. Matthews replied that although in general the editors had mainly to rely on what was provided by each department, in order to get more comparability they asked the departments to run their models with certain changes in assumptions of population, GNP and so on. This represented one small step in bringing some uniformity into the separate departmental models.

2) Dr. Norse felt that it was not necessary to introduce mechanical interfaces between sectors to look at interactions. As an example, it takes about one ton of petroleum to catch and land a ton of krill. Since krill is used for fish meal or paste

and is not a high value product, it cannot carry the cost of the oil input. The option of increasing the yield of the sea from 70 million tons to 100 million tons by the harvesting of krill is easily seen to be not viable.

3) Dr. Hrabovszky felt that because of the influence of environmental people the choice of data from the US Department of Agriculture had been limited to the most pessimistic of the alternatives. Throughout the report any problem is blown out of proportion and only the problems are emphasized. Painting such a gloomy picture will only succeed in scaring people.

Mr. Matthews replied that the main reason that the summary is more pessimistic than the analytic work of the various departments is the abundant evidence that the inclusion of linkages leads to worse results. All who worked on the report would agree that it is very tentative and that more useful models will include linkages and a multiplicity of scenarios.

Professor Mesarovic noted that although all models are criticized in one way or another, at least the report has clearly indicated the assumptions and the conclusions of those who wrote it.

There is a tendency to look at models as if they represent reality. Yet at the moment there are major changes in the world and perhaps the most difficult task is to anticipate change, such as the shift of classical industry to the developing world whereby Brazil is now a major exporter to Africa. There is a need to be more imaginative in choice of scenarios rather than too concerned with improving parameters and projecting recent trends.

4) Dr. Leach asked when the data base for the trends stopped, since the Global 2000 Report is largely an exercise in trend projection. After all there have been some very sharp changes in trends in the last 3-4 years in some important sectors such as energy.

Mr. Matthews replied that the different departments cooperating in the study selected different time periods, which created problems for the project analysts who were trying to fit the pieces together. It was not possible, for example, to convince the US Department of Energy to make estimates past 1990. The cut off for most data was about 1978.

At a recent meeting of the American Association for the Advancement of Science, several of the people from the Department of Agriculture and the Bureau of the Census who had been involved in the study gave detailed papers concerning recent changes, and felt that the changes were not significant. This is of course not so for energy. The US Department of Energy had assumed that there would be no price rise up to year 2000, which had prompted Dr. Barney to include in the report a graph showing a 5% a year price rise as his estimate.

4 Modelling in Perspective

4.1 A Review of Global Models[1]

J.M. RICHARDSON

Despite many reviews and surveys, the field of global modelling is still not well known or understood, even though global models have been built for more than a decade now. This review will concern itself with seven major points.

1) There is no strictly objective position from which to carry out or view global modelling. When developing a model a choice must be made in terms of methodology as well as the policy options to be considered. Most modellers start out with a commitment to a particular method, and this is of crucial importance in determining the output results. For users of the models this observation has very important implications. In particular a user should not attempt to find the one best model, but should insist on competing models by different groups with different perspectives, should understand why they are different, and should assess the different models from the perspective of this understanding.

It is better then to state biases clearly than to pretend that none exist.

2) The field of global models is very diverse. No brief review can do justice to it, but to give a sense of the diversity a list of some models and some exercises using models is given below.

- Forrester developed the World Dynamics model, which was followed by the Limits to Growth.
- IIASA Conference stimulated models have included

[1] More formal papers have appeared recently, as the article in the Unesco magazine "Impact", vol. 31, 1982.

Mesarovic – Pestel, Bariloche, Moira, Sarum, Fuji and the UN model of Leontiev et al.

• A new group in terms of IIASA concerns have been international economic models, including: Project Link, a collection of national economic models and an effort to link these; Dynamico, another UN model which attempts to define goals and to move towards them using optimisation procedures; Inferum, a collection mostly based on the developing world.

• Although it has in the past been taken for granted that global models could not address political concerns, two new projects have attempted to do this. Simpact, developed at the University of Geneva, treats the relations between USSR, China and USA on a global basis. Globus is a much more ambitious model with 25 nations representing 85% of the world population, with well developed economic and trade sectors, and comes from the Institute for Comparative Social Research in Berlin.

• At the recent 9th International Global Model Conference at IIASA ("Global Modelling in the Service of Decision Makers") a collection of 21 projects, some with models and some not, was presented.

• ILO basic needs models consider employment factors.

A project at the All-Union Institute for Systems Studies in India has a model based on regional and national problems.

• Sarum has been applied to Australia and New Zealand.

The Fuji model emphasises the ESCAP region of South-East Asia.

• The GPID model of Galtung has a rich and different perspective.

• The USA Global 2000 report, the OECD Interfutures project and the Common market FAST study all work from a government perspective, drawing upon models.

• Zencap is a look at energy-capital relationships, developed in Zurich.

• The IIASA energy and food-agriculture models are strongly committed to a bottom-up approach.

3) There have been few concrete, specific results of direct

relevance from model studies. Most global models have some endogenous agriculture sectors, but in many, especially the econometric models, the treatment is not very relevant to the questions of nutrition, food supply or hunger. Some models have well-developed and relevant agriculture sectors, but unfortunately they do not deal well with energy. One, for example, had in the early stages fine energy and agriculture sectors, but the two sectors were not linked. Even in models with strong agriculture sectors there tends to be either a very strong demand emphasis (if you could cure poverty, you could end hunger) or a supply emphasis (if you could produce enough food you could end hunger). There is not much evidence of models which have a sufficient linkage between supply and demand factors, or with an awareness of supply constraints such as land and soil fertility. With the exception of the Global 2000 Report there are no concrete policy recommendations directly germane to issues of food and energy.

4) However there is much to be learnt from a decade of global modelling. Some particular aspects of particular models have made a very real contribution.

Firstly, and possibly the most important, is a concern with values and a vision, provided by Bariloche. The modellers ask what kind of world is wanted, have defined that and considered ways of reaching the goal. There is a real need for this emphasis.

Global models, in particular the early models, have evidenced a genuine concern for the vital issues facing mankind and the planet: sustainability, hunger, basic values, etc, growing out of a real commitment to address the problematique of the Club of Rome. Later work in the field has retreated from this emphasis on vital issues. There is a need to return to this kind of concern.

Some models such as Forrester-Meadows and Sarum have set superb standards for documentation. It has been possible to take those models, more than elsewhere, use and apply them, analyse and criticise them. Since people can understand these models it has been possible to take them apart and to build on them.

Several models, such as the Limits to Growth, Sarum and Leontiev, have included a endogeneous treatment of the environment, which should be built upon.

A concern with the behaviour of individual producers has

appeared in the Moira model.

The representation of international trade, especially in agriculture, has improved.

With well defined agricultural sectors, there has been a linkage of field studies and global concerns.

An area in which there is a need for further attention involves the embedding of national concerns in a regional and global context.

Fuji, for example, has linked model results with UN debates, such as how the results relate to the objectives of the Third Development Decade or to the Lima target.

5) There is more consensus in the field than is commonly recognised. Members of the global modelling community have spent more time discussing and projecting to the public the 5% areas of disagreement than the very broad areas of agreement, which are as follows.

- Population and physical capital cannot grow forever on a finite planet.
- No reliable and complete information exists on the degree to which the earth's physical environment can absorb and meet the needs of further growth in population and capital. There is a lot of partial information which can be interpreted optimistically or pessimistically.
- Continuing business as usual policies through the next few decades will not lead to a desirable future, nor even to meeting basic human needs. It will instead result in an increasing gap between rich and poor, to problems with resource availability and to worsening economic conditions.
- Because of these difficulties, a continuation of present trends is not a likely future course. Over the next 3 decades the world's economic and social systems will be in a state of transition to some stage which will be not only quantitatively different from the present.
- The exact nature of this future state, and whether it will be better or worse than the present is not predetermined, but is a function of the changes being made right now.
- Owing to the momentum inherent in the world's social and physical processes, policy changes made soon are likely to

have more impact with less effort than the some set of changes made later.

- By the time a problem is obvious, it is often too late to solve it.

- The interdependence of peoples and nations over time and space are much greater than commonly imagined. Actions taken at one time at one part of the world have far reaching consequences that are impossible to predict intuitively and also impossible to predict with any precision, perhaps not at all using computer models.

- Because of these interdependencies, single simple measures to reach narrowly defined goals are likely to be counterproducttive. Decisions should be made within the broadest possible context, across space, time and areas of knowledge.

- Cooperative approaches to achieving individual or national goals usually turn out to be more beneficial in the long term than competitive approaches.

Although this list may seem to be commonsense, most policies are being made on the basis of assumptions which are diametrically opposed to these assumptions. Therefore this type of consensus needs to be expressed forcefully.

6) There is a need to communicate more effectively what is known, and to learn more about what is not known.

When a recent meeting on utopian views on the future degenerated into purposeless bickering, the participants asked themselves why they had behaved that way. They felt themselves as a group to form a microcosm of the global problematique, and set out the following thoughts.

- Why is existing knowledge not being used more effectively to end hunger, to promote development, to support peace, and to narrow the rich-poor gap? How in a scholarly community is it possible to produce information which can be used effectively?

- A great deal of knowledge is not being acted on. There is a diversity of ideas, some mutually contradictory, and a tremendous mine of information which is not sorted out, organised, synthesised or accessable.

- There is a lot of uncertainty, disagreement and conflict

which needs to be cleared up. Such disagreements confuse those who wish to act.

- Much information is tinged with hopelessness. The presentation of known complexity, of uncertainties and confusion creates an impression of struggle and impossibility.
- Often there is little sense of community, which hinders communication.
- The group was not good at listening, did not feel listened to.
- Individuals have little commitment to search for agreement. They define themselves through disagreements, see one another as opponents and obstacles.
- Individuals try to convince, persuade or overwhelm each other, to see disagreements or blockages and nuisances, as divergences from achieving goals.
- To bolster self-images, individuals mystify and put down others when communicating, use specialised languages that are efficient but not understood by others. They are validated by complexity and intellectual superiority.
- People felt part of an inbred closed loop system that trains, rewards and punished on the basis of similarities to the norm and not effectiveness.
- People are forced to operate primarily in the material domain, not in the spiritual one. No reference is made to god, to love or to mysticism. Indeed these are labelled by that most degrading word, unscientific.
- As to responsibility and commitment, the group felt that they work hard, bringing considerable talents to the work, persist against difficulties, yet believe most of the time that they cannot prevail or make any difference. Knowledge, data, reports, speeches are produced without any belief that they will be listened to or used. This may be somewhat of a comfort, for the thought that somebody will take notice of pronouncements leads to scared, uncertain and inadequate feelings.
- Solutions are framed in terms of what others should do. When they don't do what is required others are treated with disrespect and downgraded.

7) Having described what they felt to be some aspects of the

present situation, the group tried to describe a community which could achieve the necessary results. It includes the following elements.

- What is known is sorted out from what is suspected, believed or wished.
- Communications are clear, concise and compelling so that those who need to take action are acting on the best information available.
- Work is efficient, using the scientific method to clear up uncertainties, perform useful experiments and increase the base of knowledge.
- Communications are intended to illuminate issues, convey opportunities as well as obstacles, to empower people and to give them the confidence to take action and discover knowledge for themselves.
- Members identify with commonalities rather than differences.
- Disagreements are used creatively as opportunities for learning, exploring and rethinking. Those who differ are valued, and there is a search for real world evidence to resolve differences.
- Energy and concern is redirected towards achieving common purposes rather than used in establishing contrary positions and refuting others.
- All capabilities, intuitive and spiritual as well as scientific and logical, are recognised and balanced.
- Honest mistakes are welcomed, learnt from, and publicised as much as success, so that others will not repeat these mistakes.

DISCUSSION POINTS

1) Dr. Robinson pointed to a connection between the two points, that it is impossible to be completely objective, and that there is more consensus amongst global modellers than is commonly believed. A person outside the modelling community will notice that people who work on global models have quite a lot of similarities of background, and that there is a lot of thinking that does not come into the models at all. The consensus is

partly due to the fact that these are people who have been through a certain academic training and also are people whose ideas can be funded, because if you have a radical new idea you just don't get the funding. Every one who has worked with models should be well aware of this, of the restricted group who are working on models.

One important radical idea not now included is the attitude towards the natural world. Anyone who is very concerned about other life forms on earth would hesitate to suggest that we could support 6 billion people, since we know that we have already wiped out too many other life forms. Thus an extremely different outlook comes from that different philosophical background.

2) Professor Cole pointed out that throughout history there have been consensuses of ideas, such as those listed by Dr. Richardson. The problem is that responses to them are often completely different. Also if you do go down the list you will find that countries are most often not directly opposed to these principles; many countries for example have positive population policies.

As noted by Dr. Robinson, the consensus amongst modellers is a sociological phenomenon. IIASA in particular has been responsible for seeking the concensus, by imposing a concensus on the modellers, by weeding out, by funding or through ideological differences, those who do not fit into the model which IIASA and the major institutions who support these models would like to see.

In the USA and other parts of the Western World there are increasing numbers of religious broadcasts, which are part of a fundamentalist revival which is taking place in many parts of the world. In a certain sense this implied consensus is presented here as a fundamentalist revival, implying a rejection of the reality of the idea of diversity.

3) Professor Mesarovic felt that energy-agriculture linkages are treated by the models, being implicit in the economic system. If the world is considered in a hierarchical fashion, there are both the physical world around us and the economic world in which we operate. Economics is the system used to measure performance, to allocate resources and to make decisions. A physical agricultural model and a physical energy

model are fully linked and integrated through an economic system. The two models are not unlinked, but are linked in the way they are linked in the real world.

4.2 Interfutures Project of OECD

D. NORSE

INTRODUCTION

Uncertainty often feeds on itself with the result that it becomes self-fulfilling. Thus in the industrialised countries a common reaction to recurring or persistent inflation and structural adjustment problems is to live for today and let the future take care of itself. People in these countries tend to concentrate on preserving the status quo and on achieving further increases in real incomes, whilst a large proportion of the population of the Third World, who have experienced relatively constant or even declining real incomes over the past 10 years or more, concentrate on survival. By ignoring the fact that their future depends to a considerable extent on what happens in the rest of the world, the developed countries are creating greater uncertainty. They are intensifying problems of agricultural production and industrial structural adjustment, impeding the transition from petroleum to less depleted or renewable energy resources, and creating new problems.

Much of this uncertainty stems from lack of awareness of the nature of the interdependence of nations and of the complex interactions that occur between sectors like agriculture and energy. These linkages have temporal as well as spatial dimensions and hence appropriate simulation models can play a useful role in determining the policy implications of, for example, a change in energy prices on the costs of food production. But as I hope to show by reference to results from the Interfutures Project these direct micro-economic effects may be of secondary importance when compared with some indirect macro-economic effects.

THE INTERFUTURES PROJECT

It was an exercise in prospective analysis completed by an interdisciplinary group at the OECD Secretariat in Paris between 1976 and 1979.[1] The project was established to study "the future development of advanced industrial societies in harmony with that of developing countries".

The main goal was to make an objective assessment of the problems that the advanced industrial societies will face in the long term, both in their own development, and in their relations with developing countries. The project tried to identify which of the current trends or problems have long-term implications and also what are the new problems or trends that could emerge over the next two decades. The aim was to indicate where current policies need to be changed to resolve long-term problems and influence trends so that future governments might have less need to resort to crisis management and to use measures which cause severe transition problems. To accomplish the above it was necessary to:

- examine sectoral interdependence and the strategies open to the actors determining or affected by this interdependence.
- analyze possible socio-economic developments within and between advanced industrial countries (largely excluding those involving european centrally planned countries)
- obtain a clearer picture of the casual and operational features of likely Third World development.

But the world is also faced by unpredictable events, which are breaks from the trend. Whilst the timing and magnitude of such discontinuities cannot be forecast, it is possible to improve our understanding of their transmission across national and international economies by exploring alternature futures. Consequently INTERFUTURES looked at a number of scenarios of the world economy depicting alternative and not equally plausible pathways to the year 2000.[2]

---

[1] Facing the Future: Mastering the Probable and Managing the Unpredictable. Final report of the INTERFUTURES Project, OECD, Paris, 1979.

[2] D. Norse, Scenario Analysis in INTERFUTURES, Futures, Vol. II.

The approach

Contrary to what one might regard as normal OECD practice, it was not an exercise in economic forecasting and it did not use conventional econometric models. Economic forecasting is unsuitable for most of the issues examined by Interfutures, which were as much socio-political as economic, and for time frames of 10-20 years or more. Also such models are often structurally short-term, and inappropriate for examining long-term issues.

A prospective analysis approach was adopted by Interfutures because even those global models which are not based on traditional analytical or forecasting methods, still have a number of weaknesses. They only partly or unsatisfactorily capture the structural changes which may affect economic and social dynamics in the long term.

Interfutures used two principal models, SARUM (the Systems Analysis Research Unit Model)[1] and a trade model constructed by members of the Interfutures team, which focused on the international monetary system to examine balance of payment problems and long-run exhange rate movements. Of the two models, SARUM played the dominant role. SARUM was chosen after examining most of the other global models available in 1976, and after gaining operational experience with some of them. SARUM is a flexible model. The regional and sectoral composition could be changed to include the main nations and geopolitical blocks, and the industrial sectors of particular interest to Interfutures. SARUM contains clear linkages between the main sectors of the economy. For example, energy feeds into the production of fertilizer, into the production of irrigation water, into land development and into the costs of food production. It also includes a mechanism whereby substitution takes place between the factors of production. It is able to simulate the rise and fall of trade barriers and the effect of cartel action. The impact of the balance of payments on exchange rates is included.

The principal role of SARUM was the production of the numerical framework for the long term scenarios. The construction of global scenarios, though central to the prospective

[1] Departments of the Environment and Transport, Research Report 19, SARUM 76 Global Modelling Project, HMSO, London, 1977.

analysis, provided insights only on certain problems and policy issues, because of the necessary aggregation of nations and industries and the simplification of some socio-economic features. The global scenarios were complemented by more specific long-term studies: eg. on key industries and on development strategies open to Third-World countries. These specific studies were subsequently integrated into the analysis.

The scenarios were constructed in a number of stages:

- the definition of conceptual models, largely based on studies of the socio-economic and socio-political dynamics of the OECD area since the Second World War, including shifts in the balance of power of groups in the OECD, and also on changes in North-South relations,
- the translation of these conceptual models into qualitative and quantitative parameters which can fit into the mathematical model,
- the production of the qualitative framework which is the simulation phase of the exercise,
- having the numerical output, the enrichment of that output, making up for the deficiencies in the model and introducing the results of external studies, and
- analyse the political and economic content, th problems which may block the development of the scenario,

Each global scenario is an image of the year 2000, ie. an end-state scenario. The concept of an end state was used to elaborate the sequence of socio-economic and political transformations and events which could lead up to it. Based mainly on explicit sets of assumptions, the scenarios were also influenced by the assumptions explicit in the computer models used to form their macroeconomic framework.

Two aspects must be stressed. First, the scenarios are not, and should not be interpreted as, forecasts. They are a means of improving our understanding of the long-term global, regional or national consequences of existing or potential trends or policies and their interaction. Second, although global, national, and regional models were used in the construction of scenarios, the model output was the starting point and not the end point of the scenario analysis. Whilst some of these models are complex and useful analytical tools, they all rely on a core of invariant macroeconomic, or even microeconomic, structural relationships. This structural invariance is generally not acceptable for long-term scenarios because, as the time horizon recedes, the model reveals more accounting consistency than structural consistency, necessitating the introduction of analyses exogenous to the model.

The INTERFUTURES scenarios had four main functions:

- to provide coherent macroeconomic and geopolitical frameworks of the world economy over the next quarter century;
- to assess the overlap and interaction of policy issues arising from changes in relations between OECD countries and from differing patterns of North-South relations (eg. problems that could be triggered off by growing economic disparities between industrialised countries and by the growth of industrial redeployment);
- to explore the consequences of broad strategies that might be adopted by OECD governments (eg. for energy or international trade); and
- to integrate the complementary studies.

THE SCENARIOS

Interfutures constructed six scenarios. The central hypothesis and associated assumptions for some of these scenarios were extreme and not very plausible but the choice was justified because they helped to sketch out end of the century boundaries for international political and economic relations. They were also an effective way of revealing the possible strategies of different groups of actors and incompatibilities between these strategies. The scenarios examined range from a high growth, consumption oriented, non-conflictual pathway; to a slow or moderate growth future based on post-materialist values; to ones of economic and political but not military conflict either North-North or North-South. In brief they are:

Scenario A: a re-run of the 1960s.

Assumptions

- Sustained and Stable High Economic Growth
- Collective Management of OECD
- Convergence of Relative Productivities
- Social Consensus for Structural Change
- Materialist Values Dominate
- Increased Free Trade
- Progressive Integration of Third World into World Economy
- Large aid flows

Scenario B1: on the road to NIRVANA

Assumptions

- Moderate Economic Growth
- Social Consensus for less Market Oriented Growth
- Post-Materialist Values Dominate
- Faster Integration of Third World into World Economy

Scenario B2: Persistent Economic Disequilibria

Assumptions

- Moderate economic growth
- Social discord
- Materialist values dominate
- Slow structural change
- Trade barriers persist
- Limited progress in the integration of Third World into world economy
- Modest aid flows

Scenario B3: Divergent Productivity

Assumptions

- Moderate economic growth

- Wide intra-OECD country differences in the level of social concensus for structural change
- Divergence of relative productivities

Scenario C: the South Opts Out

Assumptions

- Slow Economic Growth
- Deterioration in North/South Relations in Early 1980s
- Third World Countries Cut some of their Economic Links with the North
- Collective Self-Reliance in the South

Scenario D: a Fragmented World

Assumptions

- Slow and then Moderate Economic Growth
- Mounting Protectionism in early 1980s between USA, EEC and Japan
- Preferential Aid, Private Investment and Trade Flows: North America with Latin America, EEC with Africa and South Asia, Japan with South East Asia.

NATIONAL PLANNING AND POLICY ANALYSIS ROLE OF GLOBAL SCENARIOS

The type of global scenarios explored by Interfutures can contribute in a number of ways, for example:

- indicate the global context in which national decisions may have to be formulated,
- demonstrate possible outcomes of broad strategies that might be adopted by the North or the South, such as the more open attitude to the South by OECD countries in Scenario B1, or the de-linking of Scenario C,
- determine the likely effects of gradual changes or sudden shocks to national and regional economies from resource depletion, whether it be of fossil fuels or agricultural land or both,

and how these effects interact and are transmitted.

I will try and indicate the nature of this contribution by describing in some detail the agricultural image of two of the Interfutures Scenarios. Although to describe the scenario is insufficient. The analyst or national planner must interrogate the image presented by the model output or the enriched scenario. He or she must continually challenge the underlying assumptions, and question the validity of apparent feedback responses and interactions, which may reflect a deficiency in the structure of the mathematical model rather than any mechanism apparent in the real world.

Scenario B1 – on the Road to NIRVANA or towards a post-materialist society

Assumptions: There is a more rapid adoption of post-materialist values[1] in the OECD area than in the other scenarios. This permits a consensus on moderate rates of economic growth as an objective, with less stress on material goods and more on gaining fulfillment through leisure and increased participation in political and social activities. Consequently there is a reduction in the social and political forces opposed to the integration of the developing countries into the world economy. The OECD area ODA rises to 1% of GNP as in scenario A, and trade barriers to LDC agricultural exports fall quickly. This favours the more rapid adoption of basic needs strategies and speeds up improvements in income distribution.

Macro-economic image. Over the last quarter century growth in the OECD area averages 3.4% per year, and for many countries it is 30-40% less than in Scenario A. Growth in the developing countries also slows down but not as much as in the OECD area; growth rates are 10-15% less than in Scenario A.

Agricultural image: Growing consciousness of the health risks associated with the over consumption of food, or with particular eating habits is apparent in all OECD countries, and is assumed

[1] R. Inglehart, The Silent Revolution: Changing Values and Political Styles among Western Publics, Princeton University Press, 1977.

to continue in all of the scenarios. However, it is more intense in Scenario B1 than in the others. Gross food energy consumption per capita falls in absolute terms. Governments adopt nutrition policies and there is an end to the current situation in some OECD countries where the Ministry of Agriculture employs policies to encourage consumption of dairy products, sugar etc. whilst the Ministry of Health advise lower consumption levels. The demand for meat and dairy products in the OECD area is lower than in scenario A for a given income level as people make more active responses to possible health risks. Relative to the other scenarios, there is a reduction in the emphasis placed on convenience foods, and those of little nutritional value. The fall in the demand for dairy products and greater consciousness of the ecological hazards associated with certain agricultural practices leads to a reduction in the amount of nitrogen fertilizer applied to pasture. Social demands are for progress towards the adoption of sustainable agro-ecosystems. Whilst this places upward pressures on production costs, it has no appreciable effect on retail prices, which are under downward pressures because of weak demand.

The production and processing of certain food products (like sugar, oilseeds, beef) are transferred more rapidly to the developing countries. in all, agricultural production in the OECD area at the end of the century is almost 25% lower than in the strong growth scenario. Hence big shifts in manpower in agri-food activities – shifts which affect Australia and New Zealand in particular, although they still expand their exports to developing countries in East Asia and to Japan.

At the same time, the liberalisation of trade allows the Third World to increase the value of its exports to the OECD. These are 40% higher at the end of the century than in scenario B2, which assumes similar economic growth rates but less trade liberalisation, and twice the value projected for scenario A. Furthermore, the OECD countries cooperate with the developing countries to increase their food self-sufficiency and facilitate the expansion of trade inside the Third World. For example, other developing countries are able to follow Botswana's example and establish modern extensive livestock production systems to supply OECD markets, although exports do not start to grow substantially until the late 1990s. In all, the food situation in the developing countries improves appreciably under the combined effect of exports, aid for basic needs, and internal development strategies, even if per capita supply does not increase very much. By the year 2000 their share of world food production is 7 percentage points higher than in scenario B2.

In this Scenario rising energy prices do not have a major impact on food production costs. The agricultural sector currently accounts for about 4% of total energy consumption in the developed countries and a somewhat higher proportion of commercial energy consumption in the developing countries. Increased efficiency of direct and indirect energy use, and lower energy intensities than many other goods and services limit increases of real food prices.

Plausibility and implications for North-South relations

There seems little likelihood of a broad consensus for intentionally slower growth emerging during the next decade and possibly longer. Although some minorities in the OECD area currently express this preference, its adoption on a wider scale is likely to be inhibited by the present recession and by existing institutions. Consequently a more open attitude to a New International Economic Order (NIEO) may have to be stimulated by political leaders with vision in the North, or a broader awareness of mutual interest and the potential of the South as an engine of growth for the North.

But even given a more open attitude to NIEO in the real world there are justifiable limits to the speed at which agricultural production and processing can be transferred from developed to developing countries, limits that question this assumption in the scenario. Within the OECD area there is an ageing farm population who are not very sensitive to economic forces. There are no or few alternative agricultural commodities that farmers can produce to replace those relocated to the LDCs, eg. in some states and regions of the USA and Europe sugar represents 10-30% or more of arable farm income, thus rapid restructuring is not possible. A number of other forces would no doubt oppose rapid transfers, eg. the oilseed production and cocoa processing lobbies in the USA. The new entrants to the EEC, Spain, Portugal and Greece would probably endeavour to persuade the Commission to raise rather than lower barriers to certain vegetables, fruit, tobacco and various processed agricultural raw materials. in order to achieve such a transfer governments would have to formulate policies for a 10 or more year adjustment process.

In the short-term only a few developing countries could take advantage of the greater access to OECD markets since their supply elasticities are too low to achieve the flows projected by the scenario. But in the longer term many could develop their primary production and processing sectors and the scenario becomes more feasible.

As in the North, the institutional changes in the South implied by the scenario are unrealistically fast, and hence food production for domestic consumption could not expand so rapidly. If income redistribution policies were pursued vigorously food demand would grow much more rapidly than that implied by the growth rate of average per capita incomes. Explosive growth in food demand would undoubtedly come up against domestic supply bottlenecks and hence import requirements could be higher than those projected.

Scenario C – a breakdown in North/South relations

Assumptions: North/South relations deteriorate in the early 1980s. Dissatisfied with progress towards a new international economic order, the Third World countries (excluding China and OPEC) adopt a policy of collective self-reliance. Vis-a-vis the OECD, they break the links that they consider to be responsible for their dependence on the North. In particular they nationalise the bulk of the assets owned by the OECD countries on their territories and they rapidly increase by a factor of three the trade barriers against imports from the OECD but reduce by 75% the barriers holding back trade inside the Third World. They also double the price of their minerals.

The OPEC countries remain neutral with regard to their oil supplies – they sell it to all who want it, but progressively double the real price between 1978 and 2000. ODA from the OECD area is cut in 1980 to .1% of their GNP because of the nationalisation of their assets. The OPEC countries endeavour to compensate for this loss, and their aid is three times greater than in scenario A.

The governments of the developing countries place greater

stress on agriculture and basic needs strategies to overcome the disruption caused by the de-linking.

Macro-economic image: Gross world product and average incomes are lower than in any other scenario. The Third World's share of GWP increases, but it is a larger share of a smaller cake. In the same scenario, Japan's share drops appreciably, and incomes in all OECD countries are lower. Thus nobody benefits from allowing international relations to deteriorate in the manner assumed by this scenario. The loss by the North of markets and investment revenue in the South, the reduction in the flow of low priced consumer goods from the South, higher energy and raw material prices, all severely disrupt economic growth in the OECD area.

For the people of the Third World, the greater autonomy carries with it certain costs in terms of lower incomes which are, on average, about 17% less than in scenario B2. Latin America, East Asia and certain countries of North Africa and the Middle East benefit most from the production and trade opportunities created by de-linking from the North. Consequently it is the low-income and predominantly agricultural countries of Sub-Saharan Africa and South Asia which suffer most.

Agricultural image: The initial impact is much more severe in the North than in the South. External markets are very important for agricultural price and income maintenance in countries like the USA. Farmers in the OECD area have heavy investments in fixed assets which carry annual charges which have to be met. Thus they are badly hit by the loss of markets in the Third World, and by the reduced demand at home. In contrast most of the Third World's food production takes place in the subsistence sector, which is not heavily capitalised, and subsistence farmers are generally well acquainted with instability.

After the disequilibrium of the early 1980s, the slower increases in incomes in OECD countries restricts demand for meat and for luxury goods from the Third World: tropical fruits and out-of-season vegetables. Poor industrial growth makes it easier for governments to maintain the parity of agricultural incomes, although the employment situation prompts them to maintain and even expand agricultural employment. For domestic reasons, therefore, they are tempted by agricultural protectionism. In any event, the persistence of barriers against imports of goods from the Third World (beef, sugar, oil plants, certain vegetables) and the lack of assistance to the developing countries for the production of exportable goods makes it possible by the year 2000 to increase agricultural production for domestic consumption by 20 to 30% in North America, the EEC and the other European Member countries of the OECD. For Japan, the loss of industrial markets in the developing countries and the non-liberalisation of agricultural markets mean an increase of some 5% in the proportion of consumption covered by national production (instead of a reduction of 50% in scenario A.). Australia and New Zealand lose markets both in the OECD and the Third World. The only increases are in their exports to North America which partially replace exports from Latin America. Because of the socio-political changes inherent in the logic of the scenario, domestic demand for food products falls by 20% in value compared with scenario A.

In the Third World, the low level of aid and the fall in agricultural exports earnings affect the poorer sections of the population, particularly in Sub-Saharan Africa and South Asia and make it barely possible to keep income disparities at their present level. Latin America, the Middle East and North Africa are the only regions which increase their agricultural production for domestic consumption more than in Scenario A.

Food production is reduced in South Asia and Sub-Saharan Africa in comparison with scenario A, per capita supplies are lower but 20% above the present level. However, because of the socio-political changes inherent in the logic of the scenario, distribution is more equitable and there is less malnutrition in the Third World.

As far as costs are concerned, increased production in most OECD countries results in cost levels which are 10 to 20% higher than in scenario A. In Australia and New Zealand, on the other hand, both production and costs fall. Production costs also rise in the Third World – in the region of 10% in Latin America, 20% in the Middle East and North Africa and 40% in East Asia, while the decrease in production in Sub-Saharan Africa and South Asia hardly has any effect on costs in those regions. The volume of trade in agricultural and food products falls, but above all, it changes in structure: a drop in trade within the OECD; a fall in exports from North America to the OPEC countries; a reduction in Third World exports to the OECD, not offset by the growth of trade inside the Third World. Thus, Latin America loses 90% of its exports to the OECD area, but only increases its exports to other regions of the Third World by 40-100%. The situation is even more serious for Sub-Saharan Africa which loses most of its markets in the OECD countries.

Plausibility: It is not very high in such an extreme form. Both the North and the South lose by such actions and would no doubt negotiate to avoid them. Furthermore it may be questioned whether there is currently sufficient common ground amongst Third World countries and enough socio-political cohesion at the national level to embark on such a dislocation of the world economy. If initiated, it could be a region by region act, rather than a collective one. Some countries would probably withdraw from such agreements, since the advantages to them are likely to be much smaller than the losses and internal socio-political cohesion could be greatly undermined by the economic costs. Thus it may be questioned whether many of the newly industrialising countries would accept the short-term falls in income and possible political instability even though in the longer term they could replace OECD exports.

Food production in the developing countries would be faced by two opposing pressures, one resulting from the fall in fertilizer and pesticide supplies from the North, the other the substantial increase in domestic demand caused by rapid attempts to improve income distribution. Per capita demand could grow 1 to 2 percentage points faster than average per capita incomes. The majority of developing countries do not have the supply elasticities in their fertilizer, pesticide and food production sectors to meet explosive increases in demand.

In this scenario the interaction between agriculture and energy are complex and substantial, because the reduction in world trade and reduced export earnings magnify the impact of the rise in petroleum prices. Thus both the direct (micro-economic) and indirect (macro-economic) impacts of this rise are apparent in the simulations and can be analysed.

On the direct side there is the impact on the costs of food production. In the real world this will stimulate cost reducing responses. Those aimed at increasing the efficiency of energy (by switching from petrol to diesel driven farm machinery, or from flood and overhead irrigation to trickle irrigation). And others aimed at substituting between factors of production eg. land or irrigation in place of fertilizer.

On the indirect side there are shocks that are transmitted right across the economy. Energy imports as a fraction of total exports rise, introducing balance of payments pressures that may be reflected in declining exchange rates. In such situations countries might engage in competitive devaluations of their currencies to sustain exports to pay for energy imports and create political frictions over and above those assumed as part of the central hypothesis of the Scenario. Whilst within non-oil producing countries the scenario implys frictions between food and energy producers, and possible political instability. Thus high costs of energy imports and foreign exchange shortages for food imports, result in increased domestic production of food and energy. But if the demand for energy is more price elastic than for food, and supplies of capital are limited there could be competition between these two sectors for investment, with food production loosing. Consequently gross food supply falls and prices rise.

These direct and indirect feedback responses and interactions are represented to a greater or less degree in the SARU model used by Interfutures. Hence policy analysts can interrogate the model or scenario and make iterative changes to determine possible responses to policies they may wish to consider.

Some other scenarios features

Scenario A hits a energy supply problem: it is not feasible since with the time constraint the energy demand cannot be met, the establishment of the capital structure, etc., is not realised. It also is in some ways a Marshall Plan scenario. The aid flows exceed the aid flows to Western Europe in the late 1940's. Partly because of the structure of the model and the input assumptions, particularly concerning the rate of change of institutions, the poverty and nutrition problems of the Third World are not resolved.

Scenario B2 refers to persistent economic disequilibria (where we are now) projected forward, including strengthened protectionism and very little integration of the Third World.

The scenarios can very usefully indicate possible global contexts within which national decisions will be made. They can demonstrate the possible outcome of broad strategies open to the North and the South. They can clarify the links between problem areas. They can indicate the possible impact of external shocks, such as cartel action. They can assist in the formulation of a policy response, but cannot provide precise and early policy options.

The link between energy and agriculture

Most attention have been focussed on the direct impact of rising energy prices on agricultural production, such as the impact of fuel costs on the use of tractors and on irrigation and the price of fertilisers and insecticides. Given time there will be substitution between factors of production, to less energy intensive factors, such as substitution of land for fertiliser. There is also a feedback loop in the stimulation of research and development.

Indirect links may have an even greater effect on the economic and social welfare of countries. The rise in the market price of oil may lead to an increase in the cost of imports, a balance of payments deficit, and a fall in exchange rate. Consequently, even though a country may then sell more as its exports are cheaper, the food imports will be more expensive. Furthermore, because of a balance of payments problem and dependency on foreign energy sources, investment is likely to be diverted into increasing production of fuel. In most of the low-income, oil-importing countries capital is at a premium because saving ratios are low, and so there could be competition between agriculture and energy for available capital.

In South Asia where the system of production is much less energy intensive and where the options are not foreclosed, the costs of production at the farm gate may stay relatively stable. In the more energy intensive production of East Asia, with few opportunities for substituting land for other inputs, prices rise appreciably.

If trade continues, Japan may be able to export sufficient manufactured goods to compensate for the rising cost of imported fuel. If there is some trade delinking there is a destruction of markets, they cannot do this and they devalue their currency. Numerous frictions may be created by the resultant competitive devaluation.

CONCLUSIONS

Hopefully this description of some of the Interfutures modelling and scenario analysis meets the Symposium objectives of indicating the role that global models can play in national planning. They can certainly help to:

- achieve a better understanding of the general structure and dynamics of major economic systems;
- identify or clarify the links between problem areas like energy and agriculture;
- indicate possible policy responses, although they cannot indicate the precise form of such responses.

Furthermore the Interfutures approach is not restricted to global analysis, it is equally useful for national planning, and can be adopted in constructing global models by the "bottom-up" method.

DISCUSSION POINTS

1) Professor Royer felt that in the scenarios, the strategies chosen were largely those of the North, with little room for manoeuvre for the South. No scenario considers how the South could play a greater role in the world system.

Throughout the work, better integration into world markets is taken to be synonymous with the welfare of the South. But there is disagreement on this point, and many experts hold the opposite position. The more the agriculture system is integrated into the world market, the more the poor people will suffer. In scenario C, with delinking, it is suggested that a low level of aid and a fall in agricultrual export earnings will harm the poor sectors of the population. This is not certain, possibly the rich would be the ones to suffer. Also in scenario B1 some regret is voiced that there may be some resistance to the transfer of food processing activities to the South. This may not however be a cause for concern.

2) Dr. Norse noted that export crops are labor intensive and provide employment opportunities. Also when the rich in the urban areas are affected, their demand for food from the rural sector goes down.

Dr. Herrera also felt that the scenarios are biased. For example, it was assumed that Third World countries minus China and OPEC adopt a policy of collective self-reliance. It would have been interesting to include China and OPEC in the Third World grouping, for without OPEC this group of countries is lacking in money, capital and export possibilities.

Dr. Norse noted that the structural change implied in the suggested scenario would take some time, of the order of 10 years or more, and the assumption on the neutrality of China and OPEC meant that other developing countries still had access to their capital and markets.

Dr. Papin wondered whether the scenarios are not too complex for useful model experiments. If the scenarios were simpler, with just a few changes, it might be possible to identify the underlying factors more clearly. After all, if the scenario is not accepted, the conclusions will not be accepted.

Dr. Norse agreed that it is preferable to study one factor at a time and to clearly distinguish the impact of each individual change. But lack of money and time prevented this. It was only possible to carry out a limited number of model experiments and to consider a limited number of scenarios. It would be useful to now carry out step by step model studies since in analysing the output there was difficulty in isolating the impact of the various changes made for each scenario.

3) Professor Richardson asked what were the expectations of the Interfutures project.

Dr. Norse replied that he had no very substantial expectations at the political level because of his experience with similar work for the Prime Minister's Office of the United Kingdom. The Prime Minister had asked if there was any thing in the Limits to Growth. The Minister for the Environment had been to Stockholm and asked whether environmental degradation and resource problems were serious. A group was then formed which gave the answer that the main limits to growth were institutional problems, and not physical resources requiring immediate or short-term action, so decisions could be postponed beyond the 5-year election cycle. That was the end of the exercise as far as the British government was concerned.

4) Professor Richardson asked to what degree the expectations were achieved in terms of results.

Dr. Norse felt they had not achieved as much as hoped for, but that the concept and approach of Interfutures is the correct one. A straight forward modelling approach is not very fruitful, but must be enriched from elsewhere with the addition of soft variables and socio-political issues.

Professor Lesourne noted that due to time constraints it had been necessary to limit the number of scenarios and it was not possible to examine the possibilities of discontinuities in the operation of the world system or potential crises as such. It was also not possible to link the modelling results with a number of quantitatives studies made without models, or to use several models in tandem. The effort to use adequately one model was great enough.

5) Professor Richardson asked what could be learnt from the Interfutures experiences, in particular in areas where the results fell short of the expectations.

Dr. Norse felt that they had shown that scenarios analyses have a role to play in the modelling field. The only regret is that Interfutures ended so early that it was not possible to disseminate the information and to produce the material in a form appropriate for a general audience. It would have been useful to tell others of the mistakes and successes but they were not able to do this.

Professor Lesourne added that it would have been necessary to launch a special effort, country by country on a systematic basis, in order to have the required impact. The result was different from country to country. Where there were good connecttions, where they knew the right people and had access, it was possible to discuss important issues and the influence on general attitudes was then significant. Elsewhere the influence was less, mainly through the distribution of papers. In all such studies of global issues the time spent on the problem of contact, dissemination of results and discussion is as important as the time spent on the project itself.

Dr. Robinson was of the opinion that the Interfutures project had been too large. In an organisation with many member states, the tendency is to have one or two people from each and to get a rather big team. But interdisciplinary research is best carried out by a small team of about 6 people who can then communicate, and the central team at Interfutures did that job. This had relevance to the UNU as it looks forward to other exercises of this type because if you have 20 people at your disposal, it would be best to have 3 separate teams rather than one.

Dr. Norse agreed with this suggestion. In order to capture the diversity of cultural and academic biases, there is a lot to be said for having 2 or 3 small groups rather than one big one.

6) Dr. Robinson referred to the results of model experiments carried out in New Zealand using the SARUM model. A study was made of the effects of the energy price rise of 1979 and a further doubling of energy prices in the middle of the 1980's. As mentioned by Dr. Norse, the key factor in the modelling is in the balance of payments. The New Zealand balance of payments is so unfavorable that while in the initial year of a price rise the country is paying twice as much for energy, within a couple of years the country had devalued so much that the cost is 3 to 4 times as much in New Zealand currency. That is certainly the most important feedback.

As far as the agriculture sector is concerned, since New Zealand is an exporter of food the key feedback from the model is that many buyers such as the Japanese can no longer afford to buy so much food. They produce food more expensively at home, stop buying New Zealand exports, and the terms of trade worsen for New Zealand. This is the illustration of the way in which the model describes a qualitative behaviour that one might not have thought of without model.

5 Modelling of Food–Energy Factors

5.1 The Constraints of World Markets on Basic Food Production

S. COLE

POPULATIONS VERSUS PEOPLE

Most people would agree that the "bottom up" approach to global modelling should start with people. Nevertheless in global models people are soon forgotten. Like other models, the data described in this paper was drawn from national accounts and United Nations regional compilations of national statistics. Populations are counted in millions and cash in billions. The "rounding errors" in the calculations with the model would feed Bangladesh for a year.

The late E.F. Schumacher was rather emphatic on the subject of people and macro-economic models: "If it (economic thinking) cannot get beyond its vast abstractions, the national income, the rate of growth, capital/output ratio, input-output analysis, labour mobility, capital accumulation; if it cannot get beyond all this and make contact with the human realities of poverty, frustration, alienation, despair, breakdown, crime, escapism, stress, congestion, ugliness, and spiritual death, then let us scrap economics and start afresh."

Even if we cannot incorporate people as individuals into our economic models we may, nevertheless, use these models to explore futures in which people have a greater place. For example, we may start with specific local or untypical but desirable examples of the way people live and ask what would be the overall effect if the experience was more widespread? We may ask what would be the effect on world food production if people in the richer economies consumed less meat or on energy consumption if they drove smaller cars. Would the adoption of the seemingly advantageous way of life on a wide scale create conditions under which it could not be sustained? Rightly or wrongly and in part because of Schumachers work, "appropriate technology" has often been viewed as technology with a human face. We can therefore ask the question whether appropriate technologies, used more widely, could improve our way of life.

This is the question underlying in this paper. Thus with a small sleight of hand, we can pretend that we are humanising our global model or at least going half-way. In fact what we shall do is to take data from comparative studies of so-called "appropriate" production technologies in developing countries and use the data in a global economic model to speculate about the survival chances of poor developing countries who adopt appropriate technologies on a significant scale. This data on appropriate technology is generally collected by researchers spending considerable time in factories and villages counting, for example, the number of times a vat is stirred or a field weeded. The job skills are assessed as to whether they are science based or learned by doing. One member of the project team collected data for the model in this way, so some fairly direct links between the model and the ultimate subjects of the model can be claimed, but that is as far as it goes. In order to check whether the information so collected is representative and hence "generalisable" it must be combined with other data. Thus the model finally lends itself to broad sweeping statements and simplistic solutions.

The problem addressed in the paper takes on board a number of recent proposals. The first is the principal recommendation of the Brandt Commission (1979) that "massive" development finance should be given to the least developed countries, especially to promote food (and energy) production. The second is the condition spelled out by President Reagan at the Cancun meeting of Heads of State that United States development aid would only be given to those countries who are prepared to abide by the "rules of the market place". The third starting point is the position of the World Bank which, given recent experiences in the Sahel, is proposing to support appropriate technology while simultaneously going along with the aid policies of the U.S. Administration. Put in these over-simplified terms, the prescriptions of our leading politicians and international agencies are well matched to the possibilities of our model. The model used is particularly suitable for a study of these policies since it focuses on the interrelations between technological change and international and domestic market forces.

At this point a small apology is in order. The model does not deal explicitly with "agriculture" or "energy" although a version of the model employing the usual sectors of agriculture, industry and services has been prepared. Agricultural production in developing countries in the version of the model used in this paper is largely subsumed under the heading of "basic" goods production which also includes food processing and other activities. The reason for using this version of the model here is that the associated studies of appropriate technology were largely carried out with this classification in mind. What matters in the long run, of course, is not agriculture per se but the nutritional value of the food people consume.

THE STRUCTURE OF THE MODEL

The international economy

In the model, the world economies are divided into six groups. This division is based partly on existing economic and political structure and performance (as of 1975) and partly on potential for change. Such classification is inevitably a compromise but, as far as possible, economies have been allocated on this basis. Three groups of developed economies are identified in the model:

1 *Technologically Dynamic Industrial Economies:* those OECD economies which appear potentially be front runners in any new technologies.

2 *Technologically Lagging Industrial Economies:* those OECD economies which exhibit less technological adaptation.

3 *Centrally Planned Economies:* the industrialised centrally planned industrial economies of the CMEA.

The "developing" economies are also divided into three groups:

4 *Newly Industrialised Economies:* the rapidly industrialising developing economies.

5 *Oil Exporting Developing Economies:* largely the OPEC countries.

TABLE I
The grouping of the economies.

Economic Group	1	2	3	4	5	6
Combined domestic product ($1975 billions)	2351	1102	1086	540	244	284
Combined population (millions)	414	243	361	400	148	2481
% Share of agriculture in GDP	3.5	5.0	15.9	14.6	11.8	30.0

Key to Economic Groups.

(1) Technologically Dynamic Industrial Economies: Federal Republic of Germany, Japan, Netherlands, Sweden, Switzerland, United States of America.
(2) Less Dynamic Industrial Economies: Australia, Austria, Belgium, Canada, Denmark, Finland, France, Ireland, Italy, New Zealand, Norway, South Africa, United Kingdom.
(3) Centrally Planned Industrial Economies: Czechoslovakia, Democratic Republic of Germany, Hungary, Poland, Union of Soviet Socialist Republics, Bulgaria, Yugoslavia.
(4) Newly Industrialised Economies: Argentina, Brazil, Chile, Columbia, Costa Rica, Greece, Hong Kong, Israel, Republic of Korea, Mexico, Phillippines, Portugal, Puerto Rico, Singapore, Spain, Taiwan, Turkey.
(5) Oil Exporting Developing Economies: Algeria, Iran, Iraq, Libya, Saudi Arabia, Venezuela, Kuwait, Nigeria.
(6) Least Developed Economies: Angola, Benin, Burundi, Central African Republic, Chad, Congo, Egypt, Ethiopia, Ghana, Guinea, Ivory Coast, Kenya, Lesotho, Liberia, Madagascar, Malawi, Mali, Mauritania, Morocco, Mozambique, Niger, Rwanda, Senegal, Sierra Leone, Somalia, Sudan, Togo, Tunisia, Uganda, United Republic of Cameroons, United Republic of Tanzania, Upper Volta, Zaire, Zambia, Dominican Republic, El Salvador, Guatemala, Haiti, Honduras, Jamaica, Panama, Trinidad and Tobago, Bolivia, Paraguay, Afghanistan, Burma, Bangladesh, India, Jordan, Laos, Lebanon, Malaysia, Nepal, Pakistan, Sri Lanka, Syrian Arab Republic, Thailand, Yemen P.D.R., Papua New Guinea, Cambodia, Peoples Republic of China, Vietnam, Rumania, Mongolia, D.R. Korea, Albania, Cuba, Uruguay, Peru, Indonesia, Equador, Zimbabwe, Bhutan.

6 *Least Industrialised Developing Economies:* all other developing economies whether market oriented or centrally planned.

The countries making up the six groups are listed in Table I.

We will be concerned primarily with the last group which includes more than half the countries of the world. Agricultural production accounts for about 25% of all output in these economies. In the most industrialised economies it is less than 5% of total output. The differences in the economic structures of the developed economies are considerably less than those between the developed and the developing groups taken as a whole and much less than differences within the developing group itself. For the developing countries the evaluation of potential technological performance is based largely on their present structures. UNIDO (1978) and OECD (1979) studies indicate that 80% of all capital goods production in the developing countries is in some 8 countries.

The domestic economies

The domestic structure of the model attempts to capture some aspects of the relationship between production techniques and the type of production. Especially important is the relationship between who produces which goods and who consumes them. Within each of the six groups of economies the major division of consumption and production is between "basic" and "luxury" or "non-basic" goods with a third investment goods sector.[1] This division cuts across the conventional sectors of agriculture, manufacturing and so on. For example, in the least developed group of countries "basic" goods would include staple food stuffs such as maize flour and rice, as well as many manufactured household items and everyday clothing. It also includes essential medical services and schooling. Luxury goods includes those goods which Stewart (1976) classes as having "high income characteristics", for example, caviar, whiskey, fancy motor cars and colour televisions.

The main economic "actors" in each economy are labour

[1] The equations of the model and a solution algorithm are given in Chichilnisky and Cole (1979) and Clark, Cole and Lucas (1979).

and capital. Labour is divided into "skilled" and "unskilled" workers. These two groups differ in their relative numbers in each economy, their employment in the different sectors of production, the income they receive and its sources. Their patterns of consumption of basic and non-basic goods and saving are also different. Skilled and unskilled labour are therefore taken to be identified with high and low income households respectively. Both groups own capital although this tends to be concentrated in the higher income (i.e. higher skilled) group. Government activity is not excluded explicitly so Government consumption is divided between low and high income households. Government policy changes such as redistributive tax transfers between income groups can nevertheless be simulated in the model. Therefore, stretching a point, some sixty groups of people are identified – high and low income workers and entrepreneurs in three sectors plus a Government in six economy types of economy. But this is the limit of personification in the model.

Domestic and international markets: Some of the mechanisms which affect the bargaining strengths of these actors in the domestic economies are formalised in the model through assumptions about the supply of labour and other factors. For example, it is assumed that the more people earn, the more people will work. It is also assumed that entrepreneurs (and state capitalists) will move their investment around so as to increase the rate of return.

The same assumptions are made for the centrally planned economies. While it may seem incongruous to treat these economies in the same way as the market economies, for present purposes, it may be argued that all economies, however strictly planned, are obliged to pay attention to the constraints of world markets, both in their domestic and international policy. In agriculture especially, many Eastern European Economies and also the Peoples Republic of China have private small holdings operating alongside collective farms and peoples' communes.

The exogenous data for each economy are the supply of capital, the technical coefficients for each production sector, the composition of consumption, and any domestic income transfers. The balance of payments are also taken to be exogenous, and to depend on the level of development finance and conditions on its repayment. For the base period, these data are used to calculate an equilibrium with total demand equal to total supply for each commodity worldwide. Commodity prices are uniform across all economies. Within each economy, demand and supply for each factor of production are equal and there are no excess profits. The income and expenditure of each income group are separately balanced.

When changes are made to the exogenous parameters, as in the experiments described later, a new equilibrium is calculated with worldwide demand and supply again equal. Capital is redistributed within each economy so as to achieve a uniform rate of profit. This gives new levels of output of goods by sector in each economy, employment levels for skilled and unskilled labour, factor prices and net trade by economy and sector.

Model parameters

The model is set up to match the state of the world economy in 1975 as closely as possible. This year was chosen as the best compromise given available data. The data used was drawn from a number of sources. The main problem is that only for a few countries are there consistent data which break down the income and expenditure flows for different categories of workers and households. This has been done for a number of countries in the form of social accounting matrices (see for example Pyatt and Roe 1977). For each region the social accounting matrix for an economy, typical of the economic group as a whole, has been prepared so as to provide key data on the input-output and income distribution. These data supplement other more aggregated statistics from United Nations sources.[1]

Basic and luxury goods: The starting point for data preparation is the definition of basic and luxury goods. Even the simple assumption that what the very poorest groups consume should be considered totally "basic" and what the very richest consume as totally "luxury", is difficult to apply empirically. For the present estimation, it is assumed that the proportion of goods consumed from each production sector by the unskilled (low

[1] Some of this work is described in Meagher (1980), Okolski and Karminsky (1980) and Cole and Evans (1980), Cole and Meagher (1981).

income) group, defines the proportion of basic goods production from that sector. We also assume that the basic and luxury goods, produced by the six groups of economies, are equivalent and are tradable internationally. It is therefore assumed that in the various economies, households consume different amounts rather than different types of luxury or basic goods. The shares of luxury consumption in the twelve income groups' consumption baskets, are shown in Table II. The consumption of basic goods decreases fairly systematically as the per capita income level rises.

TABLE II
Comparison of consumption and factor availability across economy groups.

Economy Group	1	2	3	4	5	6
Share of luxury Consumption:						
High income	0.82	0.74	0.70	0.69	0.59	0.54
Low income	0.69	0.63	0.62	0.65	0.53	0.45
Share of agriculture in Consumption:						
High income	.035	.036	.016	.12	.22	.35
Low income	.05	.054	.02	.15	.33	.60
Factor Availability:						
Ratio of high to low skill labour supply	3.5	2.35	0.75	0.54	0.28	0.30
Capital Stock per Capita[a]	18.26	12.28	4.74	1.77	0.47	0.32

[a] Units are thousands of $US (1975) per person.

This is also true of consumption of agricultural products. The poorest group in the least developed countries spend 60% of their income on agricultural products, whereas the highest income group in the richest countries spend less than 4%.

Technology: The differences between the way different goods are produced in the six economies are of key importance for the results of the model. Most theories of development suggest that systematic trends in factor intensities across countries, roughly in the order that regions were listed earlier, are to be expected, i.e. as we move from the technologically advanced high income countries capital/labour ratios should systematically decline. The most straightforward cross region comparison to make here is physical labour inputs (e.g. worker-years) and inputs of physical labour stock. Inputs of labour and capital by sector for each economy group are shown in Table III. Factor inputs to basic and luxury production in the more industrialised economies are roughly equal. For the least developed region basic production appears considerably more labour intensive. Generally speaking the production structure of the less industrialised countries is more heterogeneous.

There is again a parallel between basic and agricultural production. For example, capital labour ratios across the

TABLE III
Labour and capital inputs by sector.

Low skill employees/unit output[a]

Economy Group	1	2	3	4	5	6
Luxury	0.08	0.088	0.113	0.26	0.28	3.6
Basic	0.079	0.099	0.093	0.24	0.26	5.6
Investment	0.075	0.080	0.084	0.107	0.112	0.98

[a] Units are employees per $US1000 output (1975 prices).

Capital stock/unit output.

Economy Group	1	2	3	4	5	6
Luxury	2.2	1.45	0.67	0.77	0.42	2.3
Basic	2.1	1.5	0.66	0.76	0.42	2.1
Investment	0.64	0.93	0.58	0.77	0.37	1.2

economy groups differ by a factor of 200. In the least developed group of economies, at least, it may be argued that the behaviour of the agriculture sector or basic food production, at least, will parallel that of the basic goods sector in the model.

Skills and capital· The capital stock in each group of economies has been estimated using data in Doblin (1978). The capital stock per capita in each economy is given in Table IV. Combined with information on the operating surplus in member countries of each group from the U.N. Yearbook of National Account Statistics, it is possible to deduce a rate of profit for each type of economy.

The structure of production in the various economies is to some extent historically determined by factor supply conditions. Changes in the structure of production will in turn affect the composition of the workforce. The extent to which this affects the relative wages of skilled and unskilled workers will depend on their relative scarcity in the economy. The gap between the current level of employment and full utilisation and the current wage level, provide a measure of the supply elasticity of unskilled labour. Thus unlike the famous model of Arthur Lewis (1954), an unlimited supply of unskilled labour is not assumed and increasing employment for low skilled workers will increase their wages. To estimate the supply elasticity of unskilled labour, it is argued that every person of working age is potentially an unskilled worker. The supply elasticity of skilled labour is less easy to assess. For the purpose of this exercise, it is determined from the proportion of people in each economy who have received higher education or vocational training (this is described in Cole and Miles (1981). In the most advanced economy about 20% of people of working age fall into this category. In the least developed economy only 4% do so. The ratio of the supply elasticities of skilled versus unskilled labour determined in this way, are shown in Table II.

The results again show a systematic trend. In the most developed economies, skilled labour is relatively more abundant than unskilled labour, while in the less developed economies the reverse is the case. It follows that if production rises in the developed economies, with no change in the relative inputs or availability of skilled or unskilled labour, then wages will increase faster for unskilled than for skilled labour. The reverse is true for the developing economies. These differences in the relative abundance of skilled and unskilled labour again are important for an interpretation of the experiment described later.

Social accounts: The data for each economy in the model in the form of a social accounting matrix. These social accounts for the least developed group of economies to be compared with that from later experiments, is shown in Table IV. Key data for all economies for the base year (1975) are shown in Table V. In Table IV, the first three columns show the cost of each factor in the production of each commodity. The next two columns show transfers from the skilled and unskilled wage and investment income accounts to households. Purchases of luxury and basic consumption goods and investment by households, are shown in the next two columns.

Table V supplements the data of Table IV, showing the rate of profit as a percent of invested capital and the level of employment (in billions of worker-years). The other data shown, total domestic product of the region, total household income (including development aid), total sectoral output and trade levels for each commodity are measured. The 'foreign' account column shows only *net* trade and the balance of payments. The latter, whether deficit or surplus, is divided between each income group in the same proportion as their share of total domestic savings.

Experiment 1: Development finance

This part of the paper describes an experiment in which a fixed amount of development finance, $30 billion, is passed from the most industrialised to the least industrialised economies. The amount chosen is in line with the suggestion of the Brandt Commission (1979) and other studies. The Commission suggests (p228) that the aggregate needs of all the least developed countries for external capital to support a 5-6% growth rate is estimated by UNCTAD as $11 billion per year during 1980's and $21 billion per year in the 1990's. They also quote the ODC (Overseas Development Council), who consider that some $40-54 billion will be required by the low and low middle income countries to achieve a growth rate of 3.5-4%. What is of

TABLE IV
Social accounts for least industrialised developing economies – (Base year – 1975).

	Luxury	Basic	Invest.	Skill	Unskill	Capital	High inc.	Low inc.	Foreign	Total
Luxury	15	16	11	0	0	0	83	42	–11	159
Basic	15	17	9	0	0	0	71	53	5	172
Invest.	7	6	5	0	0	0	47	11	–20	57
Skill	25	16	15	0	0	0	0	0	0	57
Unskill	28	48	2	0	0	0	0	0	0	79
Capital	67	67	12	0	00	0	0	0	0	146
High inc.	0	0	0	57	0	126	0	0	18	201
Low inc.	0	0	0	0	79	20	0	0	7	107
Total	159	172	57	57	79	146	201	107	0	981

Amounts in billions of 1975 US dollars. All figures are rounded down to nearest billion US dollars.

interest here is whether such aid in absence of supporting policies is likely to stimulate food production.

TABLE V
Results of experiment base year – 1975.

Economy Group	1	2	3	4	5	6
GDP.	2351.3	1102.1	1086.2	539.9	244	283.9
High income	911.8	435.9	448.3	389	199	201.8
Low income	1363.4	686.1	619.9	187.8	56.9	107.1
Profit %	10.14	12.04	33.06	40.32	114.66	18.61
Skill employment	634.3	254	245.8	163	40	57
Unsk. employment	949.9	488	276.3	91.9	29.9	79.9
Luxury output	1887.8	921.9	1281.3	510.7	201.1	159
Basic output	1300.3	743.1	822.2	270.2	105.1	172.9
Invest. output	1009.3	568.1	527	137.9	64.7	57
Lux. exports	42.7	–11	–29.1	–21.3	28.3	–10.5
Basic exports	–42.3	–12.9	59.3	6.2	–15.1	5.4
Invest. exports	75.5	3.7	–12.3	–22.1	–25.3	–20

KEY:
1 = Technologically dynamic industrial economies
2 = Technologically lagging industrial economies
3 = Centrally planned economies
4 = Newly industrialised developing economies
5 = Oil exporting developing economies
6 = Least industrialised developing economies

World price of commodities:

Luxury = 1.0
Basic = 1.0
Invest. = 1.0

The income of the least developed economies is about 4.1% of total world income and their population is 61% of total world population. For the purpose of the experiments, it is assumed that the ultimate beneficiaries of development finance in the developing countries, are intended to be the lower income households of these economies. Thus, in the following experiments, investment income arising from the transfer is taken to accrue to this group in addition to other income arising from changes in their level of employment. It is also assumed that the full amount of the transfer is from the high income group in the most industrialised economies. This is the most favourable situation we can imagine. Clearly if the finance is transferred as direct investment or a loan, the benefits to the poorest groups are likely to be reduced.

The impact of such a gift is shown in Table VI. While the shifts from the base period are somewhat exaggerated, they nevertheless indicate the *direction* of change likely to be induced by the development aid.

TABLE VI
Results of experiment – 30 billion dollar capital gift.

Ecónomy Group	1	2	3	4	5	6
GDP	2351	1102.2	1086.1	539.6	243.9	289.5
High income	910.5	435.6	448.9	389	199	201.8
Low income	1364.5	686.5	619.1	187.5	56.8	112.7
Profit %	10.13	12.02	33.07	40.36	114.71	18.60
Skill employment	633.9	254	245.9	162.9	40	57
Unsk. employment	950.1	488.2	276.1	91.7	29.9	80
Luxury output	1860	922.9	1268.2	508.3	200.2	206.2
Basic output	1330.1	739.7	830	265.9	105.2	145.7
Invest. output	1009	572.2	533.5	144.7	65.6	39.1
Lux. exports	14.5	−10.5	−42.9	−23.7	27.4	35.6
Basic exports	−13.5	−16.7	66.8	1.9	−15.1	−23.5
Invest. exports	74.9	7.1	−6	−15.4	−24.4	−37.2

Even though no change is assumed in the production techniques, the experiment suggests certain changes in the structure of production and trade. Luxury production in the recipient economy increases by some 30% while basic goods production declines by 16% and investment goods production by 40%. Thus agricultural production seems to have been harmed rather than helped by the transfer. The relatively small shift in employment and incomes belies the larger shift in production and trade. Because incomes and hence domestic levels change little, the economy shifts from being a net exporter of basic goods to a net exporter of luxury goods and net imports of investment goods nearly doubled.

The donor country does not provide these additional investment goods imports to the recipient economy. These exports are instead provided by other economies, notably the newly industrialised group and also the centrally planned and less dynamic industrial economies. The donor economy provides instead, the additional demand for basic goods. What this result suggests, for example, is that in an unfettered world economy, if the United States should provide aid to Pakistan, the latter will produce export quality shoes and run down its agriculture, import wheat from the United States and purchase milling machines from South Korea.

By comparison with the shifts of production and trade, there are rather small changes in income. At first guess, we might expect the $30 billion new investment to raise the total income of the economy by an amount pro rata to the existing stock (i.e. about $11 billion). This does not happen because adjustment to the market leaves employment and wage levels relatively unchanged in the new equilibrium. The income of the recipient group in the least developed economies is increased only by an amount roughly equal to the increase on investment income ($30 billion x .185). The income of the high skilled group is practically unchanged. In fact, in the *donor* economy, there are more marked shifts in employment. The income of the low income group rises while that of the high income group falls. This is despite the fact that the direct transfers in this experiment are identical with those in the base year. The income distribution in other economies also changes. It improves in less dynamic industrial economies and worsens in the newly industrialised and centrally planned economies. Clearly, in obtaining the equilibrium a rather complex process of adjustment has taken place which calls for some explanation.[1]

The reason for the shifts observed in this experiment, arise from the underlying assumptions of the general equilibrium model and the precise magnitudes of the factor supply and technical coefficients shown in Tables II and III. When additional capital is introduced into an economy, additional production becomes possible. In most cases, this will require that additional labour is employed. Given the assumed labour supply equation, this causes the price of labour to rise. If this

[1] Most authors of general equilibrium models prefer not to attribute any economic significance to the *process* of adjustment, but prefer to demonstrate that the outcome has an internally consistent rationale (see for example De Melo, Dervis and Robinson 1978). The problem is one of causality. Since every variable is mutually dependent on all others and since there are no time lags, the result of the experiment has to be thought of as the full working through of all market processes.

happens, then capital will be relatively cheaper and production in the more capital intensive sector will become more favourable than hitherto.

In the new equilibrium, the employment levels of skilled and unskilled labour in the least developed economy, rises slightly, so capital does indeed become relatively and absolutely cheaper and the rate of profit falls. Because the luxury goods sector in this economy is the most capital intensive, its competitiveness in world markets for goods and domestic markets for factors of production is relatively strengthened. It should be noted that this is not automatically the case. The functioning of the model is such that it will redistribute capital and labour in each economy so as to achieve uniform factor prices across the economy. The small increase in the price of skilled and unskilled labour might be expected to promote production of basic goods and investment goods since the capital and basic goods sectors in this economy are respectively the least intensive in their use of unskilled and skilled labour. However, these tendencies are strongly offset by the larger fall in the price of capital.

It was noted earlier that capital used to produce commodities is interchangeable although the factor proportions in each sector remain constant. Capital is, therefore, not written off, or taken out of production during the adjustment. For example, if production of luxury motor cars rather than trucks becomes more favourable, then the output of the plant would be switched from trucks to private cars. Similarly, land used to produce staples such as maize is switched to production of hard wheat for export.

The Brandt Commission suggests that development finance should be directed specifically to production of food (which makes up much of basic production in the poorest economies). What the results of the model indicate is that if the new capital was restricted to the production of basic agricultural goods, then the rate of profit or wages in this sector would have to be lower than in the rest of the economy. However, it is implied by the assumptions in the model that in this situation entrepreneurs will invest in more profitable activities or labour will move to better employment prospects. This would lead, for example, to migration from rural areas to industrial townships and cities. Alternatively production of luxury crops would increase at the expense of basic agriculture.

Thus the initial attempts to tie down the investment for food production will soon be frustrated. Indeed the additional finance will lead to an acceleration in the "modernisation" of the developing economy exhibiting the common syndrome of a decline in basic agriculture and increased dependence on other countries for capital goods. The switching of production of staple foodstuffs to cash crops for export with the increased importation of staples, in particular, has characterised many developing economies in recent years. Iran and Chad, for example, have increased production of cotton for export while basic agriculture has practically collapsed. Brazil has switched from the staple bean crop to soya for export and sugar cane for production of gasohol fuel for motor cars.

The main conclusion of this first experiment, therefore, is that domestic and international market adjustments can considerably reduce the beneficial effects of development finance. Whatever the intrinsic merits of the Brandt Commission's proposals for increasing food production through massive development assistance to Third World countries, any condition that recipients of such assistance should operate as market economies, may nulify any potential gains and do nothing to improve social distribution. Furthermore, the effects of development finance appears to be to weaken those basic production sectors (such as agriculture) which the Commission hopes to strengthen.[1]

[1] One arguable limitation of the present model in evaluating Brandt Commission type proposals, is that the level of stock in each economy is fixed exogenously and full utilisation of capital stock is assumed. The Brandt Commission emphasises the IMF view that "massive" transfers are particularly important to the industrialised economies when, as in the present recession, there is underutilised capacity. It may be argued, however, that to bring unutilised capacity back into production would require an increase in the rate of profit in the same way that in the model it is assumed that an increase in the wage rate is a prerequisite for an increase in the labour force. In the experiment the rate of profit in the most industrialised economies is rather stable and only shifted by less than 1% of the rate in the base period. Consequently, unless capital supply was very elastic, it is unlikely that this particular addition to the model would greatly affect the results.

Experiment 2: Appropriate technology in the face of world markets

In this experiment the $30 billion development finance is used to promote appropriate technology in the basic goods sector of the least developed economies. The interesting question here is whether a developing economy could expect to adopt these techniques and also adopt free trade policies at home and abroad.

Definitions of what constitutes "appropriate" technology vary widely. A recent extensive survey by Vitelli (1980), shows that a principle concern of the majority of studies in this area, is to show that the more labour intensive techniques will be competitive with modern sector production. These studies also assume that factor prices will be unchanged by the introduction of the new technology. While this may true for a specific production unit in a large economy, it cannot be assumed for a strategy of development in which appropriate technology is to be introduced extensively as a matter of policy. This is especially so when the purpose of introducing these techniques is to increase the incomes of at least one of the factors, i.e. low income labour. This increase obviously could affect the economic viability of the technique. Indeed, a technology which is competitive at the prevailing wage level, may be quite uncompetitive if factor prices change.

In order to make use of existing studies as a basis for the evaluation of an appropriate technology strategy, a precise definition of the technology is required. A particular "basic needs" oriented development strategy was defined and from a secondary analysis of several studies (Cole and Nunez-Barrigga 1981), it was possible to identify techniques which are compatible with this strategy. The strategy includes the idea that the appropriate production techniques themselves should be produced in the developing economy and that the production units should typically be small. To quote Schumacher (1974) again:

> Production methods must be relatively simple, so that the demands for high skills are minimised not only in the production process itself but also in matters of organisation, raw material supply, financing, marketing and so forth.

In order to quantify the factor inputs to the techniques in a consistent fashion which can then be related to parameters in the model used here, the factor shares are compared directly with those of "modern sector" techniques used in the production of luxury goods. The resulting comparison of appropriate and modern sector techniques, is given in Table VII. As the standard errors to the results of the study indicate, there are wide variations in the factor inputs to production within and between sectors. The analysis of existing data was backed up by a second study by Date (1980) in India of the range of *appropriate* techniques which could be used in selected

TABLE VII
Estimated coefficients for appropriate technology experiment – ratio of factor inputs to basic needs versus modernisation strategies.

	Unskilled output	Skilled output	Labour output	Capital output
Food processing (sector average)	3.17	0.72	(1)	0.03
Textiles (sector average)	(1)	(1)	4.38	0.48
Housing (sector average)	1.9	0.05	3.5	0.39
Averages over sub-sector studies	2.9	0.39	4.0	0.41
Standard deviation	2.3	0.51	3.5	0.25

Source: Cole and Nunez-Barrigga (1981).
(1) No data available.

sectors. Some of Date's results for cottage or village scale techniques versus those suitable for small towns are shown in Table VIII. The possible range of factor inputs between competing *appropriate* techniques is comparable with that between modern sector techniques and the appropriate techniques as selected in the Cole and Nunez-Barrigga study.

Date's results also shed some light on the energy requirements and scale of production of appropriate techniques. Energy inputs especially do not show a clear pattern. For example, in vegetable oil extraction only a relatively small amount of (bullock power) energy is used. By contrast, in soap manufacture the cottage industry appears to be relatively inefficient in its consumption of (firewood) energy.

TABLE VIII
Relative scale and factor inputs for different 'appropriate' techniques.

Sector	Scale	Capital output	Labour output	Energy output
Tanning[a]	.005	10.4	.018	(1)
Vegetable oil extraction[b]	.06	8.9	.003	.008
Soap making[c]	.012	3.4	.08	2.8

Based on Date (1980).

[a] Cottage scale bag tanning versus small scale chrome tanning – bag tanning uses only natural heat but takes 12-13 weeks compared to chrome tanning which requires additional heat.
[b] Bullock Ghani versus 4 tonne screw expeller.
[c] Cottage scale vat versus town scale.

The results for labour inputs in Table VII are comparable to those of Bartsch (1977) for cereal production in Asia. Bartsch distinguishes between *technology* – whether a cereal is a traditional, an improved or a high yielding variety (HYV) – and *technique* – the degree of mechanisation. Each variety may be produced using hand ploughs or tractor drawn ploughs. The results of Bartsch's study are summarised in Table IX. This data for other inputs are less clear cut. The HYV's require more energy intensive fertiliser, pesticides etc. than other varieties and hence greater intermediate inputs. However, output per hectare is higher with the HYV's so, if land was the only capital input, capital labour ratios would be lower for these crops than for traditional varieties at a given level of mechanisation. In practise, whatever the possibilities, new varieties tend to be more mechanised and this increases the associated capital labour ratios.

TABLE IX
Comparison of factor inputs to rice and wheat production using different crop varieties and levels of mechanisation.

	Crop variety		
Inputs	Traditional	Improved	High yield
Land Output	4.75 (2.5)	1.7 (1.8)	1 (1)
Labour Output			
Traditional	6.25 (9.1)	5.8 (7.5)	4.4 (4.6)
Some mechanisation	2.6 (4.0)	2.2 (3.1)	1.6 (2.0)
Mechanised	1.4 (1.7)	1.3 (1.5)	1 (1)
Other inputs (number of fertiliser or other dressings)	0	4	7

Based on Bartsch (1977): data for wheat shown in parenthesis.

Obviously, on the basis of these empirical findings any generalisation to sector wide or economy wide analysis must be treated with caution. For the experiments the changes in technology will be based on Table VII. It is assumed that inputs of skilled and unskilled labour and capital to appropriate techniques for the basic goods sector are respectively 39%, 290% and 41% times those prevailing in the luxury goods sector. It is further assumed that, with the $30 billion development finance, these appropriate techniques will be added to the existing $390 billion capital stock in the basic goods sector in the least industrialised economies and owned by low income households. The aggregate factor inputs adjusted pro-rata. At prevailing factor prices this should ensure that the competitiveness of the sector is marginally increased domestically and abroad. In other words, if factor and commodity prices remained the same we should expect this sector to expand and exports of basic goods from the economy to increase.

As the results of the experiment in Table X shows, this does not happen. Instead there is a rather dramatic fall-off in basic goods (i.e. agriculture) production in the least industrialised countries and a much increased dependence on external sources for these goods.

TABLE X
Results of experiment – 'Appropriate' technology. (Open Economy)

Economy Group	1	2	3	4	5	6
GDP.	2350.4	1102.3	1085.7	539	243.7	278.1
High income	907	434.9	450.2	389	199	210.6
Low income	1367.3	687.3	617.5	186.9	56.7	92.5
Profit %	10.1	11.9	33	40.4	114.8	20.3
Skill employment	633	254	246.2	162.7	39.9	55.7
Unsk. employment	950.7	488.5	275.5	91.4	29.8	69.2
Luxury output	1784.2	923.2	1236.9	503.3	197.3	315.1
Basic output	1411.3	735.5	850.4	258.2	106.9	45.3
Invest. output	1008	579	547.1	157.5	67	6.5
Lux. exports	–62.2	–11	–75.7	–28.7	24.4	153.6
Basic exports	64.9	–21.8	86.4	–5.8	–13.4	–111
Invest. exports	73.1	12.7	7.1	–2.6	–23.1	–67.7

In equilibrium, the introduction of the new technology has triggered a reduction in employment of both skilled and unskilled labour in the least developed economies. This results in a 4% and 17% decline in their respective wage rates. At the same time the cost of capital in this economy increases by 9%. A striking feature of these results is that although the factor price movements are in the opposite direction from those in the earlier "Capital gift" experiment (Table VI) the production shifts are in the same direction.

Since a lowering of factor prices is most favourable to those sectors which are intensive in its use (or vice-versa), it is seen from Table III that a rise in the price of capital favours expansion of luxury production. Similarly, a fall in the cost of low skill and skilled labour respectively favours basic and investment goods production. Given that production of investment goods is much less capital intensive than other sectors of this economy and that the fall in the price of skilled labour is comparatively small, production of capital goods should be favoured. This does not happen. This suggests that the explanation cannot be found simply in terms of the prevailing factor inputs and that the reason must lie in the changes in production techniques or the relationship between production costs in the least developed economy and the rest of the world. The following experiment exploring the impact of appropriate technology in a closed economy will serve to justify this assertion.

The experiment indicates a tendency for the combination of policies, appropriate technology and free trade, to push the least developed economies into greater dependence on other economies for both basic commodities and investment goods. Further, the increase in the rate of profit in the least developed economy more than offsets the loss in wages of the high income group; for the low income group, even the additional ownership of the new capital stock does little to offset their loss of wage income. Hence the introduction of the appropriate technology creates pressures to added external dependence and worsens the absolute and relative incomes of the poorest group. This is quite the reverse of the intention.

Since the shifts in production in this experiment are rather large, the assumption of "putty" capital used in the model is stretched beyond credibility. Nevertheless, if we treat the result as indicative it lends support to the idea that open economies and appropriate technologies do not mix and possibly exposes a contradiction in the relationship between development agencies and the activities of the current U.S. Administration.

Experiment 3: Closing the market

The last experiment (Table X) suggested an incompatability between appropriate technology and an open economy. It should not be viewed, however, as an objection of this technology under all circumstances. The figures used for the experiment were tentative and it is reasonable to ask what amount of capital and skilled labour would have to be displaced from production in order to support the trebling of unskilled labour inputs assumed. Halving again the amount of skilled labour and capital previously assumed, for example, does little to prevent the decline of the basic goods sector.

At least as important to examine whether it is the nature

of the technology or the openness of the economy, which is responsible for the apparent failure to achieve the objectives set for the appropriate technology. As noted earlier, one aspect of the "basic needs" oriented definition of appropriate technology adopted in deriving Table VIII was that an economy should be much less dependent on other economies for critical commodities. This does not imply that it should not trade, but rather that net trade flows for each class of commodity should be relatively small compared with total domestic consumption. This is effectively the case for the more industrialised groups of economies.

The impact of "closing" the least developed economy group is demonstrated by calculating the equilibrium solution for these economies using the same adjustment to the technical coefficients as in the last experiment. The total net importation of each commodity is not permitted to exceed a nominal $1 billion and in addition, the balance of payment deficit of the economy is set to zero, i.e. the economy is no longer subsidised from outside. The "rest of the world" economy is computed separately.

The results of the experiment shown in Table XI, provide a major contrast to those of the open economy. Despite the assumed loss of $7 billion support (through the balance of payments deficit), the total income of the low income group is only some $3 billion less than in the base year experiment and some $8.2 billion less than in the capital gift experiment. Unskilled employment rises by some 3%. The level of employment of the high skilled group increases by more than 11%. The consequent rise in wages balances the decline in their investment income of 11.6%. The decline in total income of this group arises from the loss of the support through their trade deficit.

Since income levels are roughly those of the base period and net trade levels are negligible, domestic production is sufficient to satisfy all domestic consumption and investment. The introduction of more unskilled labour-intensive technology has indeed led to the desired increase in employment and wages and also enabled some redistribution. However, it has also led to an even greater demand for skilled labour. This is largely because the previously imported skill-intensive investment good, is not produced domestically. Even so, the total budget of the low income group is slightly less than the base period, while the budget of the high income group is very much less.

TABLE XI
Results of experiment – 'Appropriate' technology (Closed Economy)

Economy Group	1	2	3	4	5	6
GDP.	2351.3	1101.9	1086.2	540.1	244	288
High income	922	436.2	447.7	389	198.9	183.2
Low income	1378.3	685.7	620.4	188.1	57	104.7
Profit %	10.1	12	33	40.2	114.6	15.9
Skill employment	634.3	254	245.6	163	40	63.6
Unsk. employment	949.8	487.8	276.4	92.1	30	82.5
Luxury output	1889.3	919	1287.8	512.6	200.4	175.8
Basic output	1298.5	749.3	820.4	275.5	106.9	170.1
Invest. output	1008.9	562.5	521.3	130.7	63.4	78.9
Lux. exports	26	–13.3	–22	–19.4	27.7	.7
Basic exports	–50.3	–5.8	57.8	11.5	–13.3	–.6
Invest. exports	75.2	–1	–17.9	–29.3	–26.6	–.3

In equilibrium, relative domestic prices differ from those for the world economy. The price of basic goods is reduced by 4% and the price of luxury goods by 3%, relative to the price of investment goods, whereas in the world economy they are roughly equal. Without these price differentials the new levels of production would not be possible in the internal domestic market just as they are not viable in the international market. It should also be noted that the change in relative prices of basic goods means that the real value of incomes in terms of their purchasing power of basic goods has increased.

The experiment supports the assertion put forward earlier, that the failure of the "appropriate" technology experiment was due to international market forces rather than the structure of the domestic economy, or even the factor proportions assumed for the new technology. This last experiment indicates that a basic needs oriented strategy, with appropriate technology as a major component, by necessity must also place restrictions on trade. It also follows that when a developing economy which has previously employed appropriate technology and trade restrictions, chooses to become more integrated into the world economy, it may experience strong pressures to unbalance the production structure and worsen the income distribution. A case in point may be the Peoples Republic of China.

Experiment 4: A basic needs oriented economy

A shift in production techniques may also be associated with a change in consumption habits. As pointed out by Stewart (1976) and others, an increased use of appropriate techniques would be linked to greater consumption of basic goods. For example, more contents and less fancy packaging. This could increase the "value for money" in terms of essential nutrients. A study by Kaplinsky (see Table XII) indicates that the unit cost of food energy in commercially branded breakfast cereals is some twenty times that of maize flour. Clearly similar Tables may be constructed for most consumption goods and services. Thus even if there was no increase in income a shift in consumption habits could lead to an improved situation for key social indicators such as health, education and life expectancy. This should be born in mind in interpreting the next experiment where only economic and not social values are considered.

TABLE XII
Comparison of unit nutritionally costs for maize flour and locally manufactured breakfast cereals.

	Unit price difference	Protein	Calories	Vitamins (average)
Weetabix Maize flour	20	19	21	5
Weetaflakes Maize flour	25	30	24	7
Post Toasties Maize flour	27	36	27	10

Based on Kaplinsky (1978).

In the final experiment, it is assumed that the share of luxury goods in the consumption basket of low income groups in the least developed economy, is reduced by 30% and that for high income groups by 20%. For brevity a second change is also made in this experiment. Since it is possible that a basic needs strategy would also be reinforced by redistributive taxation, it is assumed that 13% of the high income is transferred to the low income group. This is a transfer within the least developed economies of $20 billion. This redistributive change nevertheless leaves *employment* and the wage income of the high skilled group at a higher level than in the base period. The single economy experiment is then repeated with these changes. The results are shown in Tables XIII and XIV.

TABLE XIII
Results of experiment – 'Basic Needs' technology

Economy Group	1	2	3	4	5	6
GDP.	2351.3	1101.9	1086.2	540.1	244	295.2
High income	922	436.2	447.7	389	198.9	157.8
Low income	1378.3	685.7	620.4	188.1	57	137.4
Profit %	10.1	12	33	40.2	114.6	15.7
Skill employment	634.3	254	245.6	163	40	61.9
Unsk. employment	949.8	487.8	276.4	92.1	30	87.1
Luxury output	1889.3	919	1287.8	512.6	200.4	141.9
Basic output	1298.5	749.3	820.4	27.5	106.9	199.8
Invest. output	1008.9	562.5	521.3	130.7	63.4	78.8
Lux. exports	26	–13.3	–22	–19.4	27.7	.7
Basic exports	–50.3	–5.8	57.8	11.5	–13.3	–.6
Invest. exports	75.2	–1	–17.9	–29.3	–26.6	–.3

World price of commodities:		Developing economies price:	
Luxury	= 1.000	Luxury	= .965
Basic	= 1.000	Basic	= .995
Invest.	= 1	Invest.	= 1

The introduction of the domestic transfer of $20 billion is responsible in large part, for the additional $32.7 billion increase in the low income groups budget. In addition the shift in the composition of consumption also creates a significant multiplier effect in the economy. In other words, the "exogenous" shift in consumption which may result from changes in technology (or simply a lack of luxury imports) may, nevertheless, through the market have a significant impact on income distribution. The increased consumption of low skilled labour-intensive products by both income classes, creates an additional demand for low skilled labour, which increases income and in turn raises total demand for the labour-intensive good.

In an open economy, when the composition of consumption

TABLE XIV

Social accounts for least industrialised developing countries – "Basic Needs" technology.

Luxury	Basic	Invest.	Skill	Unskill	Capital	High inc.	Low inc.	Fore
13	18	14	0	0	0	47	38	0
12	20	13	0	0	0	63	87	–1
6	7	7	0	0	0	47	11	–1
24	21	24	0	0	0	0	0	0
26	65	4	0	0	0	0	0	0
49	65	14	0	0	0	0	0	0
0	0	0	70	0	107	0	0	–20
0	0	0	0	95	21	0	0	20
132	198	78	70	95	128	157	137	–1

in billions of 1975 dollars. All figures are rounded down to nearest billion US dollars.

of the developing country population changes, the tendency is for the structure of trade to change. The multiplier effect, noted above, cannot operate since it would immediately imply a rise in the domestic price of the labour-intensive good, which would not be supportable against the world price. When the developing economy is delinked from the world economy and the composition of production changes to match local demand, then commodity and factor prices also change significantly. It is the consumption by the low income group of goods which they themselves produce which leads to the income multiplier.

With the domestic transfer included, the gain to the low income group over the base period is $30.1 billion. Without this transfer, but with the balance of payments income subsidy added in (for comparability with earlier results), the increase in income is $17.1 billions. Nevertheless, the combined income of the two income groups declines by some $13.1 billion. With the income subsidy of $25 billion from the richer economies restored, the total increase would be about $11.9 billion. This increase above the "base period" is more than the "Capital gift" experiment in Table VI. Furthermore the composition of production and consumption in the economy are better matched. This could be interpreted as implying a much lower level of external dependence, especially since the "appropriate" technology by definition implies the possibility of its production in the least developed economy.

Turning to the social accounts, Table XIV throws some additional light on the meaning of the result. As noted above, the employment and hence wage income of the skilled labour group rises even faster than for low skilled labour. This is because, in the closed economy, skill intensive production must satisfy domestic demand. The reduction of income for the high skill group arises from lost profits rather than from lost wages. Given that the high income group is certainly not homogenous in reality, the result has quite different significance for capitalists by comparison with skilled labour (including bureaucrats and management) and the workforce in general. However, while the implied "scenario" is not attractive to local entrepreneurs, it should be less adversive to development banks since profitability in the recipient economies remains significantly higher than in the donor economies.

CONCLUSIONS

This paper has used a model of the world economy to study the use of development finance for supporting the appropriate technology on the growth and distribution of basic foodstuffs production in the least developed countries. The model focuses attention on the impact of market forces and so, at a time when considerable international pressure on the recipients of development aid to respect the functioning of the market, the results of such a model are especially relevant. It has been argued that the key results presented here depend on certain assumptions. These assumptions are: (1) the existence of world markets for commodities and domestic markets for factors of production, (2) differences in the availability of factors between economies (e.g. abundant unskilled labour in the least developed economies and abundant skilled labour and capital in the most advanced economies), (3) differences in factor inputs between sectors (i.e. capital goods production is especially skill intensive in all economies, labour intensity is highest in the least developed economies).

A number of limitations of the model have been pointed out. In some cases the findings are exaggerated but this is to be expected in a model with such a simple structure. The results are found to be insensitive to some parameter values and assumptions, but rather sensitive to others, so the precise magnitudes of some parameters are important. This includes, of course, the assumptions made about the factor inputs to "appropriate" technology. However, even if the estimation and structure are not a perfect reflection of each economic group, it may well capture the situation of many economies within the groups. The model appears to simulate well, for example, the switch of agricultural production from staples to cash crops, the increase in export of luxury food crops and the accompanying increase in imports of basic foods observed in many industrialising developing economies. It can be argued, therefore, that the model may well provide a useful indicator of the direction of changes, or the *pressures for change*, which would be experienced when certain policies are enacted. In this sense, the model should be viewed as a "story-telling" model which may illustrate and sometimes guide analyses of alternative policies. Accepting this, in several respects it tells a different story from other models.

The main results are summarised in Table XV. The results from the first experiment suggest that when a developing economy receives development aid, the benefits are redistributed worldwide in such a way that the recipient economy gains relatively little additional income and little reduction in its level of poverty. This is because market forces create an economic environment in which production structures and international trade flows can change by relatively small changes in the level of income. Thus, it appears that considerable social disruption, in particular, rural urban migration could be engendered without compensating increases in material well-being or redistribution.

TABLE XV
Summary of results for income and distribution in the least developed economies.

Experiment	% Increase in low incomes	% Increase in total income	Low income share of total %	External balance %
Base Year (Table v)	0	0	34.7	11.6
Capital gift (Table vi)	5.2	2.9	35.8	30.0
Appropriate-1 (Table x)	–13.7	–1.9	30.5	0.5
Appropriate-2 (Table xi)	–2.3	–6.8	36.4	0.5
Basic Needs (Table xiii)	28.2 18.9(1) 16.1(2)	–4.4 –4.4 2.8	46.5 39.8 38.9	0.5

(1) Without domestic transfer of \$20 billion.
(2) Without transfer but retaining same external subsidies of \$25 billion as in the base year.

The functioning of markets is such that development finance in some cases may have a greater impact on the income distribution of the donor (and other) economies than on the recipient. Even so, contrary to the hopes of the Brandt Commission, development finance may do little to bring idle capacity in the

Northern industrial economises back into production.

The second set of experiments deals with questions of technical choice and the use of development finance for the promotion of appropriate technology.

The adoption of appropriate technology in a fully open developing economy appears disastrous for basic agriculture. Nevertheless, in a closed economy appropriate technology may have a very significant redistributive function. Once the destabilizing effects of external market forces have been brought under control, through a tightly regulated trade policy, for instance, then domestic market forces can reinforce social policies. Increased domestic demand leads to increased employment and wages; and so to further increases in demand. In some cases, the loss of some national growth, which would otherwise have been gained from international trade, is compensated for by "low income multiplier" effects in the domestic economy which reinforce other social objectives.

The reverse process is also suggested. When a previously closed economy, which has relied on less advanced or "appropriate" technologies for production of mass consumption goods, including foodstuffs, becomes more fully exposed to world market forces, it is likely to face extreme pressures in its production structure and income distribution. To avoid this, a very rapid process of modernisation may be called for, with consequent dramatic domestic social transformation. This appears to fit well with the present situation of China.

From these experiments it appears that economic growth through modernisation is gained at the expense of economic and ultimately political independence. Conversely, the use of appropriate technology sacrifices some growth. Nevertheless, the overall reduced capital intensity of a closed economy using appropriate technology (arising from changes in sectoral productivity and the composition of output) may offset considerably some of the apparent advantages of an open economy using more sophisticated techniques.

DISCUSSION POINTS

1) This model is of particular interest since it includes some key real world linkages such as between the income distribution

in the developing countries, technologies, the openness of the world market and the financial relationships between developed and developing countries. The scenarios considered are also linked to on-going programmes, usefully examining the recommendations of the UN agencies, Brandt Commission, World Bank, etc.

2) One participant found the results interesting, being diametrically opposed to those arrived at by many people working on dual economy models or within the classical economic tradition. It is believed that the opening of the economy tends to improve income distribution, achieve higher growth and provide more employment for unskilled workers. Similar results are derived from studies of economic history.

It is difficult to identify the particular features of the model which lead to these results. Policy makers might be able to use the models if the differences were made more explicit. What assumptions and behavioural relationships make the difference, and have others left out crucial international relationships? How important is the assumption on the wages of unskilled labour, or the degree of substitution between capital and labour within specific sectors?

Professor Cole noted that although the model defies the historical experience of certain Asian countries, there are other studies, especially by Latin America economists, which argue that terms of trade and interactions with the international market have not been to the advantage of the developing economies. The results follow directly from the structural assumptions.

In this model there are 6 technologically different types of economies, grouped by structural characteristics rather than by geographic distribution. The technologically dynamic economies of the North are best able to implement most advanced technologies, and there are the lagging ex-colonial economies and the centrally planned economies. In the South the 3 groups are the newly industrialised economies, the oil exporting economies and the least developed economies.

The emphasis of the paper is on the least developed group. It is possible that transfers to the newly industrialised countries would give quite different results, following the different definition of the production structure. There are quite big differences between the groups of economies.

The co-director, Professor Chichilinsky, and Professor Cole have reduced the model to a more orthodox two-sector, two-economy model with which one can identify differences from the Samuelson Theory. The major differences are the variable labour supply, and the different technologies between the industrialised and developing economies: one has a more abundant supply of labour, the other has very different technologies. The simplified model can be solved algebraically and the results for the terms of trade and the growth processes emerge from this analysis in line with the results presented here.

With only one economic group supplying labour and consuming an amount of production, with some being exported, then as labour increases wages go up and the consumption by labour may increase more rapidly than the rate at which new labour is demanded. Assuming fixed technologies, the demand for labour is proportional to the sum of domestic consumption plus exported consumption. There are two solutions to this model where supply and demand of labour are equal. For one an expansion of exports results in an increase in the demand for labour and an increase in domestic consumption. For the other the same changes can only be achieved by reducing domestic consumption, and overall production is reduced. Thus with a variable labour supply two model results are possible. Some of the published papers provide a more detailed analytic description of the model.

3) Dr. Leach felt that the model experiments have shattered most liberal illusions about what to do. He asked how aid flows are characterised, since in practice most aid is specific to specific projects.

Professor Cole agreed that it does make a difference how aid is transferred and to whom it is given. The Brandt Commission for example suggests that aid should be given for food production. In this model such a policy provides new investment in agriculture, but the result is to reduce the rate of profit in agriculture below the rates of profit in other sectors (or to cut wages) and thus to encourage entrepreneurs to move to other sectors (or to encourage labour to migrate). The results expose possible contradictions in internationally advocated strategies.

4) Dr. Frohberg wondered if there was not too much flexibility in the system with respect to production and the

transmission of world market prices into domestic prices, asked how the demand side of the model is built, and was puzzled by the results of international investment.

Dr. Cole emphasised that the experiment described is one of a series. In this the money transferred is to be spent on investment goods. One way to do this is to take the investment expenditure from the most industrialised group and put it into the least industrialised group. The price mechanism then ensures that the money is then spent back in the most industrialised group, the pattern is more complex. In the experiment described the investment in the least developed countries were designed to maximise the profitability of that economy.

The model assumes that entrepreneurs shift stocks so as to maximise their returns. This is followed by a change in structure – the price of capital is reduced relative to the price of labour, which favours the production of luxury goods. Then as world prices adjust to a new equilibrium, the most industrialised countries tend to export more basic goods, to fill the gap in the less developed economies. This results in a shift in income distribution in the most industrialised economies as production shifts from luxury to basic goods which are more low-skill intensive.

By contrast the extra demand for investment goods is met by the newly industrialised countries. To understand this, it is necessary to consider the parameter magnitudes, the structure of production and the labor supply elasticities in the model.

5) Dr. Frohberg enquired whether the model had anything to say about whether economies should be closed. Professor Cole explained that there can be large structural changes and there seems to be a trade off between growth, income distribution and structural dependence in the outside world. The decision cannot be made only on economic criteria, but the model can be used to study the phenomenon.

6) Professor Malu wa Kalenga queried the definition of appropriate technology. He felt that the concept of appropriate technology offered a process for choosing from a broad range of options, seeking to optimise the solution for a reliance on the problem solving capabilities of local people, suitability to environment and to cultural impact. In that sense it can be applied in a closed or open economy, can be simple, inter-

mediate or even high technology, and can be applied following a soft or hard route.

Professor Cole agreed, and explained that part of the effort was methodological, to try to combine three different types of analysis: case studies, scenario analyses and modelling. This was partly a technical exercise, exploring the relevant conditions for a particular type of technology. As an economy expands it will move from one phase with one type of strategy to another. The different regimes exhibited by a model like this would even suggest diametrically opposed strategies at different points in the history of an economy.

References

Acero L., Cole S. and H. Rush. (1982). Methods for Development Planning of Long Term Development. UNESCO, Paris.

Bartsch W. (1977). Employment and Technology Choice in Asia Agriculture. Praeger: New York.

Chichilnisky G. and S. Cole. (1978). A model of technology distribution and north-south relations. *Technological Forecasting and Social Change.*

Clark J., Cole S. and H. Lucas. (1979). Estimation and solutions of a macro-model of north-south development. *Applied Mathematical Modelling,* 3. (3).

Cole S. and I. Miles. (1980). *Labour Supply in the major world regions.* UNITAR working paper. Science Policy Research Unit, University of Sussex. mimeo March 1981.

Cole S. and A. Nunez-Barrigga. (1981). *Basic needs technology – A starting point for a macro-economic evaluation.* UNITAR working paper. Science Policy Research Unit, University of Sussex. mimeo.

Cole S. and A. Meagher. (1981). *Growth and income distribution in India – A general equilibrium approach.* La Trobe University Discussion Paper 13/81. August.

Date A. (1980). *A comparison of technological alternatives in some products and processes in India.* UNITAR Working Paper. Academy of Development Science, Taluka Karjat.

De Melo J., Dervis K. and S. Robinson. (1978). *Planning models and development policy: Computable general equilibrium models.* Woodrow Wilson School, Princeton. Discussion Paper no 79. March 1978.

Evans E. and S. Cole. (1979). *Data sets for the 3-region global model.* Department of Economics, Columbia University, mimeo.

Doblin C. (1978). Capital formation, capital stock and capital output ratios (concepts, definitions, data, 1850-1975). International Institute for Applied Systems Analysis, Laxenburg, RM-78-70.

Independent Commission on International Development Issues, (1979). *North-south – A programme for survival.* Pan World Affairs: London.

Karminsky B. and M. Okolski. (1980). *An estimation of the UNITAR model for Poland.* Faculty of Economics, University of Warsaw. mimeo July 1980.

Kaplinsky R. (1980). Inappropriate products and techniques in UDC's: The case of breakfast foods in Kenya. In Acero *et al.* (1978).

Kittler F. (1979). *An empirical investigation of international trade flows.* UNITAR working paper, Columbia University, New York.

Lewis W.A. (1954). *Economic development with unlimited supplies of labour. The Manchester School,* (May 1954).

Meagher A. (1980). An estimation of the UNITAR model for national economies. In UNITAR (1980).

OECD (1979). Capital goods: Structural evolution and world prospects. OECD Paris. FUT (79) C.4.

Pyatt G. and A. Roe. (1977). *Social accounting matrices for development planning with special reference to Sri Lanka.* Cambridge University Press.

Schumacher E. (1974). *Small is Beautiful.* Economics as if people mattered. Harper and Row: New York.

Stewart F. (1976). *Technology and Underdevelopment.* MacMillan: London.

UNIDO. (1978). *Capital Goods Industry Preliminary Study.* UNIDO/ICIS: Vienna.

UNITAR. (1980). The UNITAR macro-model. Cole S. (ed). Volume 1 – National Models in the World Economy – Model Structure and Estimation. Sussex. June 1980.

Vitelli, G. (1980). *The Chaotic Economics of Technical Change: A Survey throughout the Choice of Technique.* Institute for Development Studies. University of Sussex. mimeo.

5.2 Systems Approach to Modelling of Energy Effects on Agricultural Production

V.A. GELOVANI and L.V. TRETYAKOV

Complex dynamic systems which we have to deal with while studying contemporary problems of global development possess different kinds of interacting processes with feedback and timelags, the possibility of modifying the structure in time, discrepancy between long-term and short-term goals and effects. The models of these systems rapidly tend to become more complicated as effort is made to accomodate in them as many details as possible of these interconnected processes. The building of such highly complex models involves considerable expenditure of time and resources. In this case, heavy expenditure on the creation of a unique model may be recouped by prolonged use.

It is doubtful whether this approach to modelling of the dynamics of large-scale systems is relevant to decision-making at the present time. What is required is a pragmatically realizable series of models, making it possible to investiage the major problems for which there are formal descriptions both relevant to the research aims and satisfactory to the experts. Each specific goal calls for a model of its own which will be the result of multidisciplinary research by sociologists, scientists, engineers and mathematicians. Being problem-oriented a model at the same time must most fully reflect the specifications of the research problem and correspond with existing knowledge and data.

Both top-down and bottom-up approaches have merits and drawbacks. We use the top-down approach but feel that with this approach we sometimes lose the specificity of a country or a region when trying to imbed it in a typical basic model structure. The bottom-up approach may be of greater interest to particular governments for planning purposes. We are also developing models of selected countries which we hope to combine in a global model after finishing the more aggregated global model, hopefully by the end of this year.

Particular choice of the structure of an energy-agriculture model is related to the task set by researchers or planning bodies. Nevertheless some issues of global or universal character may be of common interest.

Though agriculture acquire a relatively small share of commercial energy supply now, the growth of agricultural production is and will be based on energy-related inputs of fertilizers, machinery, irrigation, insecticides and pesticides. The use of all or at least some of these inputs is a precondition of successful introduction of high-yielding varieties and multiple cropping. Oil and natural gas have been the principal energy sources for production of fuel and chemicals needed for restoration of soil fertility, field operations, irrigation, pest and weed controls.

In 1981 IIASA Energy Systems Program published the findings of the global and long-term examination of the energy future where strategies were identified for transition from a globe reliant on oil and gas to one served by sustainable sources of energy (Energy in a finite world, IIASA, 1981). Corresponding change in the structure of energy supply will raise the problem of technology adaptation.

Another path of development is connected with technologies enhancing self-sufficiency of agriculture including the reduction of its dependancy on commercial energy.

Increasing use of chemically fixed nitrogen does not diminish the role of biological nitrogen fixation that now is largely associated with legumes. (E.N. Mishustin "Biological nitrogen in the Soviet agriculture," Report to VIII Int. Congress on Fertilizers, 1976).

Bio-conversion of biomass and wastes in agriculture might also play an important role in technological development. (T.K. Ghose, Food and nutrition bulletin, v.2, N3, 1980, UNU). As Brazilian experience shows biomass produced in agriculture is an alternative source of commercial energy. (Dr. L. Navarro, Report to 16th Latin American Food Production Conference, 1980).

Substantial energy cost is related to development of land resources. Estimated limit of arable land expansion is some 3,2 – 3,4 bln. Ha according to: (1) "The World Food Problem" v.2, Washington D.C., 1967 and (2) Buringh et al. "Computation of the Absolute Maximum Food Production of the World," Wageningen, 1975. Though not all this area can be cultivated (N.N. Rozov et al. "World land resources . . .," 1978 give the maximum arable area of 2,7 bln. Ha) related cost must be enormous taking into account that now it takes approximately 1 tonne of coal equivalent to open 1 additional Ha to cultivation in developed agriculture.

Global trends of climatic change which have been discussed lately may have serious impacts on agricultural production if certain measures are not taken. One of the major conceivable processes influencing the climate of our planet is carbon dioxide release caused by burning of fossil fuels. If man keeps releasing carbon dioxide into the atmosphere at current rates the global average surface temperature will rise from the present global average.

Year	2000	2025	2050
t°C	0,9	1,8	2,8

Warming will change with latitude. If, for example, average warming is 1°C

Latitude	0	20	40	60	80
Local warming	0,5	0,6	1	1,7	3

M.I. Budyko *et al. Man-made global climat change* (1981). See also H. Flohn *Possible Climatic Consequences of a Man-Made Global Warming* and *Life on a Warmer Earth* IIASA.

As we see it energy and agriculture are connected by multiple direct and indirect links between various elements of these two largest production systems. An adequate choice among these elements and connections to be included in a model depends on the goal of the study and must be considered in the context of a certain country. The whole volume of work often has to be repeated all over again, even for a minor change in the initial statement of the problem.

The contradiction between the multiplicity of tasks and countries and limits on the time and resources available for construction models evoked some methodological responses. We think that between two approaches of building very complicated (universal) models and of using simplified methods there is a third way, based on the concept of a problem-oriented, man-machine modelling system. This system of mathematical, algorithmic, program and information software enables experts in conversational mode with a computer to carry out all the processes of building, reviewing and operating the model for a broad range of activities involving simulation of processes in a given problem area. The system does not initially contain a model, but instead a base that lets the expert generate the particular model structure corresponding to his research aims. The system greatly facilitates man-model contact due to accessibility of the system to the researcher at all modelling stages, possibility of obtaining data in suitable form for substantive interpretation.

Institute for Systems Research (ISR) has built a modelling

system for the study of global development processes. We assume that a long-term perspective is required for comprehensive understanding of global problems.

ISR global model has hierarchical structure and consists of interacting regional models. We speak of a region as a group of countries or a country.

Each regional model contains sectoral models of economy, demography, food and others. The main output variable of Food Model is the level of nutrition. It characterizes the degree to which requirements in nutritional elements are met by food consumption. Nutrition level influences on demography and economy. Main inputs from other models are capital investments, fertilizers, population and labour, scientific and technological progress.

Principal sources of food are grain and meat production. Fishery is now included in the model in a simplified version.

Arable area consists of nonirrigated and irrigated land. Processes of development and degradation influence land area available.

Yield is a function of several production factors: fertilizers, capital, labour, the level of scientific and technological progress. They are calculated in other submodels or given exogenously. Crops are distributed for food, feed and reserves. The distribution depends on quantity of all the food produced, preferences of people and reserves building policies. In meat production feed is utilized coming from agriculture and non-agricultural production sector.

Nutritional needs in a region depend on the total population, age-sex and professional structures.

DISCUSSION POINTS

Dr. Makhijani and Professor Lesourne asked about the connection between production and the money market, and the role of incentives and the domestic pricing system in the models. The pricing systems cannot after all be independent of the behaviour of the productive units.

Dr. Tretyakov indicated that they had not solved this problem in modelling, but noted its importance in the selection of control variables in the model. It is important to identify the points of initiative, which may be normative in centrally controlled economies, or market mechanisms in the market economies.

5.3 Food-Energy Modelling for an LDC

J.G. KRISHNAYYA

In a modelling study it is important to cope with the reality of the situation, with the environments in which decisions are made and the kinds of players in the game in LDCs. It may be distressing to realise who are making decisions and on what grounds, yet it is necessary if the modelling is to be of real value. Since the analyses must take into account specific local conditions there will be some limit on their universality.

During 1980-81, the Systems Research Institute constructed a socio-economic-demographic model of India[1] as part of a worldwide ILO project. With this model we have been able to conduct some useful policy analyses.

This is primarily a basic needs availability model, integrating demography, economics and availability of basic needs, and not directly oriented to energy and agriculture.

One of the parameters generated is the domestic demand for fuel based on rural and urban population and per capita requirements and another is the commercial availability of energy for private consumption from the input-output model. The model gives us the ratio of these figures in the form of an "Energy availability index." From 1970 to 2000, the model reports figures from 31% to 34%, with only one scenario for rapid-growth-based-on-very-high-exports giving a figure above 40%. This implies that the bulk of domestic energy needs are met from "non commercial"[2] sources. In an LDC where close to

[1] ILO Working Paper WEP 2-32/WP-25

[2] "Commercial" sources are coal, oil, electricity, hydro and nuclear power.

80% lives in rural areas, this is understandable. However, it also implies a very high rate of drawdown of India's forest resources which may not in fact be sustainable. That this is so is also confirmed by the more conventional indicator of price: The user price of fuel wood has risen faster than general prices over the last 10 years, despite the fact that virtually all forested land is in government control, and the delivered price is an administered one.

The management of forests is very antiquated and cutting schedules are fixed 20 years in advance. With cutting schedules and the amount of wood available fixed, it becomes necessary to raise prices as demand exceeds supply.

This signal from our model that the fuel needs of consumers will not be met from commercial sources, and our knowledge that conventional forestry will not be able to make up the supply, suggest that a more careful analysis of energy needs and sources is urgently needed.

The existing model, though useful in identifying a problem, was not built to provide the answers. It is limited by the number of sectors in the Input Output table (10), by the countrywide aggregation of the variables, and by the fact that neither Agriculture nor Energy is given any special place. Something more is needed, and in this Paper, I shall suggest a line of research which will, I trust, result in a model more relevant for examining the Food-Energy nexus in a fashion relevant to an LDC. Because of India's size and variety, it is likely that the resulting model will largely be useful for any other LDC.

Today in India food is available to all – and particularly to the cities – at reasonable prices because the energy available to rural areas – sunpower, fertiliser, electricity and oil for pumps, agricultural waste and fuel wood for processing and domestic needs, petroleum for transport – is adequate. What when the prices continue to rise? What when today's unpriced agri-waste, *gobar,* and forest scraps become priced? What when fuel wood is no longer available in much of the country?

Detailed study is needed of the relationships between food production (and productivity, since land is limited) and the use of different forms of energy. Since India is a large country with many different eco-climatic zones, it is necessary to do this at two levels – at the regional level, distinguishing the Punjab and the well-watered Gangetic basin from the drought-prone Deccan plateau from the hilly regions, and also at the aggregate, country-wide level. Separate consideration of each region is needed because whatever the eco-climatic conditions, the actual choice of crops is usually such as to make each area as self-sufficient food-wise as possible.

Each state in India knows that its own self-sufficiency in food gives it an enormous leverage on the central government for bargaining for money and projects, and for other decision making.

Only after the local food needs are met, does it make political and practical sense to think of the overall optimal "division of labour" as between regions, to produce cereals, legumes, vegetables, fruits, trees, etc. and commercial crops such as cotton, tobacco and jute.

In the 1950s commercial crops were taking over while food prices were being controlled, and this continued even in the 1960s when food was being imported. Subsequently the pressure for self-sufficiency has borne fruit and there is a balance between commercial and food crops.

Nevertheless we need an overall model to determine what is the optimal geographic distribution of different crops and what the opportunity cost is of the deviations from this optimal. Determining a true optimum is non-trivial, as found by the FAO and IIASA in their efforts, and it may well be that a reflection of the actual land-use pattern in different regions and evaluation of the effects of small changes from this may be easier and more productive.

Today we are concerned with the future availability of Energy and its impact on Food production. Indeed, even the technology and the type of crop may have to change so as to balance out the energy needs. For instance the "dwarf" varieties of wheat produce much less burnable waste than formerly. Is another change desirable? Already we have seen as against the general shift to mechanisation of farms, in a few reported cases, a shift back to animal power – because of the better matching up of inputs and outputs (waste = animal feed; animal waste = gobar = energy/fertilizer; animals are multipurpose and in densely inhabited areas, provide all the mobility required). In short, the number of variables is quite large, and our freedom of action similarly wide, and only modelling can help us find useful paths to follow.

For each type of eco-region, we need a model that tells us (at least) what the opportunity cost is of "doing things the same way as before," and which permits us to test for alternate ways of matching energy needs and sources. These need not be direct matches, of course, as with replacing petroleum-based fertilizer with alternative nutrients (and ways of capturing atmospheric nitrogen). Similarly, the need for a new source of energy may not have been explicit before because foods and agricultural commodities may not have been processed in that way before. Thus, solar dryers are economical today, and even solar-driven refrigerators are now being introduced. With this technology available in the producing area, how does it change energy use "down stream?"

Energy consumption in towns and cities is constrained to make use of electricity, petroleum and gas. Coal can be used only in large central stations, and wind and sun and other sources hardly at all. This is primarily because of lack of space. Therefore solar and wind energy, etc. must be implemented in the rural areas and it is best if they are used directly in agriculture and in agro-processing since that is where the needs for energy are still "soft," and in LDCs electricity and petroleum have not totally taken over. This also suggests that long term models to study this problem should incorporate demographics and migration data as well.

INTERRELATIONSHIP OF ENERGY AND THE ECONOMY

Energy enters the economy as inputs to production and service sectors and also as outputs to meet many final demands. Though all sectors in the economy use energy along with other inputs such as labour, capital and other raw materials, the degree of dependence of a sector on energy varies according to the type of output and the production process employed. The very fact that every establishment, whether productive, trading, servicing or household, uses energy makes it an integral part of the economy. Moreover, the dependence of energy on other sectors makes the system interdependent and hence sectoral output, employment prices, etc. are to be determined as a part of a general equilibrium analysis of a socio-economic system.

A simple aggregate model of energy-economy interaction is illustrated in Figure I.

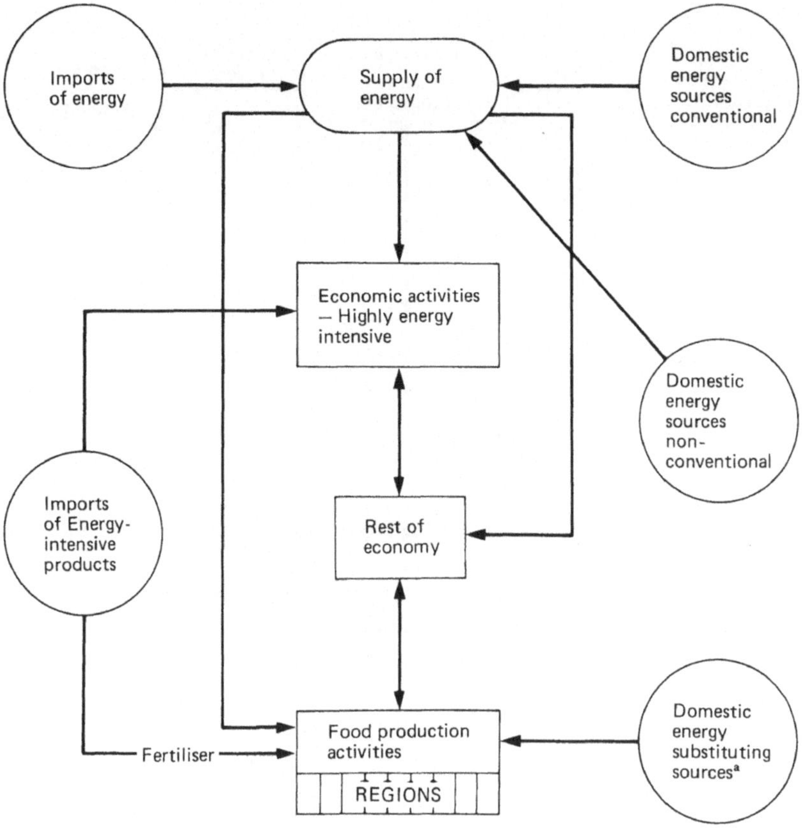

FIGURE I Energy model keeping food-agriculture separate from rest of econmy.

An alternative view is also possible in which primary focus is on Agriculture and Food-production activities, and the Energy needs are seen from that point of view (See Figure II). However, this is likely to generate sub-optimal solutions since most of the rest of the economy cannot readily use "soft energy" as conveniently as agriculture. A sectoral model is likely therefore not to measure accurately the opportunity cost of hard energy used in Agriculture. Furthermore, the changing nature of the manufacturing sector, the migration patterns and of exports (of

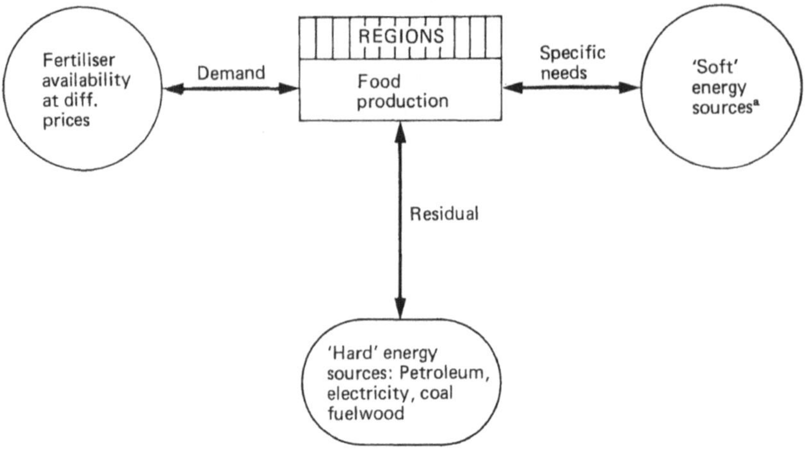

FIGURE II Alternate view, taking Agriculture-Food as the focal point.

agricultural products like cotton, jute, etc. as well as of other exports) must be taken into consideration if we want a realistic picture. Our conclusion therefore is that a model more like Figure III is needed.

The supply of energy is primarily determined by the domestic energy sources and imports. Since energy processing and transformation form part of economic activity, it also affects the supply of energy. For example, the supply of electricity depends, inter alia, on railway transportation to move coal from the mines to thermal stations.

The interdependence of energy imports and economic activity may need explanation. The effects of energy imports on economic activities can be understood via prices. Energy imports can be done in the form of primary energy (as is often done, instead of the final energy form) depending on the domestic availability of refining and processing facilities and to this extent the domestic economy influences imports. Energy may also effectively be imported in the form of final products such as fertilizer. Similarly the interdependence of domestic energy sources and economic activity can also be explained.

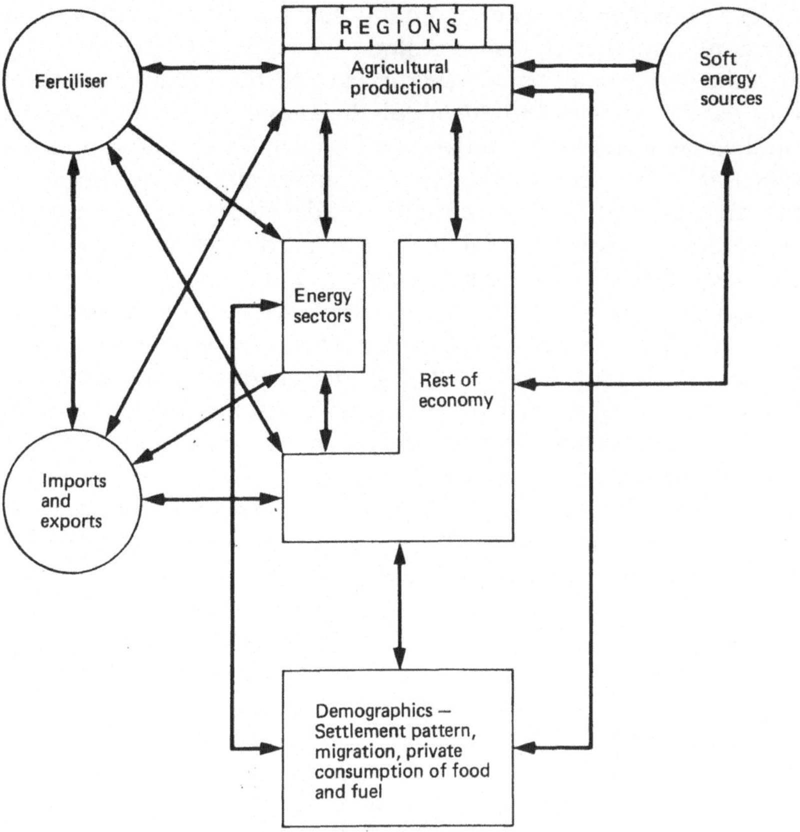

FIGURE III Proposed food-energy model.

DOMESTIC ENERGY SOURCES: DEMAND AND SUPPLY

Classification of Energy Resources

Energy Sources

- Animate
  - Human Power
  - Animal Power
- Non-animate
  - Commercial
    1. coal and lignite
    2. oil and natural
    3. hydro-electric power
    4. nuclear fuels, etc.
  - Non-commercial
    1. forests
    2. vegetable wastes
    3. dried dung, etc.
  - Renewable
    1. solar
    2. geothermal
    3. tidal
    4. wind, etc.

A comprehensive energy policy can hardly concentrate on the commercial energy sector alone which mainly caters to the manufacturing and service sectors and to the urban areas. The very fact that over 78 percent of the Indian population live in rural areas warrants formulation of appropriate energy policies specifically oriented to the local resources and energy needs of the rural people. The vast potential of developing bio-gas energy and animal power in rural India can hardly be overlooked as we have the world's largest cattle population of about 261 million, according to the 1971 livestock census. (Equally important of course is the distribution of cattle population, since economic viability of e.g. gobar gas demands certain minimum size.). The limited availability and rising prices of commercial energy resources add yet another dimension to the controversies over mechanisation in agriculture.

We can now visualize a little more closely the method by which a model can be built, which is sufficiently modularised to be buildable and yet integrated enough to be usable for policy analyses in this field.

The core of the model must be the food-agriculture-land use model, regionalised with information on animal and human population, relevant information on local "soft-energy" and fertilizer possibilities. (The model will iterate, so some starting level of non-food crops demand and of fertilizer availability will be needed.) Regional production will then be totalled and if the results require it, certain regions will be re-done. These results from the Agri model imply a certain amount of hard energy and petroleum-based fertilizer. They will be carried over to the next phase of the computation – the economic-energy module. Here we will be looking at the economy as a whole with greater interest in the energy sectors. It will be possible now to examine whether the implied level of imports and exports are feasible, whether the other economic inputs (transport, etc.) implied by the agriculture modules are feasible, and to determine income and consumption levels. Since the detail for the economy-as-a-whole is limited, we may not be able to obtain as rich information on income distribution as would be desirable. The final module would be demographic covering vital rates and migration and providing information on the resulting settlement pattern. It is hypothesised that different settlement sizes in an LDC have different energy intensities[1] and since settlement patterns are changing rapidly in LDCs, it seems relevant to keep track of them.

It is likely that the economic module will require one or more iterations through the agriculture-food module, before it balances for any one year. Consequently it may not be possible to run the model afresh for each year. It may be necessary to run it instead at intervals of 3 to 5 years and to trace out the viable trajectory for Energy and Food.

The Systems Research Institute has a serious commitment to the development and Use of models for policy analysis – and indeed for policy design – in LDCs, and we hope that we will be able to study through the model here described, the close connections between Food strategies and Energy demand in LDCs, while keeping sufficient detail about the specifics of alternate soft energy sources to be able to relate them to specific cropping patterns and development paths for rural parts of the LDC.

We invite comments and suggestions which would enable us to improve the formulation of this model.

DISCUSSION POINTS

1) Dr. Makhijani noted that although it is commonly assumed that fuel wood use is a primary cause of deforestation, two India studies have indicated this is not so. The observed price of wood went up early in the 1970s, and this was connected to the oil crisis. But why did the price of wood increase? The government owns the woodlands, but the forestry department is among the most corrupt in the government, and possibly the price rise is due to internal effects. At the same time the international corporations which were exploiting commercial timber in, for example, the Phillipines and Indonesia experienced a very severe crisis in timber exploitation and they may have used the oil crisis to force a fuel wood crisis. After all there had been only a slow growth in fuel wood use for some time.

Dr. Makhijani has investigated this question and concluded that the principle causes for deforestation were: wars, such as

[1] This is to be verified in the field in a study likely to be sponsored by ILO.

the USA move into Vietnam and the massive deforestation involved; commercial forestry such as that of German and Japanese corporations in Brazil, various Japanese corporations and the vast forest contractors in North India and Nepal; at the local level, the refusal of governments to address squarely the land distribution question. Whenever there are organised demands for land, the government gives away forest lands to groups which are organised enough to threaten the local power structure, rather than address the redistribution question. So deforestation is strongly effected by demands for agricultural land. There is a universal misunderstanding of this subject, with the problems of the poor put onto the backs of the poor by the rich.

Dr. Hrabovszky had a different view of the origin of the fuel wood crisis. The real flare up of this issue came as an aftermath of the Sahel tragedy. It was then conceived that the real fuel wood short areas of the world are either the Savannahs on the edge of rather fragile environments or the very heavily over-populated aluvial areas of Asia. The FAO has carried out a detailed survey on the imbalance of fuel wood demand and supply for practically all the developing world, including projections to the year 2000, and the results are frightening.

Dr. Krishnayya agreed that deforestation is not primarily the result of fuel wood use. A variety of reasons includes higher population densities leading to the use of more land for agricultural purposes. Since more unfavorable land is brought into cultivation, the process acclerates. Fuel wood use by tribal people has probably not affected the forest at all. However Pune City, with a population of 1.2 million people, consumes about 50 truckloads of fuel wood per day. A month ago there was no kerosene on the market, which illustrates the uncertainty in management of the system for maintaining the refineries in Bombay. It is important not to ignore such uncertainties in modelling.

2) Dr. Hrabovszky wondered if the forestry department fuel wood releases really are indicative of fuel wood supplies in India. In Bangladesh for example the department wood release is about 10-20% of total fuel wood use.

Dr. Krishayya speculated that a large proportion of traded wood comes from the forestry department, while the rest is non-commercial supplies. The price effect of the forestry wood would then be significant.

3) Dr. Makhijani felt that the changes when non-monetised fuels become monetised would be extremely important.

Dr. Spitz drew the attention of the conference to a very good article by John Briscoe in the Population Development Review of December 1979. Briscoe spent about 9 weeks in a Bangladesh village and made a very careful and detailed account of the energy use. This answers in a very detailed way the question of what happens when non-market goods enter the market. The poor people are priced out and, particularly in the village, the cost of fuel increases much more rapidly than the cost of food. This effect is particularly marked for the poor who had some elements of the biomass available to them before the charge, but none afterwards. In reference to the use of biomass there is often the assumption that biomass is a free, non-market good, but when certain pressures build up in villages biomass soon becomes priced.

4) Dr. Tims noted that the distribution of biomass availability is twice as skewed as that of land availability and thus that biomass access must be considered along with income distribution and farm size. In a model experiment it was found that as the price of fossil energy increased, farmers moved to more legumes as they require less nitrogen fertiliser; rice-wheat substitution returned since rice straw protein digestibility is about 10-15 times higher than for wheat straw as a feed for draught cattle; with substitution towards single-function cattle, as power production becomes more important, there is less milk and meat and therefore less protein available. Also since there is a need for substitute fuel, more dung must be bought, and with less dung more fertiliser (which is costly) is needed. Eventually the final cost can be greater than that of kerosene.

5) In answer to a question from Dr. Hrabovszky, Dr. Krishnayya noted that income distribution is not included in the model.

5.4 Long Term Perspectives in Energy and Agriculture within the Framework of Industrial Development, in the UNITAD Model[1] [2]

J. ROYER

BASIC PURPOSE AND SCOPE OF THE UNITAD PROJECT

The purpose of the project, at its inception in 1979, was to explore with an analytical tool – the UNITAD model or system of models – the potential impact in the long run of different development strategies on the world economy, especially in the area of industrialization, trade and growth. More generally, the project was meant to contribute to an understanding of the structural transformations which would be necessary in the North-South relations and within the domestic sector of developed and developing countries in order to attain a more equitable system of international economic relations.

The project is concentrating on three questions:

• What changes are required in the international systems (international commerce, international finance market, etc.)? The present system is not equitable to the Third World.

• What changes must the South, in its diversity and with its means, adopt in order to solve its problems?

• What are the possibilities for more autonomy in the South

---

[1] This paper is based on "the 1981 Report" on the UNITAD system, and reference is made to that Report in the text.

The UNITAD System was designed and built up from April 1979 to December 1980 by a team of research workers, with J. Royer as Project Manager, under the sponsorship of three United Nations Agencies, with the generous assistance of a Trust Fund. The model is now operated (since January 1981) under the direct authority of the United Nations Industrial Development Organization (UNIDO).

The System and its various components were designed and estimated by G.A. Cornia, A. Duval and J. Royer, who coordinated contributions ranging from two ro nine months originating from various sources, in particular the Global and Conceptual Studies Section of UNIDO.

[2] The opinions expressed in this paper are strictly those of the author and do not commit the Organisations which sponsored the project.

within the existing international system to attain its objectives? This implies a consideration of poverty, technology, the energy crisis and the planning of resource useage in the long term.

The broad development framework to be explored for the world economy was expected to follow the General Assembly resolutions and plans of action, in particular on the New International Economic Order, and the guidelines provided by the Declaration and plan of action adopted at the UNIDO Second and Third Conferences (Lima, 1975, and New-Delhi, 1979).

The need for a new world model in the UN family can find a first justification from the changes which have occurred in recent years, which call for new approaches and solutions. For instance, the now widely shared realization that industrialization in developing countries cannot expand indefinitely on the basis of growth of their export markets in the more advanced economies (which seem to have entered a long period of stagnation) has given a sounder foundation to policies of collective self-reliance. Very often, however, the implications of increased economic co-operation among developing countries have remained largely unexplored[1], owing, among other reasons, to the limited capability of existing global models[2]. Little evidence, even of a very broad nature, is then available of the real chances and limits of policies aiming at substantially changing the world industrial horizon as well as the direction and structure of world trade. Hence, a quantitative approach to these questions entails the elaboration of new tools and appropriate formalization.

Two approaches are possible: formal models alone, and systematic thinking on a broader scale which represents a certain vision of society. One example of the latter approach is the Interfutures project, which centres on North interests. The UNITAD model project intends to follow this approach, centering on the South, with the model based on a certain vision of society. A major problem to be considered relates to possible South-South cooperation, self dependence of the South and thus a study of links between the regions.

In the second place, there has been little attempt, at least in

[1] A.L. Goedz: "Beyond the Slogans of South-South Co-operation", *World Development,* Vol. 9, No. 6, June 1981.

[2] As a major exception, one should quote the evaluation of the UNITAR Project on the Future ("Technology, Domestic Distribution and North South Relations", UNITAR, New York, August 1981).

existing models, to examine the relative importance and the interactions of policies and decisions concerning the "external sector" (trade, capital flows, aid, etc.) on the one hand, and those concerning the "domestic sector" on the other. Admittedly, the theory in this area is still lagging behind. In reality, however, the external and the domestic sectors interact very closely, and to restrict the analysis to the functioning of only one of the two would be very misleading indeed. The UNITAD Project was therefore necessary to provide for an examination of the relative efficacy of domestic versus international policies in attaining given development targets.

The scope of the model has been broadened so as to take account of some crucial long-term aspects of development. Indeed, the generally acknowledged failure of traditional development policies, the poor performance of most developing countries and the persistence and increase of the most evident forms of forestry, have led both development theory and modelling to focus on a wide number of issues.[1] Among them are: the relation between income distribution and growth; the mobilization of productive resources such as labour, land and knowledge, which too often remain grossly underutilized; the balanced growth of the economy, with agriculture and industry reinforcing one another through the progressive expansion of their absorptive capacity. The importance of such issues is reflected in the text of the International Development Strategy for the Third United Nations Development Decade.

Similarly, although there has recently been a wide use of the concept of appropriate technology, the real scope for its application and its implications for trade, growth and overall development have seldom been explored systematically by means of global models. In these cases too, then, new concepts,

[1] For relatively recent reviews of global models, see S. Cole, *Global Models and the International Economic Order* (Pergamon Press, 1977) and J. Richards, "Global Modelling," *Futures,* Vol. 10, No. 5, October 1978. See also S. Gupta *et al.,* "The World Bank Model for Global Interdependence: A Quantitative Framework for the World Development Report," *Journal of Policy Modelling,* 1979, 1(2), 179-200; W. Leontief *et al., The Future of the World Economy* (New York, Oxford University Press, 1977, for the United Nations); M. Hopkins and R. van der Hoeven, "Economic and Social Factors in Development: A Socio-Economic Framework for Basic Needs Planning," ILO World Employment Programme Working Paper (WEP 2-32/WP.19), July 1979.

approaches and information need to be developed if this important aspect of development policies has to be analysed thoroughly.

To explore, even very tentatively, these broad issues it was felt that a semi-aggregated world and economy-wide model guaranteeing the consistency of all aggregates was necessary. The model specification should also explicitly include those variables representing or approximating the new issues and making it possible to consider the old ones in a novel way.

A BRIEF REVIEW OF REGIONAL MODELS AND THE LINKAGE SYSTEM

Regional models and domestic policy issues

The system includes a set of eleven regional models, of which five for developed and six for developing countries[1]. A major feature of the model is the basis of the disaggregation, which is by and large geographical vicinity. It should be noted, in this connexion, that in many cases, neighbouring countries have comparable economic structures and, to a large extent similar institutional and social conditions, hence, the geographical approach overlaps with other approaches. Geographical vicinity was used as a criteria essentially in order to provide a framework for the simulation of economic co-operation among neighbouring countries. In this respect the UNITAD system differs from other world models in which countries are grouped according to per capita GDP level, i.e. a structure implying a specific linkage to the world system, more apt to picture North-South than South-South relations.

Four different specifications of regional models have been built for Market Developing Regions, Market Developed Regions, the European Centrally Planned Economies (CPE, Europe), and the Asian Centrally Planned Economies (CPE, Asia), respectively. The four types of models, however, have

[1] The following division of countries and regions is used: North America, Western Europe, Centrally Planned Economies of Europe (CPE, Europe), Japan, Other Developed, Latin America, Tropical Africa, North Africa and West Asia, South Asia, East and South-East Asia, Centrally Planned Economies, Asia (CPE, Asia).

several common features, besides adopting to a very large extent the same production, consumption and trade breakdown. The specific features of the model for Centrally Planned Economies, Asia, will be reported in a separate section.

No attempt will be made here to describe the 3,000 equations of the system[1], though reference will be made in footnotes to relevant methodological papers. The considerable work of econometric estimation performed will not be alluded to[2]. However, bearing in mind the research objectives mentioned earlier, it may be relevant to list briefly the main policy issues addressed by regional models, with a few explanations of their potential uses in the investigation of such issues.

On the production side, every effort has been made to interrelate technology choices with other variables (employment, investment requirement, savings, and main economic equilibria). This can be done by using the coefficients of the Input/Output cores and productivity functions at a relatively disaggregated level (8 sectors for the Input/Output core and 12 for productivity functions). Thus, a broad variety of concepts of "appropriate" technologies can be reflected in the model.

Energy, further disaggregated into two primary sources and two processing sub-sectors, is one of the key policy sectors. The system can generate energy balances at the regional and world levels, with all related effects on trade and payments balances. If required, energy can be treated as a supply-constrained sector.

Agricultural production in developing countries has been

[1] A series of flow charts and the complete system of equations are available in the paper "The UNITAD Model: Main Methodological Features" (UNIDO/IS.22), May 1981.

[2] The originality of the UNITAD system lies indeed in the fact that the team had access to large data banks of the United Nations Agencies and a generous allocation of computer time. Thus, a systematic attempt was made to measure the effect of structural and institutional changes on the basis of pooled cross-section data and time series. The production subsystem was based on a series of 49 original I/O tables (UNIDO), and Enterprises Surveys for 28 manufacturing sectors in the United Nations, *Yearbook of Industrial Statistics* (UNSO and UNIDO); the agricultural production functions were derived from FAO samples of Farm Management Surveys covering 18 countries; the trade equations were derived from original country tapes (full coverage) from 1963 to 1975 (UNSO); the consumption equations were computed on the basis of the United Nations, *Yearbook of National Accounts Statistics* and an ILO Collection of Household Surveys covering 12 large developing countries, etc.

pictured as taking place in farms of different size, i.e. small, medium and large. The distinction is important since it has been observed[1] throughout the developing world that farms of different size adopt widely diverse technologies, land and labour use patterns and cropping intensities which determine, accordingly, different yields per unit of land. The introduction of farms of different sizes therefore makes it possible to test the possible effects of land redistribution policies on farm-output growth, labour absorption and income distribution. Investment in agriculture includes the capital expenditure required for land extension and increased cropping intensity.

Households allocate their income between savings and eight consumption categories (traditionally used in national accounts statistics, i.e. food, clothing, rent, furniture, health, transport, education and miscellaneous). Note that for the developing regions the household sector is broken down into rural and urban, with rural households showing higher marginal propensity to save and consumption baskets different from those of their urban counterparts. In this way the effects of exogenous changes in the rural-urban income distribution can be assessed in the system.

Government allocates its disposable income between savings and current consumption in administration, defense, health and education. Again, by changing the exogenous structure of government consumption, it is possible to reckon the effects of alternative government expenditure patterns in terms of overall growth and standard of living.

Import requirements have received an elaborate treatment in view of their economic significance. For each regional model, imports of eight groups of goods and services are computed on the basis of econometrically estimated import functions. The main arguments of the latter are an activity variable (generally the domestic production of one or more industrial sectors), the ratio of domestic versus import prices, and two policy variables, i.e. the level of protection (tariff and non-tariff) and the average domestic market size, i.e. an indicator measuring the degree of economic co-operation among neighbouring countries and/or

[1] See for instance R.A. Berry and W.R. Cline: *Agrarian Structure and Productivity in Developing Countries*, Johns Hopkins University Press, 1979. See also ILO: *Poverty and Landlessness in Rural Asia*, Geneva, 1977.

the effect of the enlargement of the domestic market. The activity variable is expected to measure complementary imports, while the other three tend to decrease import dependence through import substitution. In particular, the domestic market size is instrumental in determining the level of imports and, via the input-output framework, in apportioning final demand between domestic supply and imports. This spurs the growth of the basic products and of the capital good sectors in those regions with a large industrial base and introduces a means of measuring the long-term impact of economic groupings or income distribution policies[1].

Labour and land gaps are computed for each region by subtracting an endogenously determined demand from the respective supply given by submodels. For developing countries, the labour gap is obtained separately for urban and rural populations, so that the full impact of policy choices governing the industry-agriculture technology and interrelations can be expressed by the model.

The system as a whole and international policy issues

Three families of international issues, i.e. prices, trade and finance respectively, are embodied in the system[2].

An international price system can be superimposed on endogenous regional price-costs to simulate a world scarcity of any of the seven commodity groups, whether as a result of natural supply constraints (e.g. on the oil and gas market) or of a monopolistic or oligopolistic behaviour of a group of regions (e.g. sophisticated equipment as suggested by the product-cycle theory). As was already observed, the model can trace the effects of such scarcities on the allocation of resources at the world level, but the price increase itself is not generated by the model and must be derived from outside sources.

[1] The market size, as distinct from plant size, was underlined as an important externality in the pre-war period by Allyn Young, and more recently analysed by P.B. Rayment in a paper on "Intra-industry specialization and the foreign trade of industrial countries" to be published in a collection of essays in honour of C.T. Saunders. (Macmillan).

[2] See, for the complete set of equations, the annex entitled "Analytical Expression of the UNITAD Model", to the 1980 Report on the UNITAD System (UNIDO/IS.224), 15 April 1981.

On the trade side, any change in import requirements of a specific region influences the exports of all other regions of any specific commodity, through the trade share matrices. As was already seen, such regional changes can result from import substitution effects, which in turn can be endogenously generated by the growth process or result from a deliberate policy embedded in the parameters governing intra-regional trade or protection level. Another important group of parameters are the trade shares themselves which can be changed to simulate policies or economic processes. Policy changes are deemed to reflect decisions made by a group of countries affecting the regional level (e.g. CMEA decisions for the group of centrally planned economy countries, Europe, or the EEC for Western Europe). Economic processes can simulate the effects of export-push policies such as those observed in Japan or in East Asia; to that effect, a gravitational model was estimated on observed data, so as to provide (outside the system) values of trade shares resulting from growth assumptions.

Finally, assumptions of ODA, migrants' remittances, international banking facilities, interest rates and maturity of loans influence directly the payments and basic balances of every region. Moreover, the current payments gap affects the volume equilibria via the equality between the current payments gap and the investment-savings gap. It should, however, be observed that the financial structure of the system has been deliberately kept rudimentary and is not meant to determine reliable levels of financial variables. It should, nevertheless, convey an idea of the implications of changes in the institutional financing system for the world economy.

Centrally planned economies, Asia[1]

The regional model for Centrally Planned Economies, Asia (CPE, Asia) can be considered in many respects as a simplified version

[1] This model was built-up late in 1981 in co-operation with a group of UNIDO consultants from China. In all scenarios and sensitivity analyses reported in 1981, the CPE, Asia region played a purely passive role, i.e. importing from an exporting to other regions without any feedback to the regional economy. 1982 scenarios have, however, made use of the model described in this sub-section, with the CPE, Asia region fully interactive in the system.

of the generic model for market developing regions but its specific features, derived from an original problematic, are worth being considered.

Like for market developing regions, the model uses the Standard National Accounts (SNA) concepts and the common sectoral and commodity classifications, in particular an eight-sector Input/Output core. However, income and financial relations are expressed in real terms (except trade and external financial relations), using the same general approach as for CPE, Europe.

The original feature relates to the elaborate treatment given to the rural area. It starts with the split of the output in 12 sectors, between urban and rural activities. This is governed by a transition matrix of exogenous shares, but it might be improved, in further versions, by using a 16 x 16 Input-Output core. Productivity functions similar to those of market developing regions are then computed on the basis of exogenous figures of capital/labour ratios for rural and urban areas separately. On the basis of output and productivity figures, the model generates separate vectors for urban and rural income, consumption, employment and investment.

Although a formal labour gap can be computed for both urban and rural areas, the growth of rural income per head in the rural area is computed as proportional to a given fraction of the growth of an average "rural productivity" defined as total rural output divided by total active population in the rural area. In other words, it is assumed that the whole active population in the rural areas (computed outside the model) shares in the gains resulting from agricultural, industrial and service activities. It is worth noting that the assumptions on rural-urban migrations, together with the growth rate of agriculture and the shares assigned to rural industry are determinant in generating rural income[1].

The philosophy of the model is therefore to plan rural development *per se* so as to enable rural people (85% of total population in 1990) to satisfy most of their essential needs

[1] For the urban sectors, the growth rate of productivity is dependent not only on the shares or urban output but also on the technology used in urban activities (via the capital/labour ratios which determine the productivity functions).

from their own activities. Rural industries also contribute a significant amount of intermediate and capital inputs to agricultural output, thus feeding productivity growth in agriculture. The latter relationship might be modelled more adequately with the planned extension of the Input-Output core into 16 sectors, in which own-consumed rural inputs could be identified separately from deliveries to the urban sector.

Altogether, the model can be said to remain crude, since a number of behaviour equations were approximated by fixed proportions reflecting base year data with assumed trends. As it stands, however, it has two merits, the first being to yield 1975 figures consistent with what is known from the economic situation of the region[2], and the second, from a methodology angle, to pave the way for the modellisation of essential needs policies in the UNITAD system.

In future simulations, the real difficulty will be to use the model to project the future development pattern of the region, i.e. to translate an adequate planning strategy into the parameters of the model. This raises a number of substantive economic issues: To what extent is it possible to push the development of the rural area further without intensifying the two-way relationships with the urban sector? In such a case, what would be the level of the required infrastructure, the price system, the level of urban migrations? In turn, can the urban sector keep growing without more active trade with other regions? These are but a few complex questions for consideration if meaningful projections are to be made.

SOME POLICY CONCLUSIONS FROM THE COMPREHENSIVE SCENARIOS

Based upon the detailed analysis of the explorations achieved so far, it seems useful to take a bird's-eye view of the main policy implications; the leading thread will be the analysis of the trade and payment balances, without losing sight of the interdependence between external and domestic issues.

[1] Figures published by the government were used for most estimates. A systematic comparison with World Bank estimates convey the impression that the latter overestimate structural changes and growth in China for the period 1965-75.

Three components of trade balances, agricultural goods, energy materials and manufactures respectively should be reviewed separately. Negative balances for agricultural goods are likely to become a crucial problem in two regions, Tropical Africa and West Asia (which includes here North Africa), unless active policies succeed in reverting the low growth trends observed in the past (1.6 per cent in Tropical Africa and around 2 per cent in West Asia). In other regions, like South Asia and East Asia, high average trends leave hopes for a balanced trade of agricultural products, but such averages may hide vulnerable balances for individual countries. In the past the first impact of an acceleration of the industrialization process was often a sudden rise in food imports to ease a demand pressure resulting from high income elasticities for food in the urban areas. The Ricardian law of comparative advantages cannot justify such a rise. The truth lies in the slow or inexistent adjustment in the domestic supply of food for reasons which vary from country to country: it encompasses adverse domestic policies ranging from low internal prices for agricultural goods to institutional defects in the credit and marketing system or no adequate extension services, in other words, low priorities to the allocation of resources to the lengthy and costly adjustment of the agricultural sector. Distributive aspects also play a role, as will be seen below in the text.

The analysis of energy balances concentrates on the relation between growth process and energy demand. The slack which can be observed today in the international oil market should not deceive policy makers on potential long term tensions. Actually, the UNITAD scenarios draw attention to possible conflicting claims of developing and developed countries for scarce resources like oil and gas. A 2 per cent annual growth of the capacity of the energy sector of developed market economies for the decade such as is planned by governments in the last available reports, would require an annual growth rate of the sector output between 5 per cent and 5.5 per cent in developing market economy countries to sustain a 7 per cent GNP growth rate in the 80's and satisfy as well energy import requirements of the North. This figure should be compared with an expected 4.5 growth rate of the capacity of the sector in developing market countries. A sustained 7% growth rate for developing countries, it may be said, appears highly improbable when looked at in 1982. However, a slightly higher growth of the North would bring out the same world energy tensions. It is also noteworthy that a great deal of uncertainty attaches to the real impact, in the long run, of energy conservation policies in the North which are an essential ingredient to fill the gap. Another problem is the financing of investment required in the South to achieve the expected 4.5 growth rate of energy supply capacity. This calls for urgent consideration and decisions at the international level, taking into account the long maturation delays, ranging from 5 to 10 years, required to raise the supply capacity: in the field of energy, the year 1990 is tomorrow.

There remains the fact that negative balances of commercial energy (ignoring, for that discussion, non-commercial energies affecting the life of more than half of mankind) are still found in 1990 for South Asia, if a 5% overall growth rate is to be achieved; in other regions, the reassuring positive balances found in the scenarios hide the abysmal deficits likely to appear in the majority of individual oil importing countries. There long-term adjustments processes are required, until the time when the necessary structural changes reach a cruising speed. No short term solutions can alleviate the deficit of oil developing importing countries (and prevent a long standing halt in their growth), short of an ambitious solidarity effort of all oil exporting countries, whether developed or developing, with an adequate support of international financing institutions. It is essential here to recall that weak economies are more vulnerable to a sudden shock than advanced industrialized countries and also require a more costly process of recovery in relative terms. In this field, a specific South-South solidarity scheme might well serve as a starting point for triggering the world community effort which is called for.

The third area in which the UNITAD scenarios bring out sharp conclusions is the trade of manufactures. In order to evaluate the impact of the message, it seems essential to follow step by step the main arguments.

1) Following many authors, and in particular the conclusions of the "Interfutures team"[1], it is assumed that developed countries have now entered a new era of relatively slow progress

[1] Published by OECD.

of productivity and growth. These analyses, based on sociological, technico-scientific and political considerations, or a mix of these, refer to the long run beyond the stabilization schemes now under way. An essential issue is therefore the assessment of the implications of this radical change on the world trade equilibrium at large, and on developing countries in particular. The low exogenous growth rates assigned to market developed countries in the UNITAD scenarios, from 3 per cent to 3.3 per cent, as compared to the 4.5 per cent found in 1963-75, is meant to reveal these implications, even if they are considered over-pessimistic.

2) The second pillar of the demonstration is the time duration of the industrialization process in developing countries and its implications in terms of market power. There are obvious differences from one country to another but the magnitude of the problem can be perceived from a variety of indicators, such as the relative size of the manufacturing value-added, as a percentage of GDP, in the UNITAD scenarios for 1990: around 22% as an average for developing countries as a whole, compared with the pre-recession figures in developed countries: 25% for North America, 33% for Western Europe and 32% for Japan. Using the same yardstick, the road to industrialization will still be very long in 1990 for Tropical Africa (13%), South Asia (16%), West Asia (19%); Latin America (26%) and East Asia (27%) should aim at keeping at least the same ratio in further GDP growth.

3) The main lesson of the UNITAD scenarios is that, as long as regional markets remain fragmented, there will be a strong tendency towards increased trade dependency of the South in manufactures, but South-South economic co-operation appears to be a powerful means of reverting this tendency. In a nutshell, fragmented markets in the South will tend to decrease the self-sufficiency ratio for manufacturing from 74% in 1975 down to 68% in 1990, in spite of a growth rate of manufacturing valued added around 7% in the more optimistic scenario. In terms of trade balance, the same scenario indicates an overall deficit for manufactures as a whole (food excluded) amounting to 11.6 per cent of the GNP of the South in 1990 (10.2 per cent in 1975). The deficit in manufactures can be traced to their high import elasticities (food excluded), i.e. 1.35 per cent relative to GDP

in volume terms (1.40 in current prices). Such is the hunger for manufactures imports in the course of the industrialization process that exports cannot catch up with them in a South-North oriented trade with a sluggish northern economy, in spite of the generous assumptions on the penetration of exports into developed markets[1]. On the other hand, integrated regional markets in the field of machinery and equipment alone (admittedly a large area) can reduce by 30% and probably more the overall trade deficit in manufactures, and increase the self-sufficiency rate in manufactures up to 77%. The merit of the UNITAD scenarios, if any, is to illustrate the enormous magnitude of potential gains to be expected from South-South co-operation, without neglecting, altogether, the policies to improve North-South balances of manufactures.

4) To achieve such results, it would not be sufficient for governments to establish free trade areas for machinery and equipment among neighbouring countries, but action would be required to curb restrictive trade practices of world oligopolies and ensure an equitable share of industrialization benefits between more industrialized and less industrialized countries of the same region. As was perceived by the 1975 UNIDO Lima Conference, the small share of manufacturing value-added of market developing countries in the world manufacturing value-added (8% in the mid seventies, 12% in 1990 according to the UNITAD scenarios) can be interpreted in terms of a weak market power. A major effort of market operators and governments is therefore called for in groups of countries to counterbalance the influence of oligopoles and monopoles dominating international trade and controlling the allocation of investment in major fields. To quote a recent UNCTAD publication[2] "possibly 30-40 per cent of all international trade is on an intra-firm or related-party basis [i.e. transnational private corporations], another 30 per cent is likely to constitute state trading, a further share of international trade is captive in nature . . ." Two conclusions emerge: "it is apparent that the concept of a freely competitive international trading market is illusory" and "the barriers to new entrants in the international trade market are formidable and could well be growing greater, especially as

[1] The annual growth rate of exports of manufactures from the South appears to be 12.1 per cent in the scenario.

[2] UNCTAD/ST/MD25.

a result of concentration of market power . . . ." In view of these impressive obstacles, rapid progress towards South-South co-operation appears as a major challenge for the world community at large, worth mobilizing the energies at the international, regional and subregional levels.

5) The extension of the market for manufactures should not be sought only through co-operative arrangements among neighbouring countries but also inside each country through the harmonious development of urban and rural areas. In this respect, the analysis of employment figures of the scenarios showed the small impact of overall GNP growth and of industrial growth on rural unemployment (which should actually be interpreted in terms of underemployment). The problem is acute in Tropical Africa, South Asia and East Asia, where 70% or more of the population is living in rural areas, and also in large parts of Latin America, North Africa and West Asia. This raises two related problems, i.e. the development of agriculture, already mentioned, and that of rural industries. It should be observed that the agricultural growth process was achieved in all scenarios without changes in land distribution nor in technology. In other words, the bulk of the additional output was produced by large farms or plantations with "modern technology", i.e. with a high capital-labour ratio, and costly inputs of fertilizers. The employment effect, as was seen, was naturally minor, leaving more than two hundred million people unemployed for the South as a whole. The struggle against poverty in rural areas as well as a sustained growth of internal markets therefore point to the same direction, i.e. special policies and programmes to raise the standard of living of small and medium farmers. This should be completed by the development of rural industry which can both improve the average productivity and income in rural areas and foster productivity progress in agriculture itself. Beyond the obvious social benefits of such policies, the experience of several countries is most convincing in demonstrating the long-term impact a balanced rural development may have on further growth: in brief, a large food industry and consumption goods industry sector provide the market required to develop a number of capital goods and intermediary products, thus avoiding the difficult search for a sizeable export component for infant industries.

The international side of such domestic policies are multifold. One of them is that food aid, useful as it should be in emergency situations on the humanitarian side, calls for parallel and sustained action to alleviate the financing burden of costly and time consuming plans for rural development. The integration of rural and urban economies, in turn, offers a large area for financial co-operation, especially with regard to costly infrastructure network. Perhaps the fast technology progress in decentralized energy sources (biomass, solar) or in modern telecommunication framework (satellite) would permit, if made available to developing countries, to delay, or even to avoid, the early development of some national wide infrastructure networks. More generally, the doubling of ODA to least developed countries, as recommended by the 1981 Paris Conference might act as a catalyst in accelerating the pace of rural development in this group of countries.

This brief review of major long-term development issues cannot be concluded without turning attention to the international financial problem, which appears both as an immediate and as a long-term challenge for the world community. It is not relevant here to elaborate on the magnitude of present indebtedness[1] which is outside the scope of this study. However, as was well perceived by the Brandt Commission, the present difficulties in the recycling of short-term balances of capital surplus countries call for an imaginative solution to take into account long-term development requirements. The simplified treatment of the debt problem in the UNITAD scenarios appears quite sufficient to demonstrate the pervasiveness of financial requirements of industrializing countries over the next decades and the need for the World financial system to find a response commensurate to the dimension of the development process, whether in the agricultural, the energy or the manufacturing sectors. The improvement of the monetary and financial world system cannot therefore but loom large in forthcoming negotiations on a North-South solidarity framework.

[1] According to the President of the IBRD: "In 1970, the debt owed to private lenders by developing countries amounted to only 45% of the total. By 1980, the percentage had burgeoned to 65%: $284 billion out of $349 billion," 13 January 1982, Press Release.

DISCUSSION POINTS

1) Professor Mesarovic asked if there was any plan to use the different UN models in a comparative analysis, and whether it is intended to use the various models as a check of one against another.

Dr. Royer answered that there was some partial simultaneous use of UN models. There is an obvious interest in using models which are sectoral in character (such as the FAO agricultural model, the ILO employment model) as checks on a world system of this type. It is however very difficult to try to compare global models which are of a very different character. The Dynamico model for example is simulating how the market will govern the world economy for the next two or three years, and the UNITAD system is not constructed with that question in mind. Probably one may try to match together the results of the individual country models, with which the UN is now well endowed, and the global models. Many interesting interactions can take place, and there is a need to check the data of the large models with the country models.

Dr. Hrabovszky spoke of the integration of the FAO and other world models. One of the fundamental features of these models is that, being target oriented, normative models, the same 7% GDP growth rate is used throughout, and the population assumptions are identical.

Dr. Royer added that for consumption and savings, the large household surveys collected by the ILO and the World Bank have been used. As for the different treatments of the technologies in rural areas, and especially rural industries, this has so far been carried out only on the advice of colleagues from China who came to UNIDO last year. Data on rural vs. urban technologies has been gathered, and it is intended to extend the experiment to other regions such as India – using the Indian experience with small scale industries.[1]

2) It was pointed out that many models are based on the systems and concerns of the North, lack flexibility, and often treat the South as an appendage or residue. This model is more concerned with the preoccupations of the South, but what

[1] This was actually achieved in 1983, by the time this paper was printed.

is the operative value as concerns the strategy which the countries of the South may adopt in particular to resolve the problem of interactions between energy and agriculture?

Dr. Royer noted that although the complexity of the system can cause problems, it is not possible to deal with problems of energy and agriculture in isolation. The detailed local studies need to be integrated into a global system. A model of such complexity is not transparent to politicians, and the question of communication between modellers and politicians is a difficult one.

3) Professor Mesarovic raised the question of consistency between the models.

Dr. Royer indicated that systems can be checked for consistency in two ways, by a check on behavioural equations or by a check of the consistency of submodel data with the world economy at large. Thus, for example, there is a selection of FAO data which can be used as inputs, and the check consists in comparing the output from this agricultural system with the FAO estimates.

4) Dr. Hrabovszky pointed out that a number of model experiments show the difficulty of providing employment in rural areas. Given the ceilings on agriculture growth rates in many countries (usually 3-3.5%), with an agriculture sector population growth rate of 1.5%, there is about a 2% per capita growth rate in incomes. These incomes are already much lower than in the rest of the economy, so built into those policies is a further pulling apart of rural and urban incomes. The FAO feels that the opposite is needed. A much more careful exploration of labour-intensive, non-agricultural development is required in order to build a development model which will take care of intersectoral income distribution problems. The model system must include such considerations at a high priority level, and search across sectors for technological systems which can help to create a better income distribution between sectors.

5) Professor Mesarovic asked how much effort is needed to initialise the system. For example, if projections are to be made again in 1985, how long will it take to update the input data? It seems that in the Leontiev system, it is a considerable task to gather all the input-output data.

Dr. Royer recognised this as a difficult problem as a system of this type includes a considerable amount of endogenous data which needs to be fed into the model. One comparative advantage of the present system over Leontiev is that there is a full command of a number of information subsystems in the UN data base, but even so it is a considerable task.

5.5 The Food and Energy Programme of IIASA

W. TIMS and K. FROHBERG

Sketching out some of the highlights of the IIASA food and energy programme implies an attempt to represent the activities of a very large number of scientists in any number of different disciplines and institutions around the world which are co-operating in this effort. Notwithstanding the fact that we have been engaged in this effort for the past 5 years, we are not yet in a position to give complete results from the research, and it may take another 5 years before we are in a position to provide results of interest or to demonstrate the system which we are trying to build.

In constructing a world system which is addressed to the problem of world hunger and which therefore contains reference to the development of agriculture and the distribution of food in the world, both between countries and within countries, we start from the presumption that the world consists of countries, that policies are made by countries and country governments, and that the international market is an interplay of policies and actions which emanate from those individual countries. It is therefore a bottom-up approach which requires first of all a study of the centers where policies relevant to the food-energy situation and its long term development are made. Then those policy centers must be modelled in a way in which the actors and their endowments are properly represented in the overall system.

A particular selection of countries has been made, taking those which are important in terms of the effect which their policies have on the international system. Those countries are therefore selected which are large in terms of agricultural output, agricultural consumption or share in international trade in agricultural products. About 25 countries and country groups have been selected, which together cover about 80% of agricultural production, consumption and trade, and also about 80% of world population. These countries vary considerably in their level of development, ranging from North America to USSR, China and a number of the larger developing countries. The smaller developing countries are modelled as a group or as regional groups. This is a closed world system, with the remaining 20% of world population and agriculture in a large number of countries taken to be dependent on the world market; that is, they do not have an independent effect through their policies on what happens in the world market.

A second consequence of this approach to the world system is the effect on the modelling, in that a system of networking is required. Modelling country policies, to make a model which is relevant and has a coherent representation of relationships and definition of policy instruments and constraints, is not something that can be done at IIASA in Luxembourg. This requires the full cooperation of people involved at policy and research levels in individual countries. Thus in order to achieve something worthwhile the activity needs to be spread around the world, and a number of specific prerequisites which originate at the national level must be taken into account. It is not possible to develop a methodology and software and to suggest that the various institutions simply apply this in their differing situations. A dialogue develops in which scientists indicate the factors of greatest local interest, and these determine to a large extent the contents of the model.

In order to make sure that the national models are suitable for inclusion in an international system, it is necessary to specify a set of conditions which apply to each so they are internationally linkable. There are then two major component parts, national models constructed locally with a significant input of methodology and software from IIASA and the Center for Food Studies in the Netherlands, and the international linkage system which runs on the IIASA computer.

It is clear that even with a limited number of major countries, all the models will not be constructed at the same time. The first in fact was started in 1975, while the last will hopefully be completed in 1985. Therefore one is gradually moving from one country to another and this creates a number of problems.

The major intended characteristic of the models is to distinguish the major actors in each economy. Actors may be farmers, landless laborers, consumers, government, urban workers. In each case, in order to catch the full dimension of the problem of hunger and malnutrition, the actors to be depicted in the model must be very carefully selected. For each group of actors it is necessary to represent their endowments, capital, land, available labour, how those factors of production are used and in what kind of hierarchies the actors act with respect to each other.

As well as hierarchies between consumers, producers and governments, there are also hierarchies within communities. Certain people dominate the trade sector, credit sector and input supply sectors, and this may have a significant impact on the factor earnings that other groups of people such as the landless or small farmers may get from their endowments. The distribution of food and income depends on those hierarchies, both at a national and local level, and it is hoped to include these effects in the models.

Since national institutions have different kinds of interests, the models, which fit into the total system, are quite different. The models for the western economies are largely economic, and do not move far beyond the specification of economic relationships. For the East European countries, USSR and probably for China, there is more emphasis on the role of technology, agronomy and life styles. In the models for developing countries the use of physical resources is added to the economic and technological components.

There are many ways in which the low yields of developing countries can be increased. It will be interesting to identify the various possibilities for resource use in agriculture and to determine the choices which are available to governments and farmers.

Since the economic models apply generally, all models have similar economic components. The developing countries have a large component of agronomic modelling, starting with physical factors such as climate and soil and investigating various possibilities for crop and livestock development. In models of this type it is essential to include the full economy and to include, for example, the constraints on government action, which depend on development in non-agricultural sectors. Since it is important that the model not become over-complicated, non-agricultural activity is aggregated into one sector in the model, and some of the criticism of this extreme level at aggregation is valid.

At least the inclusion of a full economy picture avoids hidden sinks in the model, where things which cannot be allocated simply disappear, or where deficits and shortages are not noticed. It also forces the modellers to take full account of financial balances – each actor has certain resources available and cannot spend more.

For developing countries the approach is to analyse the agricultural sector not only in some detail as far as commodities are concerned but also to consider the physical factors of crop and livestock production. After examining factors such as soils and climate, choices available to farmers are identified, along with economic and physical constraints.

It is clear that energy plays a very important role in the system, and this becomes more evident as experience is gained in constructing this type of model. The crops grown and the livestock produced for food probably represent only one third of the total calories produced in some countries. The rest of the plants and animals form a very basic input into the energy cycle at the local level.

Although we do not expect to achieve the completeness of the ideal as described by Dr. Sachs, we hope at least to include the key factors and to represent the main elements of the energy cycle. It is essential to do so – in some countries which have been modelled it seems that agriculture strategies have not succeeded in the past because the energy consequences have not been fully taken into account.

This is a long, hard project. It has been in operation for 5 years, some of the models have been completed, others are in a fairly advanced stage, while others are about to be started or still to be started. It takes about 3 years for a good research team to construct a national model. One problem is that by the time the last model is completed, the first may be out of date. In order to run the world system the models must be kept up to date, and this is a considerable problem. Thus it is necessary not only to construct the models but also to document all stages of their development.

It is important to start from a listing of the basic statistics, a clear list of sources, then to go through all the steps, listing every manipulation and to describe the model fully. If the documentation is well done and put into a computer program it should be possible for a new researcher to find out what the original data were, to go to the sources and find the data, to enter them into the computer file and to run a check. It is not easy and can be boring, but it is necessary if the models are to be updated from time to time.

Setting up this kind of system is a part of the methodological work undertaken. Some significant advances have been made and it will be possible to take a local model which is running now and update it in a few year's time with new census data, new budget surveys, new production accounts and so on. In terms of the investment in those models it would be a waste to use them only once or twice. This system is intended to remain operational world-wide, and to be improved over time.

One of the major outputs of the system so far is not the running of the models, the policy alternatives at a local level or the setting up of the simplified international system, but that gradually there is a growing group of scientists around the world who think along the same lines, who have become familiar with a particular way of looking at how an economy functions and at how agriculture and energy interact, and who ask about the effects of policies. This may in the future form a suitable basis for a discussion of international policy issues, since this will be a group which will think more or less along the same lines and speak the same language.

DISCUSSION POINTS

1) Professor Meadows enquired whether in designing the world market interactions the role of the multinational corporations and the large trading corporations had been considered. Do they merely implement actions within the market system, or do they have a distorting effect on the market? Should they be integrated into the representation of the world market?

Dr. Robinson insisted that when building a model the first step must be to identify the major actors and the factors which will dominate the system. In the present day world the multinational and transnational corporations are separate entities and extremely important actors with separate behavioural patterns from those of countries. Having considered simply countries as basic units a very important qualitative feature has been left out of the system. After all such large companies are important in both the energy and agriculture sectors. This point is so important that perhaps it invalidates the modelling.

Dr. Tims did not think that the modelling of cartels or multinational corporations presented any particular problem. Given that the commodity breakdown is uniform for all national models, it is possible to study, for example, wheat in both the countries and at the world market level and to insert cartel arrangements (which is basically what multinational corporations are).

This has not yet been done but one of the basic considerations of this approach is that the models should be capable of carrying out this kind of analysis. One of the main aims of the research is to find out exactly what are the effects of multinational corporations across borders on the international market.

2) Dr. Glaeser felt that IIASA and UNU followed two different paradigms. IIASA is primarily concerned with the monetary economy and the world market whereas the UNU would stress a needs orientation.

There is a significant difference in energy needs in different types of agricultural systems. While developed countries need about 10 calories of input energy to produce one calorie of food, in developing countries the ratio is the other way round, with one calorie of non-solar energy used to produce 70 calories of foodstuffs. The problem is that in most of the models presented so far a certain technological system is proposed and a certain production function.

Dr. Tims doubted that there was a clear difference between IIASA and UNU. As a step towards our understanding of interactions between countries a number of national models are being constructed, and such national models have a needs dimension. There is of course a need to convince governments that they should extend their horizons beyond simple agriculture, energy or economic growth dimensions and to insist on a needs dimension in the modelling.

The educational element is quite interesting. A lot is learnt during the modelling exercise of the factual situation in each country, and this helps to open the eyes of some people to the real development problems which they face.

3) Professor Meadows asked if there were a significant dynamic difference between the national models which are primarily economic and those with a strong agronomic physical resource base.

Dr. Tims could not give a clear response based on experience to date using the models. It could be assumed however that the flexibility in the more extended models would be considerably less. In a model which has the endowments of the different groups in it (in addition to purely economic factors), and where both constraints on groups and hierarchies within and between groups are included, government policies would probably be found to be less effective.

4) In answer to a query by Dr. Glaeser, Dr. Tims described how by specifying the production functions in considerable detail in terms of physical factors and available technology an effort was made to determine the impact of national policies. The choices open to farmers are modelled in such a way that at the next step it is possible to examine alternative government policies in terms of a choice of resources, technologies and so on.

5) Dr. Glaeser wondered if it might not be worthwhile to break up the technological system and the production function, to use a different notion of productivity, replacing yield per hectare by yield per energy unit.

Dr. Hrabovszky pointed out that energy units per unit of output are very different, as are the costs and uses of energy. It would thus be very difficult to deal with yield per energy unit.

Dr. Tims agreed that yield per hectare was not the only useful measure. For example the consideration of income levels, the possibility to influence and change them by various policies goes well beyond the yield per hectare notion.

6) Professor Richardson felt that the project had four important characteristics: quality of documentation, honesty, humility and clear acknowledgement of all contributions. He asked what more can be learned from a project of such diversity and continuity.

Dr. Tims replied that one can learn that this networking approach is aggravating, rather like pulling teeth. It is far from easy to get 25 countries to all agree on one methodology, to stay with it and to model accordingly. In a number of cases people have gone off and constructed their own models which then could not be used. This process takes a tremendous amount of coordination, discussion and getting people together to talk about what is being attempted. A large part of that takes place at IIASA, which is the main center of exchange on the modelling system.

One should also try to contribute as little as possible to the models. The central group needs to do a little, but it is much more important to have a significant group of people who are purely working on the methodology, so that they can be referred to each time a new country joins in. For then new questions are raised – somebody wants an informal sector, a split in the non-agricultural sector or addition of more commodities to the standard list. This can be done but it needs thinking through, setting up the system, constructing the software and delivering it to the user. A methodology team is needed to step in when a new question arises, to meet the needs of the users – who are the only ones able to construct good national models and to make them policy relevant.

7) Since models can soon become outdated, Dr. Glaeser asked if it would be really necessary for the UNU to use a modelling approach. After all modelling is a tool, a means and not an end in itself.

Dr. Tims was of the opinion that models are helpful, particularly to show the way in which a economy works. Markets are linked, price movements in different markets are linked, each actor is restricted by his endowments and the economy as a whole has to stay within particular constraints. It might be thought, for example, that a stabilisation of the international wheat price would be beneficial to all participants. If such a partial analysis is extended it can be shown that changes in the wheat market will effect the rice and maize markets. A change in area restrictions in the USA will effect Thailand and China, and it is very difficult to handle all of this in a verbal form. It is not possible to really tackle such problems without models.

Models are not the final answer and need a lot of interpretation to handle them properly, but they can play an important role in the analysis.

5.6 The Japanese Experience of Global Modelling: Implications for Energy and Agriculture Futures

AKIRA ONISHI

INTRODUCTION

The purpose of this paper is two fold; the one is to introduce a short history of development and the current status of the FUGI model which might be deemed as a Japanese experience of global modelling. The other is to show projections of the world economy in 1983-1990 under alternative oil price scenarios by using the FUGI model, in particular, global macroeconomic model (type IV 011-62) and suggest its policy implications for energy and agriculture futures.

The project FUGI (Future of Global Inter-dependence) was launched in 1976 by three co-leaders such as Y. Kaya, A. Onishi and Y. Suzuki who jointly constructed a new global model complex under the sponsorshipof National Institute for Research Advancement of Japan, where three models were interlinked with each other. They are, a global macroeconomic model (GMEM type I and II), a global input-output model (GIOM), and a global metallic resource model. A complete report of the project at this stage was presented at 5th IIASA Global Modelling Conference in 1977 and published in its proceedings.

Upon the request by UN ESCAP, Development Planning Division in 1978, in order to envisage the future of ESCAP member countries the author and Kaya reconstructed GMEM (type III) and GIOM respectively. The final report was presented in March, 1980 to the ESCAP, Development Planning Division where reproduced it entitled "Long-term Projections of the Economies of ESCAP Countries" in March 1980.

The current FUGI model system is composed of two major sets of models which are interlinked with each other. The first one is a dynamic global macroeconomic model (GMEM) developed by the author at the IAER, Soka University, and the second is a static global input-output model (GIOM) developed by Kaya, the University of Tokyo.

The FUGI global macroeconomic model has been used as tools for forecasting as well as planning purposes, such as the Ministry of International Trade and Industry, the Economic Planning Agency of the Japanese Government, the World Economic Information Services (Tokyo) and the United Nations, Projections and Perspective Branch.

Based on the experience of global modelling in the 1970s, the fourth generation type of GMEM was developed by the author during 1980-1981. The fourth generation GMEM (type IV 011-62), first presented at the CEME Seminar, 25-26 June, 1981, Stanford, links 62 areas of the world, consisting of approximately 12,700 equations and 25,000 steps of computer software.

This paper provides with recent information on the FUGI global macroeconomic model type IV 011-62 and shows its simulation results of the world economy for the period 1983-1990 under alternative oil price scenarios.

THE SCOPE AND STRUCTURE OF THE FUGI MODEL

Current FUGI model is a hybrid of two models, each of which has a different role in investigating the future of global interdependence. The global macroeconomic model (GMEM) provides scenario forecasting and policy simulation of the world economy from a macroeconomic view point. The global input-output model (GIOM) is used to investigate feasibility of various patterns of agro-industrial development of the world within the framework of the macroeconomy given by GMEM.

Global Macroeconomic Model (GMEM)

The FUGI-GMEM type IV 011-62 covers the entire world with three major groupings such as (1) developed market economies (AME), (2) developing market economies (DME) and (3) centrally planned economies (CPE). The AME grouping encloses the OECD, EC and other AME sub-groupings. The DME grouping also includes sub-groupings (not necessarily mutually exclusive) such as oil exporting countries, non-oil exporting countries, newly industrializing countries (Asian NICS and Latin American NICS), Asia, Middle East, Africa and Latin America-Caribbean countries. The CPE grouping has two sub-groupings, i.e., USSR and East Europe as well as China and other Asian CPE.

1) The AME grouping encloses 26 countries/regions economic models for 25 leading developed market economies and the rest of AME region.

2) The DME grouping with 31 countries/regions economic models for 26 selected developing market economies and 5 other DME groupings.

3) The CPE grouping, with 5 countries/regions economic models for 3 selected centrally planned economies and 2 CPE groupings.

For further details of country and regional classifications, see Appendix 1.

Thus, the FUGI-GMEM type IV 011-62 represents the system linking altogether 62 countries/regions economic models.

While GMEM type IV 011-62 represents the system integrating "national/regional models" which reflect the special characteristics of 62 countries/areas in the world as a whole, each national/regional model, in turn, consists of 11 sub-systems as follows; (1) Production with energy constraints; (2) Expenditure on GDP at constant prices, (3) Profits and wages, (4) Prices, (5) Expenditure on GDP at current prices, (6) Money, (7) Interest rates, (8) Government finance, (9) International balance of payments, (10) ODA and private direct foreign investment to LDC, (11) Foreign exchange rates, as shown in the Appendix 2.

One of the most important distinguished features of this model is the systems structure for determining international trade and financial flows. The system at work is one whereby economic growth in each country/region does not take place completely on a basis of self-reliance, but rather with a framework of interdependent relationships among 62 countries/regions characterized by trade and financial flows.

As recognition of this close interdependence prevails, the UN member countries of the world, both North and South, having better access to information on economic conditions in developed nations as well as the developing world, have come to a better mutual understanding of both. Under these circumstances, the FUGI global macroeconomic model seems likely to play a significant role in envisaging the future of global interdependence through trade, investment and aid.

The current purpose of the FUGI macroeconomic model is to provide alternative long-term projections of the world economy under different scenarios regarding national and international policies designed to promote the North-South dialogue and cooperation. The model is expected to play as a major instrument in projecting the current and long-term trends of the world economic activities as well as in assessing the feasibility of the goals and investigating alternative world development strategies and policies relevant to a New International Economic Order and the United Nations' International Development Strategies for the 1980s.

Global input-output model (GIOM)

The FUGI-GIOM covers the world except Asian centrally planned countries. The world is divided into 14 regions of which 7 belong to Asian developing countries, but the grouping of the countries in the model may be changed according to the purpose of investigation. There are 15 industrial sectors in each region.

The GIOM is a static input-output model which gives sectoral production and trade patterns of the world within the framework of the world economy provided by GMEM in each year. GIOM is a 14-region 15-sector model and can serve the following two purposes. The first is a simple forecast based on the statistical analysis of the past data, and the other is an optimization scheme in which sectoral outputs, exports and imports are simultaneously determined so as to satisfy a given criterion.

The model is composed of four modules connected with each other, i.e. (1) trade submodule in which the interval estimates of exports and imports are generated, (2) final demand submodule in which the final demand pattern of each region is forecast, (3) input-output submodule in which input-output coefficients are forecast, and (4) LP submodule in which the optimization scheme is to be executed.

In short, GIOM provides two kinds of computation schemes, i.e. simulation and optimization ones. The aim of the former is to disaggregate outputs of GMEM into sectoral ones, or, in other words, to forecast sectoral development patterns of the world. GIOM is used as a submodel linked to GMEM, providing more detailed information on the scenarios constructed by GMEM.

The aim of the latter, on the contrary, is not to simply forecast the future but to investigate feasibility of various industrial development target taking various constraints into account. For further information on GIOM, see Reference I.

IMPACTS OF OIL PRICE RISE AND FUTURE OIL PRICE SCENARIOS

Although an attempt was made above to explain some of the FUGI global macroeconomic model's structural characteristics, the above explanations remain rather general and causal relationships of impacts of the OPEC oil pricing on the world economy in the model system are not clear. Thus, we should like to illustrate using the model for "scenario forecast simulation," how the model can be used for an analysis of impacts of oil price setting on the interdependent world economy.

Impact of oil price rise

One possible type of oil price rise is that of sudden substantial increases such as took place in the "first and second oil crises," while another would follow a "long-term strategy" of gradual increases as advocated by Saudi Arabia and the "moderate group" within OPEC.

There is not very much that needs to be explained about this "Long-term strategy" other than to say that it envisages setting oil prices according to a formula that allow forecasts of future conditions, taking into account such factors as inflation rates, exchange rates, and economic growth rates in the major industrially developed countries. This formula was originally conceived at a time when the oil price stood at $28 per barrel. Present conditions are such that it is difficult to apply directly the formula as originally drawn up.

A slight modification will enable our present moel to carry

out simulations to yield calculations based on the future introduction of an OPEC "long-term strategy," or, alternatively, calculations when oil prices are simply linked with inflation rates in the industrially developed countreis.

For example, the OPEC "long-term strategy" relies on indicators composed of combined data from 10 major industrially developed countries. The present model treats each of these 10 countries as a country unit for purposes of calculating each country's rates of inflation, currency exchange, and economic growth. But indicators of the type used by OPEC can also be calculated by means of weighted averages.

Here, we would like to consider briefly the routes by which the impact from a rise in the OPEC oil prices (linked to rates of inflation, currency exchange, and economic growth in the 10 industrially developed countries) reaches the global economy. At such a time, the problem arises of what to take as an indicator of inflation rate. The OPEC formula uses an inflation indicator made up in 2/3 part by a weighted average among the 10 countries of dollar-base export prices and in 1/3 part by weighted average of consumer price indices.

According to the above formula, the previous year's inflation rates, exchange rates and economic growth rates for the 10 industrially advanced countries are first given, and on that basis rates of OPEC price rises for the current year are automatically calculated. (See Model equation IV-2 II-2).

In a "baseline projection," a "modified OPEC formula" is used to forecast future oil prices. In making such a projection, movements in OPEC prices over the 1972-81 period are estimated in relation to rates of inflation, exchange rates, and economic growth rates in the 10 industrially developed countries in question as follows;

OPEC oil price function: Model equation IV.II-1

$$\begin{aligned} DOT(PEO) = &-0.0884 + 0.8803*DOT(GIFI10_{-1}) + 4.1046*DOT(GDP^{\#}I10_{-1}) \\ &(-0.5146) \quad (0.7923) \qquad\qquad (1.5748) \\ &+2.5543*DOT(FERSII10_{-1}) + 1.9239*FS01 + 0.4551*FS02 \\ &(2.7426) \qquad\qquad (11.7263) \quad (3.8663) \end{aligned}$$

SAMPLE : 1972-1981 $\bar{R}^2=0.9897$ S.E.=0.1031 DW=2.4074

Where DOT (PEO) denotes a percentage change in the oil price index (1975=1). GIFI10 is a weighted average inflation rate index (1975=1) of the group 10 countries including Japan, United States, Belgium, France, West Germany, Italy, Netherland, Sweden, Switzerland and United Kingdom. It is a composite inflation rate index derived from a weighted average export price index, PE10, and consumers price index, CPI10. (GIFI10 = 2/3 * PE10 + 1/3 * CPI10). GDP#I10 denotes real GDP index (1975=1) of the group 10. FERSII10 denotes a weighted average exchange rate index of the group 10 (Inverse index, 1/FERSI10, 1975=1). FS01 and FS02 are dummy variables of the first (1973/74) and the second (1979/80) oil shock respectively, (where 1974=1, 1980=1, and other years are given a "zero" value respectively).

Rises in OPEC oil prices influence the "supply side" (i.e., the production side) in industrially developed countries which consume large amounts of energy, since a negative impact on productivity is caused by relative rises in energy cost.

Wage costs will rise with slowed-down increase in productivity. Wage cost hikes will push up wholesale prices, which in turn push up consumer prices and nominal wages, resulting in a "vicious circle" of inflation.

Interest rates will go up. In other words, as inflation becomes more severe, short-term interest rates rise, and long-term prime rates show a tendency to rise also.

As a result of rises in interest rate in the industrially developed countries — and especially so in the case of the US — Euro-dollar interest rates rise, having an influence on the currencies of Europe and Japan. In other words, as a result of the current "system" for determining exchange rates, US inflation and interest rates come to have an impact on currency exchange rates in Europe and Japan. Through such a route these influences impose restrictions on the freedom of choice in selecting monetary policies in the latter two regions.

It should be kept in mind that the advance of inflation does not in all countries take place at the same rate. According to different countries' economic structures and practices, there are indeed significant differences. For example, inflation rates are likely to be relatively low in countries which skillfully control moneysupplies and maintain an appropriate balance between growth in productivity and rises in nominal wages;

while inflation rates are likely to be relatively high in countries where labor-management agreements decide on wage rises only in accordance with consumer price rises, without taking due account of increases, or the lack thereof, in productivity.

The currency of countries whose domestic price rises are small relative to the US will be considered "strong," while the currency of countries with a high tempo of inflation will become weaker, thus tending further to accelrate domestic price rises. Accelerated domestic inflation will raise export prices, so that in the absence of an appropriate devaluation of the exchange rate, there seems likely to be a loss of international competitiveness.

A number of countries have experienced a "vicious circle" in which nominal increases in imports due to oil price hikes are not covered by exports, with the result that the deficit in current balance of payments expands and the exchange rate further depriciates.

Under the various influences mentioned above, the inflation and economic growth rates in the developed countries are determined and these in turn have repercussions on the developing countries. If the developed countries should suffer from "stagflation," needless to say, developing countries which purchase expensive capital and other goods from them will also be affected. Thus, the non-oil developing countries in particular, must import under the handicaps of both oil price rises and inflation in the developed countries.

If the real economic growth rate in the developed countries should stagnate, prices for primary products are likely to fall, the value of developing countries' exports is likely to decrease, and their import capabilities are likely to decrease as a result of this worsening in the terms of trade. Such conditions will have repercussions on the real economic growth rate in the developing countries. The model is able to show how this, in turn, can threaten the developed countries' exports to developing countries and can give rise to worldwide "stagflation" with a shrinking of real world trade.

To cope with inflation, most of the developed countries should largely rely on the restrictive monetary policy. Its economic consequences are very clear now: Such a restrictive monetary policy seems likely to introduce not only temporary reduction in inflation, but also recession of economic activities.

Then, oil situation will turn out to be reversed, because, under such circumstances, it should be impossible of maintaining higher oil prices, which in turn tend to fall down by shrinkage of the world demand for oil. The model is able to show how this, in turn, can threaten the oil exporting countries. Therefore, the model is so designed as to reflect the varying characteristics of the world economic situations in responding alternative oil price scenarios.

Future oil price scenarios

To give a further illustration, in regard to the prospects for oil prices, such factors as the activities of OPEC, the tempo of development of substitute energy sources, and progress with energy conservation, are all closely interlinked, and scenarios will differ according to the way one sees the relative weights and dynamic trends of each factor.

In this exercise, however, we have introduced the following three scenarios;

Scenario A: Baseline projection In the Baseline projection, it is assumed that the OPEC oil prices will be to a large extent indexed with inflation rates, economic growth rates and exchange rates of the selected 10 industrial countries as adopted in the OPEC long-term strategy for oil pricing. It is worth noting that the OPEC oil prices can be statistically explained by the above-mentioned "modified OPEC formula" during the successive oil crisis periods 1972-1981. This is a "modified OPEC formula" distinguished from a genuine "OPEC long-term strategy," which assumes downward rigidity of the oil prices. According to the Baseline scenarios which adopted a modified OPEC formula, as shown above, the weighted average oil price will temporarily go down by 5 dollars per barrel in 1983 and gradually recover thereafter in accordance with economic recovery of the developed countries.

Scenario B: Higher oil price scenario – The OPEC long-term strategy for oil pricing† In the OPEC long-term strategy for oil pricing simulation, "indexation" of the oil prices with respect to the 10 major industrial nations' export prices, consumer prices, exchange rates and real GDP growth rates will be taken as a policy scenario, starting from 1983 and thereafter.

This scenario seems likely to be a case of *the maximum permissible limit* of setting the OPEC oil prices in the futures.

Scenario C: Lower oil price scenario – a 5 dollars per barrel fall down of the oil prices in 1983 compared with the Baseline projection. In this scenario, it is assumed that the oil prices will further go down by 5 dollars per barrel in 1983 compared with the Baseline projection.

This is a scenario of "inverse oil shock" which might be caused by a sudden further fall down of the oil prices in 1983, although it is assumed that the oil prices will gradually recover from 1984 in accordance with the oil price formula as adopted in the Baseline projection. This scenario seems likely to be a case of the lower limit of setting the oil prices in the future.

A REVIEW OF PROJECTION RESULTS: POLICY IMPLICATIONS FOR ENERGY AND AGRICULTURE FUTURES

First of all, we may ask what will be the oil prices for the period 1983-1990 under the scenarios of A, B and C.

As shown in Table I and Figure I, the weighted average oil price temporarily goes down from 32 dollars per barrel ($34/barrel in Arabian standard oil) in 1982 to 27 ($29/barrel in Arabian standard oil), 30 and 22 dollars per barrel in 1983 and is expected to recover 1982 level by 1985, 1984 and 1988 respectively according to scenario A, B and C.

In 1990, the weighted average oil price will be 45 dollars per barrel in scenario A, 56 dollars per barrel in scenario B and 38 dollars per barrel in scenario C.

† The OPEC's long-term strategy is supposed to undergo adjustment each quarter, while the FUGI-GMEM type IV 011-62 is functioning as an annual model. Thus, this means a restraint that the oil prices will be indexed to the lagged *annual* increase in the indices of the 10 industrial countries.

TABLE I
Projections of the oil prices, 1983-1990.

									(dollar/barrel)		
	1980	1981	1982	1983	1984	1985	1986	1987	1988	1989	1990
	Actual		Estimate								
Scenario A	32.0	35.4	32.0	27.4	28.3	31.5	34.9	37.0	39.8	42.5	45.0
Scenario B	32.0	35.4	32.0	30.8	33.0	36.6	40.3	43.8	47.7	51.7	56.0
Scenario C	32.0	35.4	32.0	22.4	23.3	26.0	29.1	31.0	33.4	35.6	37.8

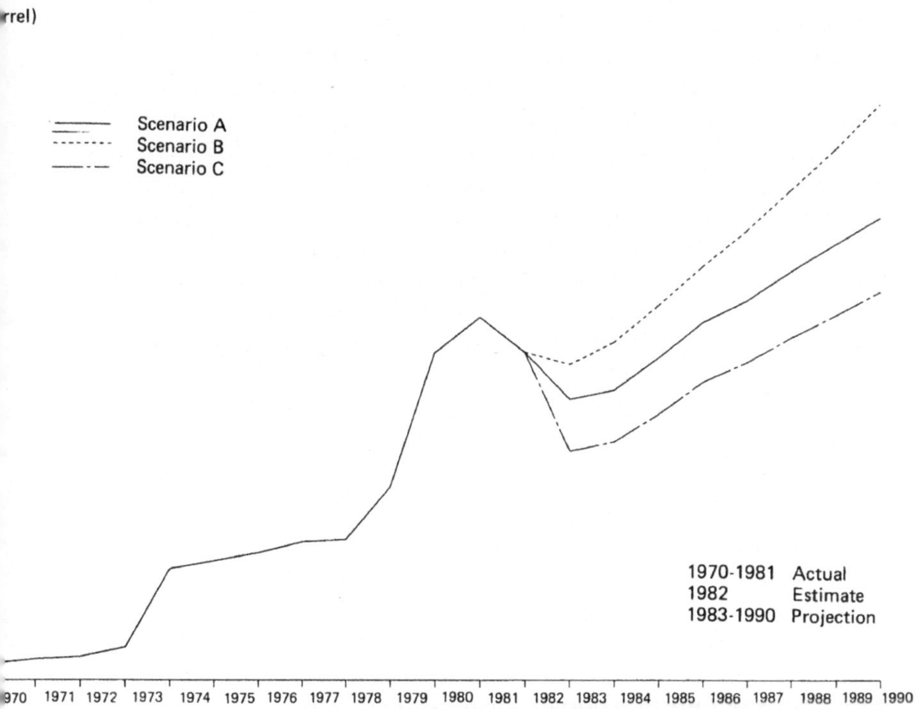

FIGURE I The oil price trends, 1970-1990.

Tables IIA, IIB and IIC indicate that, according to the projections based on scenario A, B and C, the developed countries (OECD) will maintain a 2.2, 2.1 and 2.3 per cent annual growth rate of real GDP in 1980-1990. Under this low pace of economic growth of the developed countries, it is expected that the annual growth rate of real GDP of the developing countries (non-oil exporting countries) will be 4.0 (3.8), 3.9 (3.6) and 4.1 (4.0) per cent respectively for the period. As shown in the Table II, the projection results based on scenario A, B and C envisage that relatively higher rates of real economic growth are expected for the high and medium income nations in Asia such as the Asian NICS (6.2, 6.1, 6.3 per cent) and ASEAN (6.0, 6.0, 6.1 per cent), although the Middle East (3.7, 3.8, 3.6 per cent) and Latin America (3.8, 3.5, 4.1 per cent) will decrease their economic growth rates in the 1980s. On the other side, continued lower rates of economic growth are expected for the low income nations during the period. This being the case, the North-South per capita income gap in terms of 1975 dollars seems likely to be 11.0 times in 1980 and 11.0 times in scenarios A, B and C in 1990. Therefore, it can be realistically anticipated that the main objectives of the International Development Strategies for the United Nations Third Development Decade should be directed at decreasing the North-South income gap with a special reference to the economies of the low income and most seriously affected countries.

It is then worth noting that the lower oil price will be likely to increase the real economic growth of oil importing developing countries and tend to restrain the economic growth rate of oil exporting developing countries. As a result, higher oil prices based on the OPEC strategy for oil pricing will tend to induce a larger per capita income gap between the OPEC and non-OPEC developing countries although it will be unlikely to reduce per capita income gap between the North and South as a whole.

According to the GMEM calculations, the annual average increasing rate of real exports of OPEC in the Middle East was 0.4% in 1970s, while it will lower further to an almost levelling-off –3.0% (A), –3.1% (B), –2.7% (C), although its annual average increasing rate of nominal exports will be 0.7% (A), 2.8% (B), –0.8% (C) in the 1980s. On the other side, although its annual average increasing rate of nominal imports will be 10.0% (A), 10.7% (B), 9.5% (C) in the 1980s, its trade

TABLE IIA

Projections of the World Economy, 1983 — 1990 (Baseline: Scenario A)

Annual average growth rates of real GDP

	70-80	1980	1981	1982	1983	1984	1985	80-85	85-90	80-90
WORLD	3.9	2.1	1.6	.7	2.3	3.2	3.5	2.2	3.2	2.7
WORLD (EXCLUDING CPE)	3.7	1.9	1.3	.3	2.2	3.1	3.5	2.1	3.0	2.5
AME	3.3	1.3	1.4	-.2	1.8	2.8	3.2	1.8	2.6	2.2
OECD	3.2	1.3	1.3	-.2	1.8	2.8	3.2	1.8	2.6	2.2
EC	2.9	1.3	-.6	.2	.6	2.0	2.6	1.0	2.5	1.7
OTHER AME	4.4	4.6	4.2	1.6	2.1	2.4	2.6	2.6	3.6	3.1
DME	5.7	4.5	.7	2.4	3.6	4.6	4.6	3.2	4.8	4.0
OIL EXPORTING COUNTRIES	5.9	4.0	-.7	2.6	3.9	4.7	5.1	3.1	5.4	4.3
OPEC IN MIDDLE EAST	6.7	.9	-10.1	4.2	4.5	5.2	5.4	1.7	5.4	3.5
NON OIL EXPORTING COUNTRIES	5.6	4.7	1.5	2.3	3.5	4.6	4.2	3.2	4.4	3.8
NICS	7.8	6.8	1.8	1.2	3.4	5.5	5.4	3.5	5.7	4.5
ASIAN NICS	9.0	3.4	7.7	5.6	6.0	6.6	5.8	6.3	6.1	6.2
LATIN AMERICA NICS	7.5	7.8	.1	-.1	2.6	5.2	5.3	2.6	5.5	4.0
ASIA	5.4	4.4	5.1	4.0	4.5	4.5	4.3	4.5	4.5	4.5
EAST ASIA	9.0	2.6	7.5	5.5	6.0	6.7	5.6	6.3	6.0	6.1
ASEAN	7.4	7.7	6.8	4.5	5.9	5.8	5.9	5.8	6.3	6.0
OTHER ASIA AND PACIFIC	2.8	2.9	2.5	2.9	2.6	2.1	2.1	2.4	1.9	2.1
MIDDLE EAST	6.7	1.1	-8.3	4.2	4.5	5.3	5.4	2.0	5.4	3.7
AFRICA	4.6	4.8	3.3	3.0	3.5	3.9	3.8	3.5	4.0	3.7
LATIN AMERICA AND CARIBBEAN	6.1	5.9	.4	.3	2.7	4.8	4.8	2.6	5.1	3.8
CPE	5.1	3.4	3.0	2.4	2.6	3.4	3.5	3.0	4.1	3.5
USSR AND EAST EUROPE	4.6	2.7	2.8	1.8	1.9	2.6	2.7	2.4	3.0	2.7
CHINA AND OTHER ASIAN CPE	6.3	4.9	3.7	3.8	4.2	5.3	5.4	4.5	6.2	5.3

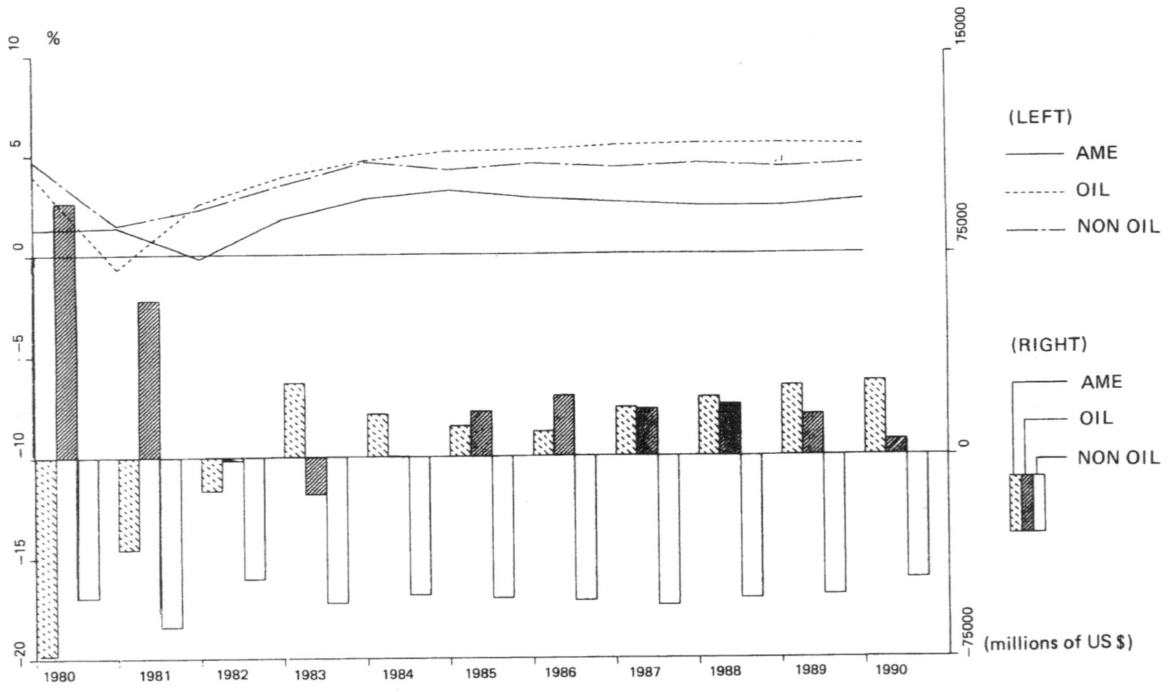

FIGURE IIA Annual average growth rates of real GDP and current balance of payments (Baseline: Scenario A).

TABLE IIB

Projections of the World Economy, 1983 – 1990 (Scenario B)

Annual average growth rates of real GDP

	70-80	1980	1981	1982	1983	1984	1985	80-85	85-90	80-90
WORLD	3.9	2.1	1.6	.7	2.2	3.0	3.4	2.1	3.1	2.6
WORLD (EXCLUDING CPE)	3.7	1.9	1.3	.3	2.1	2.9	3.3	2.0	2.9	2.4
AME	3.3	1.3	1.4	-.2	1.8	2.6	3.1	1.7	2.5	2.1
OECD	3.2	1.3	1.3	-.2	1.7	2.6	3.1	1.7	2.5	2.1
EC	2.9	1.3	-.6	.2	.7	2.0	2.5	.9	2.4	1.7
OTHER AME	4.4	4.6	4.2	1.6	2.1	2.3	2.5	2.5	3.3	2.9
DME	5.7	4.5	.7	2.4	3.6	4.1	4.4	3.0	4.7	3.9
OIL EXPORTING COUNTRIES	5.9	4.0	-.7	2.6	3.9	4.8	5.2	3.1	5.4	4.3
OPEC IN MIDDLE EAST	6.7	.9	-10.1	4.2	4.5	5.5	5.6	1.8	5.6	3.6
NON OIL EXPORTING COUNTRIES	5.6	4.7	1.5	2.3	3.5	3.6	3.8	3.0	4.3	3.6
NICS	7.8	6.8	1.8	1.2	3.4	3.8	4.7	3.0	5.6	4.3
ASIAN NICS	9.0	3.4	7.7	5.6	6.0	6.4	5.6	6.2	5.9	6.1
LATIN AMERICA NICS	7.5	7.8	.1	-.1	2.6	3.0	4.4	2.0	5.4	3.7
ASIA	5.4	4.4	5.1	4.0	4.5	4.4	4.2	4.4	4.5	4.5
EAST ASIA	9.0	2.6	7.5	5.5	5.9	6.5	5.5	6.2	5.9	6.0
ASEAN	7.4	7.7	6.8	4.5	5.9	5.8	5.9	5.8	6.2	6.0
OTHER ASIA AND PACIFIC	2.8	2.9	2.5	2.9	2.6	2.2	2.1	2.5	1.9	2.2
MIDDLE EAST	6.7	1.1	-8.3	4.2	4.5	5.5	5.6	2.1	5.5	3.8
AFRICA	4.6	4.8	3.3	3.0	3.5	3.9	3.8	3.5	4.0	3.7
LATIN AMERICA AND CARIBBEAN	6.1	5.9	.4	.3	2.7	3.3	4.2	2.1	4.9	3.5
CPE	5.1	3.4	3.0	2.4	2.6	3.4	3.5	3.0	4.1	3.5
USSR AND EAST EUROPE	4.6	2.7	2.8	1.8	1.9	2.6	2.7	2.4	3.0	2.7
CHINA AND OTHER ASIAN CPE	6.3	4.9	3.7	3.8	4.2	5.3	5.4	4.5	6.2	5.3

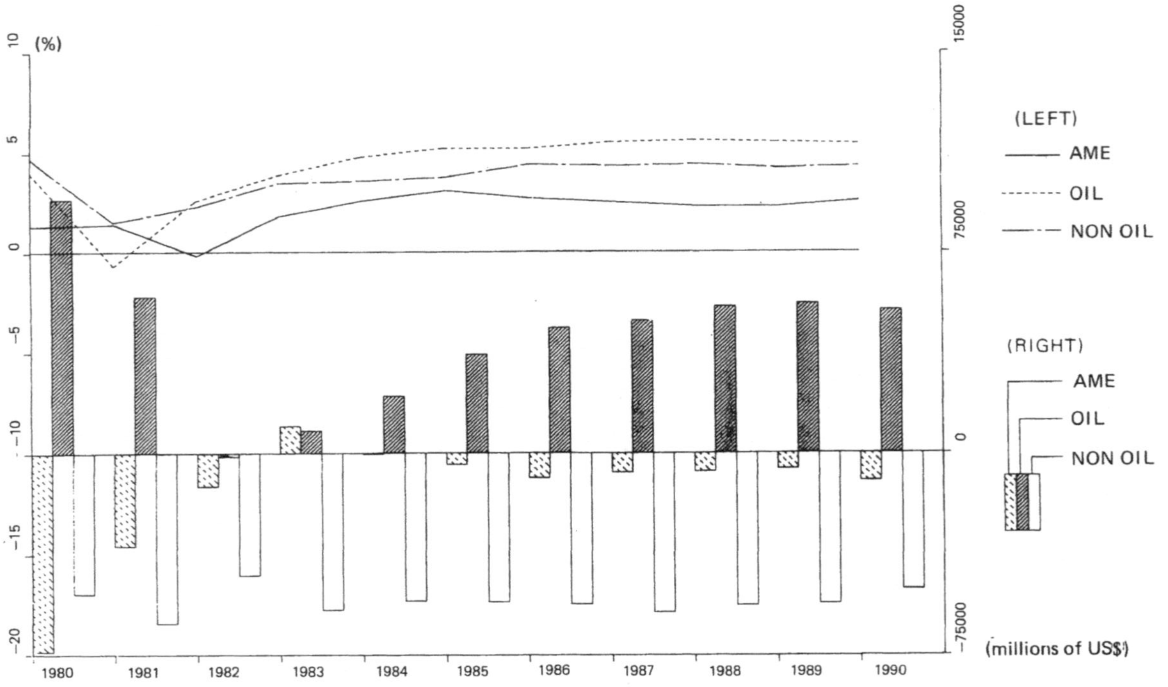

FIGURE IIB Annual average growth rates of real GDP and current balance of payments (Scenario B).

TABLE IIC

Projections of the World Economy, 1983 — 1990 (Scenario C)
Annual average growth rates of real GDP

	70-80	1980	1981	1982	1983	1984	1985	80-85	85-90	80-90
WORLD	3.9	2.1	1.6	.7	2.3	3.5	3.6	2.3	3.3	2.8
WORLD (EXCLUDING CPE)	3.7	1.9	1.3	.3	2.3	3.5	3.6	2.2	3.1	2.6
AME	3.3	1.3	1.4	-.2	1.9	3.0	3.4	1.9	2.7	2.3
OECD	3.2	1.3	1.3	-.2	1.9	3.0	3.4	1.9	2.6	2.3
EC	2.9	1.3	-.6	.2	.6	2.0	2.9	1.0	2.6	1.8
OTHER AME	4.4	4.6	4.2	1.6	2.1	2.5	2.8	2.6	3.8	3.2
DME	5.7	4.5	.7	2.4	3.7	5.6	4.6	3.4	4.8	4.1
OIL EXPORTING COUNTRIES	5.9	4.0	-.7	2.6	3.9	4.5	5.3	3.1	5.4	4.3
OPEC IN MIDDLE EAST	6.7	.9	-10.1	4.2	4.5	4.8	5.4	1.6	5.3	3.4
NON OIL EXPORTING COUNTRIES	5.6	4.7	1.5	2.3	3.5	6.2	4.2	3.5	4.4	4.0
NICS	7.8	6.8	1.8	1.2	3.4	8.4	5.3	4.0	5.6	4.8
ASIAN NICS	9.0	3.4	7.7	5.6	6.1	7.0	6.0	6.5	6.2	6.3
LATIN AMERICA NICS	7.5	7.8	.1	-.1	2.6	8.8	5.1	3.2	5.4	4.3
ASIA	5.4	4.4	5.1	4.0	4.5	4.5	4.4	4.5	4.6	4.5
EAST ASIA	9.0	2.6	7.5	5.5	6.1	7.1	5.8	6.4	6.1	6.2
ASEAN	7.4	7.7	6.8	4.5	5.9	5.9	6.0	5.8	6.3	6.1
OTHER ASIA AND PACIFIC	2.8	2.9	2.5	2.9	2.6	2.0	2.1	2.4	1.9	2.1
MIDDLE EAST	6.7	1.1	-8.3	4.2	4.5	4.9	5.5	2.0	5.3	3.6
AFRICA	4.6	4.8	3.3	3.0	3.5	4.0	3.8	3.5	4.0	3.8
LATIN AMERICA AND CARIBBEAN	6.1	5.9	.4	.3	2.7	7.4	4.7	3.1	5.1	4.1
CPE	5.1	3.4	3.0	2.4	2.6	3.4	3.5	3.0	4.1	3.5
USSR AND EAST EUROPE	4.6	2.7	2.8	1.8	1.9	2.6	2.7	2.4	3.0	2.7
CHINA AND OTHER ASIAN CPE	6.3	4.9	3.7	3.8	4.2	5.3	5.4	4.5	6.2	5.3

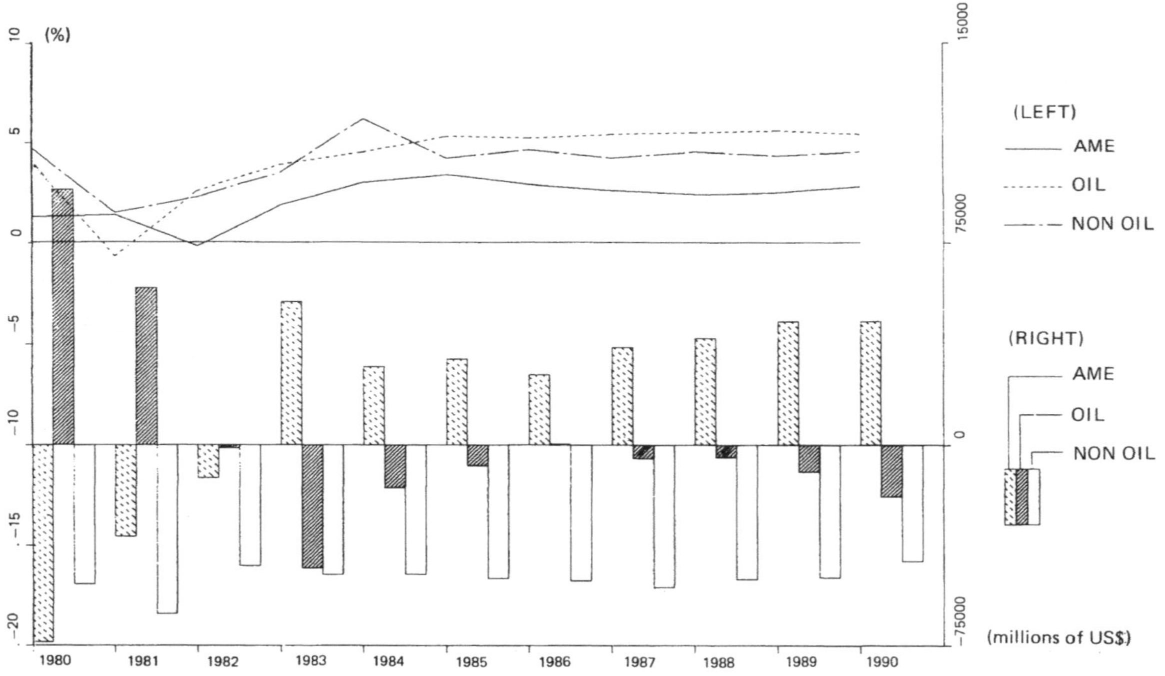

FIGURE IIC Annual average growth rates of real GDP and current balance of payments (Scenario C).

surplus is estimated to be getting smaller in its absolute terms. The ratio of a surplus of current account balance to GDP for the OPEC in the Middle East in nominal terms will be inclined to lower as 1.6% (A), 5.8% (B) and –0.9% (C) in 1990. In the absolute amount, its current balance of payments in 1990 is estimated to be 13.1 billion dollars (A), 49.8 billion dollars (B), and minus 6.9 billion dollars (C).

As a result, the oil exporting countries' (OPEC + Mexico) current balance of payments will be 5.7 billion dollars (A), 53.4 billion dollars (B) and minus 19.4 billion dollars (C) respectively in 1990.

Then, the deficit of current account balance in oil-exporting countries is likely to reflect on the developed countries in the form of a surplus of current balance of payments.

Therefore, worsening of current balance of payments and difficulties of speeding up development pace in the developing countries will likely be an important problem in the 1980s international society. Thus, an increase in financing to developing countries will provide an emergent task, together with energy conservative pattern of growth and steady development of alternative energy sources.

Second, using FUGI-GIOM developed by Kaya, we may ask what will be implications for energy conservation in terms of changes in the agro-industrial structures of the developed and the developing regions in the world under the suppositions of minimizing domestic oil consumption in the oil importing countries within a given framework of macroeconomic figures derived from FUGI-GMEM projection.

In this exercise, the domestic oil consumption in the oil importing countries in East and Southeast regions is supposed to be minimized under the constraints of global interdependence.

According to the GIOM calculations, share of energy consuming heavy and chemical industries will be shrinked while the energy conservative agriculture and services will be enlarged, although optimal changes in the agro-industrial structures may differ from country to country because of differences in the inter-industrial structures and final demand patterns.

It is, however, interesting to note that energy conservation effects by the agro-industrial changes alone may not be so large in the developed industrial countries as well as in the developing countries (around 5-10% conservation rate).

In order to conserve energy and reduce oil consumption, development of alternative energy and appropriate energy conservative technology will be likely to play much greater role even in the energy conservative agriculture.

CONCLUSION

There would be a dilemma between development of alternative energy coupled with energy-saving technology and the oil pricing. The higher will be the oil prices, the more rapid development of alternative energy and conservative technology will be expected. On the other side, too high oil prices will induce stagflation in the developed industrial countries and increase difficulties of the oil importing developing countries. Therefore, a *moderate* and *foreseeable* increase in the OPEC oil prices would be a precondition for the development of alternative energy and conservative technology based on new life style.

On account of high pricing of the OPEC oil during 1979-80, rapid replacement of oil with alternative energy sources coupled with energy conservative technology and consumer behaviours has taken place in the oil importing developed industrial countries, in particular, like Japan.

In the developing countries, energy conservative strategy for development will be emphasized and agro-based industrial development with appropriate technology will be highlighted in the 1980s. In order to achieve self-sufficiency of foods, new strategy for agricultural development with soft energy path will be a prerequisite condition from global ecological view point. Therefore, appropriate technology transfer coupled with own research and development of domestic technology in the developing countries should be incorporated into the policy scenario of the North-South coexistance and cooperation in the 1980s.

Acknowledgement

The author is grateful to his project team members of global modelling, in particular, Prof. E. Gamo, Mr. K. Aoki, Mr. T. Imoto, Mr. H. Ishii, Mr. O. Nakamura, Mr. A. Kosaki, Dr. I. Takahashi and Mr. K. Tamashiro for their cooperation.

References

Related to the whole project FUGI.

1. Kaya, Y., Onishi, A., and Kohno, A. (1974). North-South Economic Relations and Industry Transfer. Presented to the Club of Rome Annual Meeting at West Berlin (October).
2. Kaya, Y., Onishi, A. and Ishitani, H. (1977). A two-lvel multi-nation model for development planning. *MOIRA: Food Agriculture Model.* Proceeding of the Third IIASA Symposium on Global Modelling. pp. 85-149 (February).
3. Kaya, Y., and Onishi, A. (1980). FUGI-Future of Global Interdependence. *Input-Output Approaches in Global Modelling.* Proceedings of the Fifth IIASA Symposium on Global Modelling, Gerhart Bruckmann, editor. IIASA Proceedings Series 9: Pergamon Press, pp. 91-357.
4. Onishi, A. (1982). *The Australia – Japan Economic Relations in the Global Perspectives,* 1981-1990. A Report to the Australia – Japan Foundation. Tokyo (March).

Related mainly to GMEM

1. Onishi, A. (1977). Global economic model for new international order. *Hiroshima Peace Science.* Institute for Peace Science, Hiroshima University. pp. 205-233 (January).
2. Onishi, A. (1977). Using a multi-nation economic model: Projections of economic relations between Japan and developing countries in Asia (1975-1985). *Technological Forecasting and Social Change.* pp. 121-142 (October).
3. Onishi, A. (1981). Projections of alternative paths of development of the ESCAP countries in the Third United Nations Development Decade, 1981-1990. The Soka Economic *Studies Quarterly,* X No. 2, pp.175-227 (March).
4. Onishi, A. (1981). Economic forecasts for Japan and other Asian countries in the global perspective 1981-1985 – A scenario forecast of the OPEC long-term strategy for oil pricing. – *Prevision et analyse economique,* 2, No. 2, pp. 7-37 (avril-juin).
5. Onishi, A. (1982). The FUGI Global Macroeconomic Model and World Trade to 1990. (To be published by *Futures,* April 1983.)
6. Onishi, A. (1982). Methodology on Global Economic Modelling – Global Economic Modelling System –, The Institute of Applied Economic Research, Soka University, *Discussion Paper No. 82-01,* p. 47 (September).
7. Onishi, A. (1982). Long term projections of the Japanese economy in the global context, 1981-1990. The Institute of Applied Research, Soka University. *Discussion Paper No. 82-04,* p. 40 (September).
8. Onishi, A. (1982). The Theoretical Stucture of the FUGI Global Macroeconomic Model Type V 011-62, *Discussion Paper No. 82-02,* p. 57 (October).

Appendix

I. Country and regional subgroupings of the global macroeconomic model

World
World (excluding CPE)
AME
  OECD
    EC
  Other AME
DME
  Oil exporting countries
    OPEC in Middle East
  Non oil exporting countries
  NICS
    Asian NICS
    Latin America NICS
    Asia
    East Asia
    Asean
    Other Asia and Pacific
  Middle East
  Africa
  Latin America and Caribbean
CPE
  USSR and East Europe
  China and other Asian CPE

1 Japan
2 Australia
3 New Zealand
4 Canada
5 United States
6 Belgium and Luxembourg
7 Denmark
8 France
9 Germany, Federal Republic of
10 Ireland
11 Italy
12 Netherlands
13 United Kingdom
14 Greece
15 Austria
16 Finland
17 Norway
18 Portugal
19 Spain
20 Sweden
21 Switzerland

22 Turkey
23 Yugoslavia
24 Iceland
25 South Africa
26 Other Industrial Countries
27 China (Taiwan)
28 Hong Kong and Macao
29 Korea, Republic of
30 Singapore
31 Indonesia
32 Malaysia
33 Philippines
34 Thailand
35 Afghanistan, Bhutan and Nepal
36 Bangladesh and Pakistan
37 India
38 Sri Lanka and Maldives
39 Other Asia and Pacific countries
40 Iran
41 Iraq
42 Kuwait
43 Saudi Arabia
44 Other OAPEC
45 Other Middle East countries
46 Algeria
47 Egypt
48 Gabon and Nigeria
49 Libya
50 Other African countries
51 Argentina
52 Brazil
53 Colombia
54 Ecuador
55 Mexico
56 Venezuela
57 Other Latin Americ & Caribbean
58 USSR
59 East European countries
60 China
61 Korea, D.P. Republic of
62 Other socialist countries

IIB Model Specification

```
**************************************************************
* FUGI GLOBAL MACRO ECONOMIC MODEL (GMEM) TYPE IV 011-62 *
**************************************************************

                                              BY AKIRA ONISHI
                                              JANUARY 3, 1982

+----------------------------------------+
: I. PRODUCTION WITH ENERGY CONSTRAINTS :
+----------------------------------------+

  1-1.1. LOG(GDPP#/LCLF@) <AME> = F( (+) LOG(NHFCS#/LCLF@),
                                     (+) LOG(SUMT5(RD#)/LCLF@),
                                     (+) LOG(SUMT5(NHI#)/NHFCS#),
                                     (-) LOG(PEO.1*FERSI.1/PGDP.1) )
  1-1.2. LOG(GDPP#/CEMP) <AME> = F( (+) LOG(SUMT5(NHI#)/CEMP),
                                    (+) LOG(SUMT5(RD#)/CEMP),
                                    (+) LOG(SUMT5(NHI#)/NHFCS#),
                                    (-) LOG(PEO.1*FERSI.1/PGDP.1) )
  1-2.   LOG(GDPP#/CEMP) <DME> = F( (+) LOG(SUMT5(NHI#)/CEMP),
                                    (+) LOG(SUMT5(GED#.1)/CEMP),
                                    (-) LOG(PEO.1*FERSI.1/PGDP.1) )
 1-3.    LOG(GDPP#/CEMP) <CPE> = F( (+) LOG(SUMT5(NHI#)/CEMP) )
 2-1. RD#        = F( (+) (OS.1-TYC.1)/PNHI.1, (+) (GDF 1+GES.1)/PNHI.1 )
 2-2. RD#        = F( (+) GDP#.1, (+) (GDF.1+GEF.1)/PNHI.1 )
 3. DNHC#        = F( (+) NHFCS#.1 )
 4. UNEMPR       = F( (+) LOG(WSEI.1/CPI.1), (-) LOG(NHI#.1/GDP#.1),
                      (+) LOG(LCLF@), (-) LOG(GDP#.1/GDP#.2) )
 5. LW           = F( (+) CEMP, (+) LW.1 )
 6. LOG(IPI)     = F( (+) LOG(GFCF#), (+) LOG(E#), (+) LOG(GDP#-GFCF#-E#),
                      (+) LOG(SUMT5(GDP#/NP)/5) )
 7. LOG(ENGYR#) = F( (+) LOG(GDP#), (-) LOG(PEO*FERSI/PGDP) )
 8. ENGYS#       = ME#OIL + DS#OIL@ + AENGY#@
 9. ENGYR#       = MIN(ENGYR#,ENGYS#MIN@)
10. LAMBDAHAT    = ( SUMT3(DOT(ENGYS#.1) / DOT(GDP#.1))) / 3        <FOR AME>
11. LAMBDA       = DOT(ENGYS#) / DOT(GDP#)                          <FOR AME>
12. DOT(GDPC#)   = (1 - (LAMBDAHAT - LAMBDA)) * DOT(GDPP#)          <FOR AME>
13-1. GDPP#      = (GDPP#/LCLF)*LCLF                               <FOR AME>
13-2. GDPP#      = (GDPP#/CEMP)*CEMP                               <FOR AME>
14. GDPC#        = P1*(GDPP#/CEMP) * CEMP                     <FOR DME & CPE>
15. GDP#         = GDPC#                                            <FOR CPE>
16. CUR          = GDP# / GDPC#
17. CEMP         = (1 - UNEMPR) * LCLF@
18. GED#         = GED / PCG
19. NHFCS#       = NHFCS#.1 + NHI# - DNHC#

            ** <AME> : DEVELOPED MARKET ECONOMIES. **
            ** <DME> : DEVELOPING MARKET ECONOMIES. **
            ** <CPE> : CENTRALLY PLANNED ECONOMIES. **

+-----------------------------------------------+
: II. EXPENDITURE ON GDP < AT CONSTANT PRICES > :
+-----------------------------------------------+

  1-1.1. E#MAT <AME,AME> = F( (+) GDP#<J>.1, (-) PES.1<I>/PESAME.1,
                              (-) PES<I>.1*FERSI<J>.1/PGDP<J>.1 )
  1-1.2. E#MAT <ANICS,AME> = F( (+) GDP#<J>.1, (-) PES<I>.1/PESAME.1,
                                (-) PES<I>.1*FERSI<J>.1/PGDP<J>.1 )
  1-1.3. E#MAT <DME,AME> = F( (+) GDP#<J>.1, (+) E#MAT<I,J>.1 )
  1-1.4. E#MAT <OPEC,AME> = F( (+) GDP#<J>.1, (-) PES<I>.1/PESAME.1 )
  1-1.5. E#MAT <CPE,AME> = F( (+) GDP#<J>.1, (+) E#MAT<I,J>.1 )
  1-2.1. E#MAT <AME,ANICS> = F( (+) GDP#<J>.1, (-) PES<I>.1/PESAME.1,
                                (+) CAPM<J>.1 )
  1-2.2. E#MAT <ANICS,ANICS> = F( (+) GDP#<J>.1, (-) PES<I>.1/PESAME.1,
                                (+) CAPM<J>.1 )
  1-2.3. E#MAT <DME,ANICS> = F( (+) GDP#<J>.1, (+) CAPM<J>.1 )
  1-2.4. E#MAT <CPE,ANICS> = F( (+) GDP#<J>.1, (+) CAPM<J>.1 )
  1-3.   E#MAT <WORLD,DME> = F( (+) GDP#<J>.1, (+) CAPM<J>.1 )
  1-4.   E#MAT <WORLD,CPE> = F( (+) GDP#<J>.1, (+) E#<J>.1*PES<J>.1/PMS<J>.1 )
            **  E#MAT<I,J> RELATES TO EXPORTS FROM REGION <I> TO REGION <J>  **
  2-1.1. CP# <AME> = F( (+) (DFI#-(TPI+TYC)/CPI), (-) CPI/((CPI+CPI.1)/2),
                        (-) IC.1, (+) CP#.1 )               < JAPAN TYPE >
  2-1.2. CP# <AME> = F( (+) (COMPE#-(TPI/CPI)), (+) (OS#-(TYC/CPI)),
                        (-) (1+DOT(CPI)-CPI/((CPI+CPI.1)/2)),
                        (-) (IC.1-DOT(CPI.1)*100), (+) CP#.1 )  < USA TYPE >
  2-2.1. CP# <ANICS> = F( (+) DFI#, (-) CPI/((CPI+CPI.1)/2), (-) IC.1, (+) CP#.1 )
  2-2.2. CP# <DME> = F( (+) GDP#, (-) CPI/((CPI+CPI.1)/2), (+) CP#.1 )
```

```
2-2.3. CP# <CPE> = GDP# - E# + M# - CG# - GFI# - NHI# - HI# - IIS#
3-1.1. NHI# <AME> = F( (+) OS#.1, (+) OS#.2, (-) IP.1, (+) (NHI#.6+NHI#.7)/2,
                       (-) PEO.1*FERSI.1/WPI.1 )                 < JAPAN TYPE >
3-1.2. NHI# <AME> = F( (+) (OS#.1-TYC.1/PNHI.1), (+) (OS#.2-TYC.2/PNHI.2),
                       (-) IP.1, (+) (NHI#.6+NHI#.7)/2, (-) PEO.1*FERSI.1/WPI.1 )
                                             < USA. , WEST GERMANY TYPE >
3-2.1. NHI# <ANICS> = F( (+) OS#.1, (+) OS#.2, (-) IP.1, (+) (NHI#.6+NHI#.7)/2,
                         (-) PEO.1*FERSI.1/WPI.1 )
3-2.2. NHI# <ANICS> = F( (+) (OS#.1-TYC.1/PNHI.1), (+) (OS#.2-TYC.2/PNHI.2),
                         (-) IP.1, (+) (NHI#.6+NHI#.7)/2,
                         (-) PEO.1*FERSI.1/WPI.1 )
3-3.1. NHI# <OPEC> = F( (+) GDP#.1, (+) E#.1*PE.1/PM.1, (+) FCI.1 )
3-3.2. NHI# <ODME> = F( (+) GDP#.1, (+) CAPM.1 )
3-4.   NHI# <CPE> = F( (+) GDP#.1, (+) E#.1*PE.1/PM.1 )
4. HI#    = F( (+) (DFI#-TPI/CPI), (+) (NP@-NP.1), (-) IH.1, (-) PHI.1, (+) HI#.1 )
5-1. IIS# <AME> = F( (+) GDP#, (-) IP.1, (+) WPI/((WPI+WPI.1)/2), (-) IS#.1 )
5-2. IIS# <ANICS> = F( (+) GDP#, (-) IP.1, (+) WPI/((WPI+WPI.1)/2), (-) IS#.1 )
5-3. IIS# <DME> = GDP# - E# + M# - CP# - CG# - GFI# - NHI# - HI#
6. E#     = F( E#MAT<SUMJ>+(SCTI+SCTR)/PES )
7. M#     = F( E#MAT<SUMI>+(SDTI+SDTR)/PMS )
8. CG#    = CG / PCG
9. GFI#   = GFI / PGFI                    < ONLY FOR JAPAN >
10. GDP#  = E# - M# + CP# +CG# + GFI# + NHI# + HI# + IIS#   <FOR AME & ANICS>
**  GDP#  = MINMAX(GDP#,GRMIN@,GRMAX@)
11. IS#   = IS#.1 + IIS#
12. GFCF# = GFI# + NHI# + HI#
13. CAPM  = (E#*PES - Z*ME#OIL*PEO + FCI ) / PMS
                                    < IN CASE OF   OPEC & MEXICO  Z = 0 >
                                    < IN CASE OF   OTHER DME  Z = 1 >
```

```
+--------------------+
: III. PROFIT - WAGE :
+--------------------+
```

```
1-1. OS#        = F( (+) (GDP#-COMPE#), (-) IP.1, (+) WPI/PM )
1-2. OS#        = F( (+) GDP#, (-) IP.1, (+) WPI/PM )
1-3. OS#        = F( (+) GDP#, (-) IP.1, (+) PE/PM )
2.   HPI        = F( (+) OS, (+) HPI.1 )
3.   IPC        = F( (+) OS )
4.   HEI        = F( (+) OS )
5-1. DOT(WSEI)  = F( (+) DOT(CPI.1), (+) DOT(LPI), (+) DOT(OS.1/GDP.1),
                     (-) UNEMPR.1 )
5-2. WSEI       = F( (+) CPI.1, (+) LPI, (+) OS.1/GDP.1, (-) UNEMPR.1 )
**   WSEI       = MINMAX(DOT(WSEI),WSEIDOTMIN@,WSEIDOTMAX@)
6.   LPI        = (GDPC#/CEMP) / (GDPC#.0/CEMP.0)
7.   OS         = OS# * PNHI
8.   COMPE      = WSEI * LW * (COMPE.0/LW.0)
9.   COMPE#     = COMPE / CPI
10.  DFI#       = COMPE# + OS#
11.  DFI        = COMPE + OS
12.  STDC       = GDP - DFI - DFC - TI + SUB
```

```
+------------+
: IV. PRICES :
+------------+
```

```
1. DOT(WPI)  = F( (+) DOT(PM), (+) DOT(WSEI/LPI), (+) DOT(IV#.1),
                  (+) DOT(WPI.1) )
2. DOT(CPI)  = F( (+) DOT(WPI.1), (+) DOT(WSEI), (+) DOT(PM),
                  (+) DOT(IV#.1), (+) DOT(CPI.1) )
3. DOT(PCP)  = F( (+) DOT(WPI.1), (+) DOT(WSEI), (+) DOT(PM),
                  (+) DOT(IV#.1), (+) DOT(PCP.1) )
4. DOT(PCG)  = F( (+) DOT(WPI.1), (+) DOT(WSEI), (+) DOT(IV#.1),
                  (+) DOT(PCG.1) )
5. DOT(PNHI) = F( (+) DOT(WPI), (+) DOT(NHI#.1/GDP#.1), (+) DOT(PNHI.1) )
6. DOT(PGFI) = F( (+) DOT(PNHI.1), (+) DOT(PGFI.1) )    < ONLY FOR JAPAN >
7. DOT(PHI)  = F( (+) DOT(PNHI), (+) DOT(WSEI), (+) DOT(PHI.1) )
8-1.1.  PE <AME> = F( (+) WPI.1, (+) WSEI/LPI, (+) PESAME.1, (+) FERSI )
                                                          < JAPAN TYPE >
8-1.2.  PE <AME> = F( (+) WPI.1, (+) WSEI/LPI, (+) PEC, (+) FERSI, (+) PE.1 )
8-2. DOT(PE) <ANICS> = F( (+) DOT(WPI.1), (+) DOT(WSEI/LPI), (+) DOT(PEC),
                          (+) DOT(PM.1) )
8-3. DOT(PE) <OPEC> = F( (+) DOT(PEO) )
8-4. DOT(PE) <ODME> = F( (+) DOT(PEC), (+) DOT(PE.1) )
8-5. DOT(PE) <CPE> = F( (+) DOT(PEC), (+) DOT(PEO), (+) DOT(PE.1) )
          ** AME,DME,CPE & OPEC RELATE TO EXPORTING REGION<I>. **
9. DOT(PENA) = F( (+) DOT(PE) )
10. PEC        = F( (+) PESAME.1, (+) GDP#AME.1/((GDP#AME.1+GDP#AME.2)/2),
                    (-) ICUS.1 )
```

```
11-1. DOT(PEO) = F( (+) DOT(GIFI10,1), (+) DOT(GDP#I10,1), (+) DOT(FERSII10,1),
                    (+) FS01, (+) FS02 )          < MODIFIED OPEC FORMULA >
          ** FS01 : DUMMY VARIABLE ,  1974 = 1 ,  REST OF THE YEARS = 0
          ** FS02 : DUMMY VARIABLE ,  1980 = 1 ,  REST OF THE YEARS = 0
11-2. DOT(PEO) = DOT(GCI10,1)                     < OPEC LONG TERM STRATEGY >
      GCI10      = GIFI10 * (FERSII10) * GDP#I10
      GIFI10     = 2/3*PE10 + 1/3*CPI10
12. PES          = PE / FERSI
13. PMS          = ESMAT<SUMI> / E#MAT<SUMI>
14. PM           = PMS * FERSI
15. DOT(PMNA) = F( (+) DOT(PM) )
16. PGDP         = GDP / GDP#
17. IV#          = (M2/GDP#) / (M2,0/GDP#,0)

+--------------------------------------------+
: V. EXPENDITURE ON GDP < AT CURRENT PRICES > :
+--------------------------------------------+

  1. E      = PENA * E#
  2. M      = PMNA * M#
  3. CP     = PCP * CP#
  4. NHI    = PNHI * NHI#
  5. HI     = PHI * HI#
  6. IIS    = WPI * IIS#
  7. DNHC   = PNHI * DNHC#
  8. NHFCS  = NHFCS,1 + NHI - DNHC
  9. IS     = IS,1 + IIS
 10. GFCF   = GFI + NHI + HI
 11. GDP    = E - M + CP + CG + GFI + NHI + HI + IIS

+-----------+
: VI. MONEY :
+-----------+

  1-1. M1   = F( (+) GDP, (+) (DLB+DLG), (+) CBT, (-) IN,1 )       < JAPAN TYPE >
  1-2. M1   = F( (+) GDP, (+) (DLB+DLG), (+) CBT, (-) (IB,1-ITD,1)/ITD,1 )
                                                                  < USA. TYPE >
  ** M1   = MINMAX(DOT(M1),M1DOTMIN@,M1DOTMAX@)
  2. MTD  = F( (+) LB,1, (+) (GDP-CP-CG), (-) (IB,1-ITD,1)/ITD,1, (+) MTD,1 )
  3. DLB  = F( (+) (NHI+HI+IIS), (-) IN,1, (+) MTD,1/GDP,1,
               (-) (IB,1-IP,1)/IP,1 )
  4. DLG  = F( (+) (GE-R) )
  5. DM1  = M1 - M1,1
  6. DMTD = MTD - MTD,1
  7. DM2  = DM1 + DMTD
  8. M2   = M2,1 + DM2
  ** M2   = MINMAX(DOT(M2),M2DOTMIN@,M2DOTMAX@)
  9. LB   = LB,1 + DLB
 10. LG   = LG,1 + DLG

+--------------------+
: VII. INTEREST RATES :
+--------------------+

  1. IP     = F( (+) IC, (+) (LB+LG)/MTD, (+) IP,1 )
  2. IH     = F( (+) IP, (+) IH,1 )
  3-1. IC   = F( (+) IN, (+) DOT(CPI), (-) M1/GDP, (+) IC,1 )
                                            < JAPAN , WEST GERMANY TYPE >
  3-2. IC   = F( (+) IN, (+) DOT(CPI), (+) (LB+LG)/MTD, (+) IC,1 )
                                                     < USA. , UK. TYPE >
  4. ITD    = F( (+) IP, (-) MTD/GDP, (+) ITD,1 )
  ** ITD    = MINMAX(ITD,ITDMIN@,ITDMAX@)
  5. IB     = F( (+) IP, (+) LG/GDP, (-) (IB,1-ITD,1)/ITD,1, (+) IB,1 )
  6-1. IN   = F( (+) ( 1+DOT(CPI)-CPI/((CPI+CPI,1)/2)),
                 (+) ( 1+DOT(GDP#)-GDP#/((GDP#+GDP#,1)/2)),
                 (+) FERSI, (+) IN,1, (+) INUS )
  6-2. IN   = F( (+) ( 1+DOT(CPI)-CPI/((CPI+CPI,1)/2)),
                 (+) ( 1+DOT(GDP#)-GDP#/((GDP#+GDP#,1)/2)),
                 (+) 1/FERSIAME, (+) IB,1 )          < ONLY FOR USA. >
  7. IEURO = F( (+) ICUS )


+-------------------------+
: VIII. GOVERNMENT FINANCE :
+-------------------------+

<< REVENUE >>
  1. TPI  = F( (+) (COMPE,1 + HEI,1 + HPI,1) )
  2. TYC  = F( (+) IPC,1 )
  3. SSC  = F( (+) COMPE,1, (+) IPC,1 )
  4. TP   = F( (+) GDP,1, (+)  TP,1 )
```

```
  5. TDGS = F( (+) GDP.1, (+) TDGS.1 )
  6. TITT = F( (+) (M.1 + E.1), (+) TITT.1 )
  7. TIR  = F( (+) GDP.1, (+) TIR.1 )
  8. RR   = F( (+) GDP.1, (+) RR.1 )
  9. DGD  = F( (+) (GE + GLR@ - RG) )              ** GDOS<0
 10. TD   = TPI + TYC + TOU@ + TEPM@ + TP
 11. TI   = TDGS + TITT + TIR
 12. R    = TD + TI + SSC + RR
 13. RG   = R + RC@ + GR@
 14. GD   = GD.1 + DGD

<< EXPENDITURE BY FUNCTION >>
 15. GE   = GPS + GDF + GED + GH + GSW + GHC + GSS + GES + GEOP@
    15-1     GPS = F( (+) R, (-) LG.1, (+) GPS.1 )
    15-2     GDF = F( (+) R, (-) LG.1, (+) GDF.1 )
    15-3     GED = F( (+) R, (-) LG.1, (+) GED.1 )
    15-4     GH  = F( (+) R, (-) LG.1, (+) GH.1  )
    15-5     GSW = F( (+) R, (-) LG.1, (+) GSW.1 )
    15-6     GHC = F( (+) R, (-) LG.1, (+) GHC.1 )
    15-7     GSS = F( (+) R, (-) LG.1, (+) GSS.1 )
    15-8     GES = F( (+) R, (-) LG.1, (+) GES.1 )
 16. CG   = F( (+) GE )
 17. GFI  = F( (+) (GHC+GES) )            < ONLY FOR JAPAN >
 18. GDOS = RG - (GE + GLR@)

<< EXPENDITURE BY ECONOMIC TYPE >>
 19. EGS  = F( (+) GE )
 20. GIP  = F( (+) GD.1*IB.1/100 )
 21. CAE  = F( (+) (GHC+GES) )
 22. CUE  = GE - CAE
 23. SUB  = CUE - EGS - GIP

+---------------------------------------+
: IX. INTERNATIONAL BALANCE OF PAYMENTS :
+---------------------------------------+

  1. ETFOB     = F( EDTFOB )
  2. MTFOB     = F( MDTFOB )
  3. SCTI      = F( (+) E#MAT<SUMJ>*FRI )
  4. SDTI      = F( (+) E#MAT<SUMI>*FRI )
  5. SCII      = F( (+) DIAO, (+) (POINA.1+POINA.2)*IEURO.1 )
  6. SDII      = F( (+) DILO, (+) (POINL.1+POINL.2)*IB.1 )
  7. SCTR      = F( (+) FERSI, (+) DOT(GDP#.1), (-) DOT(CPI),
                    (+) GDPS.1/NP.1 )
  8. SDTR      = F( (-) FERSI, (+) GDPS.1/NP.1, (+) DFI#.1*CPI.1/FERSI.1 )
  9. DOT(FRI) = F( (+) DOT(PEC) )
 10. DIA       = F( (+) GDPS, (+) (WSEI/LPI) / (WSEIAME/LPIAME), (+) OS/GDP )
 11. DIL       = F( (+) GDPS, (-) (WSEI/LPI)/(WSEIAME/LPIAME),
                    (+) (OS/GDP)/(OSAME/GDPAME)
 12-1. POINA     = F( (+) GDPS, (+) (ICAME-IC)/IC, (+) CBP, (+) OS/GDP,
                      (+) FV20 )                            < JAPAN TYPE >
 12-2. POINA     = F( (+) GDPS, (+) (IBAME-IB)/IB, (+) CBP, (+) OS/GDP,
                      (+) FV20 )                            < NETHERLAND TYPE >
 13-1. POINL     = F( (+) GDPS, (-) (ICAME-IC)/IC, (-) CBP, (-) FV20 )
                                                          < JAPAN TYPE >
 13-2. POINL     = F( (+) GDPS, (-) (IBAME-IB)/IB, (-) CBP, (-) FV20 )
                                                          < NETHERLAND TYPE >
 13-3. POINL     = F( (+) (DOT(GDP#)-DOT(GDP#AME)), (-) (ICAME-IC)/IC, (-) FV20 )
                                                     < USA., WEST GERMANY TYPE >
     ** FV20 = (1+DOT(FERSI/FERSIAME)
              -(FERSI/FERSIAME)/(((FERSI/FERSIAME)+(FERSI.1/FERSIAME.1))/2))
 14. ESMAT     = PES<I> * E#MAT
 15. EDTFOB    = ESMAT<SUMJ>
 16. MDTFOB    = ESMAT<SUMI>
 17. TB        = ETFOB - MTFOB
 18. SC        = SCTI + SCII + SCTR + SCOT@
 19. SD        = SDTI + SDII + SDTR + SDOT@
 20. CBT       = TB + SC - SD
 21. CBP       = CBT + UTPB + UTGB
 **  CBP/GDPS  = MINMAX(CBP/GDPS, CBP/GDPS.MIN@, CBP/GDPS.MAX@)
 22. UTPB      = UTPC@ - UTPD@
 23. UTGB      = UTGC@ - UTGD@
 24. BBP       = CBP + LTCB
 25. LTCB      = DIB + POINB + OLCB
 26. DIB       = DIL - DIA
 27. POINB     = POINL - POINA
 28. DIAO      = DIAO.1 + DIA
 29. DILO      = DILO.1 + DIL
 30. OLCB      = OLCL@ + OLCA@
 31. DIOS      = DIOS.1 + PFDI - FD@
 32. GDPS      = GDP / FERSI
```

```
+-----------------------------------------------+
: X. OFFICIAL DEVELOPMENT ASSISTANCE AND         :
:      PRIVATE DIRECT FOREIGN INVESTMENT TO LDC :
+-----------------------------------------------+

  1. ODA      = F( (+) GDPS )
  2. ODAB     = F( (+) ODA )
  3. ODAM     = ODA - ODAB
  4. ODAMAT   = OMEGAMAT * ODAB
  5. AM       = F( (+) SUM(ODAM) )
  6. PFDI     = F( (+) DIA )
  7. PFDIMAT  = KAPPAMAT * PFDI
  8. FCI      = ODAMAT<SUMI> + PFDIMAT<SUMI> + AM + AC@

+---------------------------+
: XI. FOREIGN EXCHANGE RATE :
+---------------------------+

     DOLLAR RATE
     -----------
1-1. FERSI = F( (+) PE/PEUS, (+) CBPUS.1/GDPUS.1, (-) CBP.1/GDPS.1,
                (+) (IBUS.1-IB.1)/IB.1, (+) FERSIAME.1 )  < JAPAN TYPE >
1-2. FERSI = F( (+) PE/PEUS, (+) CBPUS.1/GDPUS.1, (-) CBP.1/GDPS.1,
                (+) (INUS.1-IN.1)/IN.1 )
  ** FERSI = MINMAX(FERSI,FERSIMIN@,FERSIMAX@)

          +++  NOTES +++


  #          : VARIABLE AT CONSTANT PRICES ( EX. GDPP# )
  @          : EXOGENOUS VARIABLE ( EX. LCLF@ )
  .1         : ONE YEAR TIME LAG ( EX. GDP#.1 )
  .0         : VALUE IN 1975 ( EX. CEMP.0 )
  MAT        : MATRIX VARIABLE ( EX. E#MAT ODAMAT )
  LOG        : NATURAL LOGARITHM
  SUMJ       : SUMMATION OF ROW ELEMENTS IN MATRIX
  SUMI       : SUMMATION OF COLUMN ELEMENTS IN MATRIX
  SUMT5      : SUMMATION OVER FIVE YEARS ( EX. SUMT5(NHI#)=NHI#+NHI#.1+.. +NHI#.4 )
  AME        : REGION  1 TO REGION 26
  DME        : REGION 27 TO REGION 57
  CPE        : REGION 58 TO REGION 62
  ANICS      : REGION 27 TO REGION 30
  OPEC       : REGION 31,40-44,46,48,49,54,56
  ODME       : DME EXCLUDING ANICS & OPEC


                                ================================
                                 GMEMIV 011-62 - VARIABLE LIST
                                ================================

          ** EXPLANATORY NOTES **

      -FLAG.1-
         M : MATRIX
         N : SEMI-MATRIX
         V : VECTOR
         S : SCALAR

      -FLAG.2-
         T : TIME SERIES DATA
         N : NON TIME SERIES DATA

      -FLAG.3-
         D : ENDOGENOUS VARIABLE
         X : EXOGENOUS VARIABLE

      -FLAG.4-
         F : FLOW
         R : RATIO
         P : PRICE INDEX
         S : STOCK

      -UNIT-

         MD   : MILLIONS OF 1975 US DOLLARS AT CURRENT PRICES ( NATIONAL CURRENCY
                BASE)
         MD75 : MILLIONS OF 1975 US DOLLARS AT 1975 CONSTANT PRICES
         MDS  : MILLIONS OF CURRENT US DOLLARS
         75=1 : INDEX NUMBER BASED 1975
         %    : PERCENTAGE
         TP   : THOUSANDS OF PERSONS
         MT   : METRIC TON
```

VARIABLE	FLAG	UNIT	
AC	VTXF	MDS	ODA RECEIVED FROM CENTRALLY PLANNED ECONOMY ZONE AT CURRENT US DOLLARS
AENGY#	VTXF	MT	ALTERNATIVE ENERGY SUPPLY (METRIC TON)
AM	VTDF	MDS	ODA RECEIVED FROM MULTILATERAL AGENCIES AT CURRENT US DOLLARS
BBP	VTDF	MDS	BASIC BALANCE OF PAYMENTS AT CURRENT US DOLLARS
CAE	VTDF	MD	GOVERNMENT CAPITAL EXPENDITURE
CAPM	VTDF	MD75	CAPACITY TO IMPORT OF DEVELOPING COUNTRIES AT CONSTANT PRICES
CBP	VTDF	MDS	CURRENT BALANCE OF PAYMENTS AT CURRENT US DOLLARS
CBT	VTDF	MDS	CURRENT BALANCE OF TRADE AT CURRENT US DOLLARS
CEMP	VTDS	TP	CIVILIAN EMPLOYMENT
CG	VTDF	MD	GOVERNMENT FINAL CONSUMPTION EXPENDITURE AT CURRENT PRICES
CG#	VTDF	MD75	GOVERNMENT FINAL CONSUMPTION EXPENDITURE AT CONSTANT PRICES
COMPE	VTDF	MD	COMPENSATION OF EMPLOYEES AT CURRENT PRICES
COMPE#	VTDF	MD75	COMPENSATION OF EMPLOYEES AT CONSTANT PRICES
CP	VTDF	MD	PRIVATE FINAL CONSUMPTION EXPENDITURE AT CURRENT PRICES
CP#	VTDF	MD75	PRIVATE FINAL CONSUMPTION EXPENDITURE AT CONSTANT PRICES
CPI	VTDR	75=1	CONSUMER PRICE INDEX
CPI10	STDR	75=1	AVERAGE CONSUMER PRICE INDEX OF GROUP 10 INDUSTRIAL COUNTRIES
CUE	VTDF	MD	GOVERNMENT CURRENT EXPENDITURE
CUR	VTDR		CAPACITY UTILIZATION RATE
DFC	VTDF	MD	CONSUMPTION OF FIXED CAPITAL AT CURRENT PRICES
DFI	VTDF	MD	DOMESTIC FACTOR INCOMES AT CURRENT PRICES
DFI#	VTDF	MD75	DOMESTIC FACTOR INCOMES AT CONSTANT PRICES
DGD	VTDF	MD	INCREASE OF GOVERNMENT DEBT
DIA	VTDF	MDS	PRIVATE DIRECT INVESTMENT ABROAD AT CURRENT US DOLLARS
DIAO	VTDS	MDS	PRIVATE DIRECT INVESTMENT ABROAD AT CURRENT US DOLLARS, OUTSTANDING
DIB	VTDF	MDS	PRIVATE DIRECT INVESTMENT, BALANCE AT CURRENT US DOLLARS
DIL	VTDF	MDS	PRIVATE DIRECT INVESTMENT IN COUNTRY AT CURRENT US DOLLARS
DILO	VTDS	MDS	PRIVATE DIRECT INVESTMENT IN COUNTRY AT CURRENT US DOLLARS, OUTSTANDING
DIOS	VTDS	MDS	AME'S PRIVATE FOREIGN DIRECT INVESTMENT TO DME AT CURRENT US DOLLARS, OUTSTANDING
DLB	VTDF	MD	INCREASE OF CLAIMS ON PRIVATE SECTOR
DLG	VTDF	MD	INCREASE OF CLAIMS ON GOVERNMENT SECTOR, NET
DM1	VTDF	MD	INCREASE OF MONEY SUPPLY (M1)
DM2	VTDF	MD	INCREASE OF MONEY AND GUASI-MONEY (M1+MTD)
DMTD	VTDF	MD	INCREASE OF QUASI-MONEY (TIME AND SAVING DEPOSITS)
DNHC	VTDF	MD	DEPRECIATION OF NON-HOUSING CAPITAL AT CURRENT PRICES
DNHC#	VTDF	MD75	DEPRECIATION OF NON-HOUSING CAPITAL AT CONSTANT PRICES
DS#OIL	VTXF	MT	DOMESTIC SUPPLY OF OIL (METRIC TON)
E	VTDF	MD	EXPORTS OF GOODS AND SERVICES AT CURRENT PRICES
E#	VTDF	MD75	EXPORTS OF GOODS AND SERVICES AT CONSTANT PRICES
E#MAT	MTDF	MD75	EXPORTS(MERCHANDISE,FOB) FROM REGION (I) TO REGION (J) AT CONSTANT PRICES
EDTFOB	VTDF	MDS	EXPORTS(MERCHANDISE,FOB) AT CUR. US DOLLARS, SUMMATION OF ROW ELEMENTS IN ESMAT
EGS	VTDF	MD	GOVERNMENT CURRENT EXPENDITURE ON GOODS AND SERVICES
ENGYR#	VTDF	MT	ENERGY REQUIREMENTS (METRIC TON)
ENGYS#	VTDF	MT	ENERGY SUPPLY (METRIC TON)
ESMAT	MTDF	MDS	EXPORTS(MERCHANDISE,FOB) FROM REGION (I) TO REGION (J) AT CURRENT US DOLLARS
ETFOB	VTDF	MDS	MERCHANDISE, EXPORTS(FOB) AT CURRENT US DOLLARS
FCI	VTDF	MDS	FOREIGN CAPITAL INFLOW TO LDC AT CURRENT US DOLLARS
FD	VTXF	MDS	FADE OUT OF AME'S PRIVATE FOREIGN DIRECT INVESTMENT TO DME AT CURRENT US DOLLARS
FERSI	VTDR	75=1	FOREIGN EXCHANGE RATE INDEX (PER US DOLLAR)
FERSII10	STDR	75=1	AVERAGE EXCHANGE RATE INDEX OF GROUP 10 INDUSTRIAL COUNTRIES (INVERSE)
FERSIAME	STDR	75=1	AME'S WEIGHTED AVERAGE FOREIGN EXCHANGE RATE INDEX
FRI	STDR	75=1	FREIGHT RATE INDEX IN TERMS OF US DOLLAR
GCI10	STDR	75=1	GENERAL COMPOSITE INDEX OF GROUP 10 INDUSTRIAL COUNTRIES
GD	VTDS	MD	GOVERNMENT DEBT, OUTSTANDING
GDF	VTDF	MD	GOVERNMENT DEFENSE EXPENDITURE
GDOS	VTDF	MD	CURRENT GOVERNMENT DEFICIT OR SURPLUS
GDP	VTDF	MD	GROSS DOMESTIC PRODUCT AT CURRENT PRICES
GDPAME	STDF	MD	TOTAL GROSS DOMESTIC PRODUCTS OF AME REGION AT CURRENT PRICES
GDP#	VTDF	MD75	GROSS DOMESTIC PRODUCT AT CONSTANT PRICES
GDP#AME	STDF	MD75	TOTAL GROSS DOMESTIC PRODUCTS OF AME REGION AT CONSTANT PRICES
GDP#I10	STDR	75=1	AVERAGE REAL GDP INDEX OF GROUP 10 INDUSTRIAL COUNTRIES
GDPC#	VTDF	MD75	CAPACITY GROSS DOMESTIC PRODUCT AT CONSTANT PRICES
GDPP#	VTDF	MD75	POTENTIAL GROSS DOMESTIC PRODUCT AT CONSTANT PRICES
GDPS	VTDF	MDS	GROSS DOMESTIC PRODUCT AT CURRENT US DOLLARS
GE	VTDF	MD	TOTAL GOVERNMENT(CENTRAL) EXPENDITURE
GED	VTDF	MD	GOVERNMENT EDUCATION EXPENDITURE AT CURRENT PRICES
GED#	VTDF	MD75	GOVERNMENT EDUCATION EXPENDITURE AT CONSTANT PRICES
GEOP	VTXF	MD	GOVERNMENT EXPENDITURE ON OTHER PURPOSES
GES	VTDF	MD	GOVERNMENT EXPENDITURE ON ECONOMIC SERVICES
GFCF	VTDF	MD	GROSS FIXED CAPITAL FORMATION AT CURRENT PRICES

GFCF#	VTDF	MD75	GROSS FIXED CAPITAL FORMATION AT CONSTANT PRICES
GFI	STDF	MD	GOVERNMENT FIXED INVESTMENT AT CURRENT PRICES, ONLY FOR JAPAN
GFI#	STDF	MD75	GOVERNMENT FIXED INVESTMENT AT CONSTANT PRICES, ONLY FOR JAPAN
GH	VTDF	MD	GOVERNMENT HEALTH EXPENDITURE
GHC	VTDF	MD	GOVERNMENT EXPENDITURE ON HOUSING AND COMMUNITY AMENITIES
GIP	VTDF	MD	GOVERNMENT CURRENT EXPENDITURE ON INTEREST PAYMENTS
GLR	VTXF	MD	GOVERNMENT LENDING MINUS REPAYMENTS
GPS	VTDF	MD	GOVERNMENT EXPENDITURE ON GENERAL PUBLIC SERVICES
GR	VTXF	MD	GRANT REVENUE OF GOVERNMENT
GSS	VTDF	MD	GOVERNMENT EXPENDITURE ON OTHER COMMUNITY & SOCIAL SERVICES
GSW	VTDF	MD	GOVERNMENT EXPENDITURE ON SOCIAL SECURITY & WELFARE
HEI	VTDF	MD	ENTREPRENEURIAL INCOME IN HOUSEHOLDS
HI	VTDF	MD	HOUSING INVESTMENT AT CURRENT PRICES
HI#	VTDF	MD75	HOUSING INVESTMENT AT CONSTANT PRICES
HPI	VTDF	MD	PROPERTY INCOME RECEIVABLE IN HOUSEHOLDS, NET
IB	VTDR	%	GOVERNMENT LONG TERM BOND YIELD
IC	VTDR	%	CALL MONEY RATE OR FEDERAL FUND RATE
IEURO	STDR	%	EURO-DOLLAR RATE
IH	VTDR	%	HOUSING LOAN RATE
IIS	VTDF	MD	INCREASE IN STOCKS AT CURRENT PRICES
IIS#	VTDF	MD75	INCREASE IN STOCKS AT CONSTANT PRICES
IN	VTDR	%	CENTRAL BANK'S OFFICIAL DISCOUNT RATE
IP	VTDR	%	PRIME RATE
IPC	VTDF	MD	INCOME OF PRIVATE CORPORATION
IPI	VTDR	75=1	INDUSTRIAL PRODUCTIVITY INDEX
IS	VTDS	MD	INVENTORY STOCK AT CURRENT PRICES
IS#	VTDS	MD75	INVENTROY STOCK AT CONSTANT PRICES
ITD	VTDR	%	TIME DEPOSIT RATE
IV#	VTDR	75=1	MONEY SUPPLY (M2) / REAL INCOME INDEX
KAPPAMAT	NNXR		PFDI PARAMETERS FROM AME(I) REGION TO DME(J) REGION
LAMBDA	VTD		ENERGY SUPPLY ELASTICITY WITH RESPECT TO GDP#
LAMBHAT	VTD		ENERGY SUPPLY ELASTICITY WITH RESPECT TO GDP#(3 YEARS AVERAGE)
LB	VTDS	MD	CLAIMS ON PRIVATE SECTOR, OUTSTANDING
LCLF	VTXS	TP	CIVILIAN LABOUR FORCE
LG	VTDS	MD	CLAIMS ON GOVERNMENT SECTOR (NET), OUTSTANDING
LPI	VTDR	75=1	LABOUR PRODUCTIVITY INDEX
LTCB	VTDF	MDS	LONG-TERM CAPITAL BALANCE AT CURRENT US DOLLARS
LW	VTDS	TP	WAGE EARNER AND SALARIED EMPLOYEES IN ALL ACTIVITY
M	VTDF	MD	IMPORTS OF GOODS AND SERVICES AT CURRENT PRICES
M#	VTDF	MD75	IMPORTS OF GOODS AND SERVICES AT CONSTANT PRICES
MDTFOB	VTDF	MDS	IMPORTS(MERCHANDISE,FOB) AT CURRENT US DOLLARS, SUMMATION OF COLUMN ELEMENTS IN ESMAT
ME#OIL	VTXF	MD	OIL IMPORTS FROM OIL EXPORTING COUNTRIES (OPEC+MEXICO) AT CONSTANT PRICES
M1	VTDS	MD	MONEY SUPPLY (M1), OUTSTANDING
M2	VTDS	MD	MONEY AND QUASI-MONEY (M1+MTD), OUTSTANDING
MTD	VTDS	MD	QUASI-MONEY (TIME AND SAVING DEPOSITS), OUTSTANDING
MTFOB	VTDF	MDS	MERCHANDISE, IMPORTS(FOB) AT CURRENT US DOLLARS
NHFCS	VTDS	MD	NON-HOUSING FIXED CAPITAL STOCKS AT CURRENT PRICES
NHFCS#	VTDS	MD75	NON-HOUSING FIXED CAPITAL STOCKS AT CONSTANT PRICES
NHI	VTDF	MD	NON-HOUSING INVESTMENT AT CURRENT PRICES
NHI#	VTDF	MD75	NON-HOUSING INVESTMENT AT CONSTANT PRICES
NP	VTXS	TP	NUMBER OF POPULATION
ODA	VTDF	MDS	EACH AME REGION'S ODA(NET) AT CURRENT US DOLLARS
ODAB	VTDF	MDS	BILATERAL ODA(NET) AT CURRENT US DOLLARS
ODAM	VTDF	MDS	CONTRIBUTIONS TO MULTI. INSTITUTIONS(NET) AT CURRENT US DOLLARS
ODAMAT	NTDF	MDS	ODA(NET) FROM AME REGION (I) TO DME REGION (J) AT CURRENT US DOLLARS
OLCA	VTXF	MDS	OTHER LONG-TERM CAPITAL, ASSETS AT CURRENT US DOLLARS
OLCB	VTDF	MDS	OTHER LONG-TERM CAPITAL, BALANCE AT CURRENT US DOLLARS
OLCL	VTXF	MDS	OTHER LONG-TERM CAPITAL, LIABILITIES AT CURRENT US DOLLARS
OS	VTDF	MD	OPERATING SURPLUS AT CURRENT PRICES
OS#	VTDF	MD75	OPERATING SURPLUS AT CONSTANT PRICES
OSAME	STDF	MD	AME TOTAL OF OPERATING SURPLUS AT CURRENT PRICES
PCG	VTDR	75=1	IMPLICIT DEFLATOR OF GOVERNMENT CONSUMPTION EXPENDITURE
PCP	VTDR	75=1	IMPLICIT DEFLATOR OF PRIVATE CONSUMPTION EXPENDITURE
PE	VTDR	75=1	EXPORT UNIT VALUE INDEX
PE10	STDR	75=1	AVERAGE EXPORT PRICE INDEX OF GROUP 10 INDUSTRIAL COUNTRIES
PEC	STDR	75=1	WHOLESALE PRICE INDEX OF ALL COMMODITIES IN TERMS OF US DOLLAR
PENA	VTDR	75=1	IMPLICIT DEFLATOR OF EXPORTS GOODS AND SERVICES
PEO	STDR	75=1	AVERAGE CRUDE PETROLEUM EXPORT PRICE INDEX OF OPEC IN TERMS OF US DOLLAR
PES	VTDR	75=1	EXPORT UNIT VALUE INDEX IN TERMS OF US DOLLAR
PESAME	STDR	75=1	AME EXPORT UNIT VALUE INDEX IN TERMS OF US DOLLAR
PFDI	VTDF	MDS	EACH AME'S PRIVATE FOREIGN DIRECT INVESTMENT TO DME(NET) AT CURRENT US DOLLARS
PFDIMAT	NTDF	MDS	PRIVATE FOREIGN DIRECT INVESTMENT FROM AME(I) TO DME(J) AT CURRENT US DOLLARS
PGDP	VTDR	75=1	IMPLICIT DEFLATOR OF GROSS DOMESTIC PRODUCT
PGFCF	VTDR	75=1	IMPLICIT DEFLATOR OF GROSS FIXED CAPITAL FORMATION
PGFI	STDR	75=1	IMPLICIT DEFLATOR OF GOVERNMENT FIXED INVESTMENT, ONLY FOR JAPAN

```
PHI      VTDR 75=1  IMPLICIT DEFLATOR OF HOUSING INVESTMENT
PM       VTDR 75=1  IMPORT UNIT VALUE INDEX
PMS      VTDR 75=1  IMPORT UNIT VALUE INDEX IN TERMS OF US DOLLAR
PNHI     VTDR 75=1  IMPLICIT DEFLATOR OF NON-HOUSING INVESTMENT
POINA    VTDF MDS   PORTFOLIO INVESTMENT, ASSETS AT CURRENT US DOLLARS
POINB    VTDF MDS   PORTFOLIO INVESTMENT, BALANCE AT CURRENT US DOLLARS
POINL    VTDF MDS   PORTFOLIO INVESTMENT, LIABILITIES AT CURRENT US DOLLARS
R        VTDF MD    CURRENT REVENUE OF GOVERNMENT(CENTRAL)
RC       VTDF MD    CAPITAL REVENUE OF GOVERNMENT
RD#      VTDF MD75  RESEARCH AND DEVELOPMENT EXPENSE AT CONSTANT PRICES
RG       VTDF MD    TOTAL REVENUE AND GRANTS OF GOVERNMENT(CENTRAL)
RR       VTDF MD    NON-TAX REVENUE OF GOVERNMENT
SC       VTDF MDS   OTHER GOODS, SERVICES & INCOME, CREDIT AT CURRENT US DOLLARS
SCII     VTDF MDS   INVESTMENT INCOME, CREDIT AT CURRENT US DOLLARS
SCOT     VTXF MDS   OTHER IN SC, CREDIT AT CURRENT US DOLLARS
SCTI     VTDF MDS   TRANSPORTATION AND INSURANCE, CREDIT AT CURRENT US DOLLARS
SCTR     VTDF MDS   TRAVEL IN SC, CREDIT AT CURRENT US DOLLARS
SD       VTDF MDS   OTHER GOODS, SERVICES & INCOME, DEBIT AT CURRENT US DOLLARS
SDII     VTDF MDS   INVESTMENT INCOME, DEBIT AT CURRENT US DOLLARS
SDOT     VTXF MDS   OTHER IN SD, DEBIT AT CURRENT US DOLLARS
SDTI     VTDF MDS   TRANSPORTATION AND INSURANCE, DEBIT AT CURRENT US DOLLARS
SDTR     VTDF MDF   TRAVEL IN SD, DEBIT AT CURRENT US DOLLARS
SSC      VTDF MD    SOCIAL SECURITY CONTRIBUTIONS
SSCEE    VTXF MD    SOCIAL SECURITY CONTRIBUTIONS FOR EMPLOYEES
SSCER    VTXF MD    SOCIAL SECURITY CONTRIBUTIONS FOR EMPLOYERS
STDC     VTDF MD    STATISTICAL DISCREPANCY IN COST-STRUCTURE OF GDP AT CURRENT
                    PRICES
SUB      VTDF MD    GOVERNMENT CURRENT EXPENDITURE ON SUBSIDIES & OTHER TRANSFERS
TB       VTDF MDS   TRADE BALANCE(FOB) AT CURRENT US DOLLARS
TD       VTDF MD    DIRECT TAXES
TDGS     VTDF MD    DOMESTIC TAXES ON GOODS AND SERVICES
TEPM     VTXF MD    EMPLOYERS PAYROLL OR MANPOWER TAXES
TI       VTDF MD    INDIRECT TAXES
TIPC     VTDF MD    TAXES ON INCOME AND PROFIT, AND CAPITAL GAIN
TIR      VTDF MD    OTHER INDIRECT TAXES
TITT.    VTDF MD    TAXES ON INTERNATIONAL TRADE AND TRANSACTIONS
TOU      VTDF MD    OTHER UNALLOCABLE TAXES ON INCOME
TP       VTDF MD    TAXES ON PROPERTY
TPI      VTDF MD    TAXES ON INCOME OF PROPERTY AND COMPENSATION OF EMPLOYEES
TYC      VTDF MD    TAXES ON INCOME OF PRIVATE CORPORATE
UNEMPR   VTDR %     UNEMPLOYMENT RATE
UTGB     VTDF MDS   UNREQUITED TRANSFERS OF GOVERNMENT, BALANCE AT CURRENT US DOLLARS
UTGC     VTXF MDS   UNREQUITED TRANSFERS OF GOVERNMENT, CREDIT AT CURRENT US DOLLARS
UTGD     VTXF MDS   UNREQUITED TRANSFERS OF GOVERNMENT, DEBIT AT CURRENT US DOLLARS
UTPB     VTDF MDS   UNREQUITED TRANSFERS OF PRIVATE, BALANCE AT CURRENT US DOLLARS
UTPC     VTXF MDS   UNREQUITED TRANSFERS OF PRIVATE, CREDIT AT CURRENT US DOLLARS
UTPD     VTXF MDS   UNREQUITED TRANSFERS OF PRIVATE, DEBIT AT CURRENT US DOLLARS
WPI      VTDR 75=1  WHOLESALE PRICE INDEX
WSEI     VTDR 75=1  INDEX OF AVERAGE WAGE AND SALARY PER EMPLOYEE
WSEIAME  VTDR 75=1  AME'S INDEX OF AVERAGE WAGE AND SALARY PER EMPLOYEE

              DATA SOURCE
              -------------

 A) MT

   IMF    : DIRECTION OF TRADE
   IMF    : INTERNATIONAL FINANCIAL STATISTICS
   IMF    : BALANCE OF PAYMENTS STATISTICS
   IMF    : GOVERNMENT FINANCE STATISTICS
   OECD   : NATIONAL ACCOUNTS STATISTICS OF OECD MENBER COUNTRIES
   OECD   : LABOUR FORCE STATISTICS OF OECD MENBER COUNTRIES
   WORLD BANK  : WORLD TABLES 1980 FOR NATIONAL ACCOUNTS STATISTICS OF DEVELOPING
                 COUNTRIES
   NIKKEI  : NIKKEI DATA BANK - NEEDS

 B) CARD
   UN     : YEARBOOK OF NATIONAL ACCOUNTS STATISTICS
   ILO    : YEARBOOK OF LABOUR STATISTICS
   UN     : STATISTICAL YEARBOOK
   UN     : MONTHLY BULLETIN OF STATISTICS
   OECD   : MAIN ECONOMIC INDICATORS
   OECD   : FINANCIAL STATISTICS
   OECD   : GEOGRAPHICAL DISTRIBUTION OF FINANCIAL FLOW TO DEVELOPING COUNTRIES
   OECD   : DEVELOPMENT CO-OPERATION
   OTHERS  : OFFICIAL STATISTICS IN EACH COUNTRY


EOF:693
0:
LINES:693 FIELDATA

@BRKPT PRINT$
```

6 A Total Resource Management Project

D.H. MEADOWS

INTRODUCTION

This paper defines a vision and proposes some steps to translate it into action and results. The vision is of a set of internationally-coordinated institutions and processes that will enable people in all parts of the globe to manage the earth's resources to meet human needs sufficiently and sustainably. I am writing this paper as a single individual, but I am speaking for an increasingly large group of people, most of whom have helped me shape this vision and who are continuously working to define it more carefully and to bring it about. I am writing both to and for them, summarizing the state of our conversations so far, and inviting them to react to the vision in its current form, to criticize and participate, so that as it comes into being it expresses all of our talents, ideas, and visions.

The proposal begins with the ultimate purpose of the project – the intangible, unmeasurable, but vital final goal it serves. It then spells out the specific intended results to be produced. Comparing the intended results with the ultimate purpose reveals some basic assumptions about the way those results are presumed to serve that purpose – the rationale for the project. Next come the action steps and timeline that describe what has already been done and what has been planned to achieve the intended results. Then there is a brief elaboration of one of the action steps, a sample model to illustrate what sorts of concepts and results the project will generate. Finally comes a list of concrete products – what some U.S. government contracts call "deliverables" – measurable checkpoints and results from different stages of the project.

ULTIMATE PURPOSE

The ultimate purpose of a project is the true goal, the end to which the project is a means. It is not necessarily precise or measurable or even fully attainable. It is a value statement, a guiding direction, like "east" or "truth". No matter how far east you go, there may always be east ahead of you. No matter how truly you state things, there may be more truth to say, or discover. But the goal of "east" or "truth" helps to clarify where the next step lies.

Although the ultimate purpose may be intangible and not quite reachable, it should be expressed with great care. It is the beacon of the project, the statement to look at when difficult decisions must be made or when conflicting choices present themselves. If it does not state completely or accurately the real goal or vision, the purpose statement will mislead and confuse instead of guiding and focusing.

In my experience statements of purpose have to be reworked, honed, and re-examined many times before they are exactly right. This one has been worked on for some time, and I think it is very close to right.

The ultimate purpose of the project is *to provide the information necessary to create the conditions in which people in all parts of the globe can and will manage their resources to meet their own needs:*

- *sufficiently* – so that there is no scarcity or deprivation
- *reliably* – every day without fail
- *equitably* – so that everyone is included fairly
- *sustainably* – over the long term
- *appropriately* – to their own cultural and environmental conditions.
- and *beautifully.*

Two words in this statement need to be defined broadly: *resources* refers to all natural, economic, human, and information resources; *needs* refers to all human needs, not just survival needs or material needs, but also psychological, spiritual, and esthetic needs.

INTENDED RESULTS

That purpose is one that guides many people, and there are probably many paths that contribute to its achievement. The particular path proposed here, when fully and successfully implemented, would look like this:

- There will be a set of *resource analysis centers* in many parts of the world, including the parts we now label North, South, East, and West.
- Each center will provide *continuing, top-quality information* about the condition, structure, and potential of the resource base in *its own region.* It will establish the concepts, methods, data, and personnel that will allow people in the region to perceive their own resources as a total system, to understand the dynamics of that system, and to manage it sustainably.
- The analysts in each center will be in regular contact with the important *policymakers* for that region, with the *resource experts* in the universities and elsewhere, and also with the *resource managers* – the farmers, businessmen, laborers, households whose daily decisions affect the use of resources and provision of needs.
- The centers will be *internationally linked* to exchange ideas, methods, data, training techniques, and insights about resource management and about the condition of global resources. They will compare and contrast their findings to understand the similarities and differences inherent in different resource endowments, degrees of industrialization, and types of economy.
- Each center would be *partly endowed* to ensure long-term base-level funding and *partly financed by contracts* to work on current problems deemed important in its region.
- People in the centers would be *skilled at communicating* (putting information out and taking it in) *on three levels:* scientific, political, and popular.
- The centers would not only do analysis but also *train analysts and resource managers.*
- They would use *state-of-the-art quantitative and systematic techniques* and do a *variety of studies and models,* including

in-depth studies of specific resources and comprehensive studies of the broad interactions among all resources.

- Each center would be guided in its operations and decisions by the statement of ultimate purpose in the previous section, or a statement very much like it, agreed upon by all.

BASIC ASSUMPTIONS

How does a set of resource analysis centers, no matter how excellent, connect with the purpose of creating the conditions in which people will be able to manage their resources to meet their needs? I think there is a direct connection and that the centers I have described are at least necessary, if not sufficient, to that ultimate purpose. And that reveals some of my most basic assumptions about systematic analysis, about resources, and about the factors that currently prevent people from managing resources to meet their needs.

One can never quite see all one's own assumptions. Here are some of the crucial ones I see that I must have made in order to link those intended results to the ultimate purpose.

1) Resources are vitally important – they are the source of life and wealth and the realization of human potential, and they must be protected and enhanced.

2) Resources are not being managed well – their use is in many cases unsustainable and in many places does not meet human needs.

3) One important reason for the misuse of resources is the absence of widespread comprehensive understanding of what is happening to them and of how they can be managed better.

4) There is a wealth of partial, unorganized information about specific resources.

5) Systematic analysis and modelling are necessary to sythesize that wealth of information into comprehensive understanding. Resources are so interlinked, and their dynamic behaviour is so complex that neither intuition alone nor detailed knowledge of individual sectors is sufficient to manage them efficiently and sustainably.

6) Modelling and analysis are means, not ends. They must be combined with effective communication, administration, and human interaction before the lessons learned in modelling become internalized and result in action.

7) People learn from models they have developed for themselves – they distrust models made at a distance.

8) The *process* of modelling produces most of the learning, not the finished model.

9) No person, institution, or nation fully understands how to manage the total resource base to meet human needs. This is not an area where North can instruct South or East can teach West. We all need to learn together and to proceed in a mode of experimentation, not exhortation.

10) In order to stay in touch with reality, to muster relevant facts around real problems, analysts need regular contact with policymakers and with the people – identifying with the concerns of both.

11) Analysts need some insulation from short-term political pressure – hence the need for endowment and international support.

12) Asking a good question can have impact long before the answer is found. The presence of analysts asking questions about resources, equity, and sustainability will have a salutory effect from the very beginning of the project.

13) Resources are best managed locally, with detailed and timely information about their state – but with an awareness of the larger system.

14) Interdisciplinary, problem-oriented research can be done and can be effective.

15) "Experts," "ordinary people," and "politicians" can serve each other in partnerships that are mutually empowering and that contribute to the good of the whole.

16) Development and progress need to be viewed as complex, organic, total-system phenomena, not only as parts such as "agriculture," "energy," "education," or even "liberation".

17) Lessons learned abour river basins or energy conservation or mining technology in one part of the world are at least partially transferable to other parts of the world.

18) As information and systematic understanding begin to be organized around the central purpose of people meeting their own needs sustainably, and as that purpose and that information permeate all levels of society, *either* the social and institutional changes necessary to achieve that purpose will become obvious and will be brought about, *or* it will become obvious that no momentous social and institutional changes are necessary.

ACTION STEPS AND TIMELINE

How can this interacting network of centers be made to happen? How exactly would it operate? What kinds of analyses would it do? How would it be funded?

I have only partial answers to these questions, since many of them will be determined by discussion and agreement among the actual centers involved. I would suggest the following general operating principles.

Each center would coordinate and fund its own activities and make its own models addressing the interests of its region. Analyses would include not only formal computerized models, but any gathering of information that would be useful, including perhaps data banks, maps, diagrams, and resource accounts.

The centers as a group would agree on standards of practice and documentation, and on some common elements of research design, units of measurement, time horizon, and organization of data. Perhaps one overview model would be constructed in a similar way for each region. Ideally there would be sufficient uniformity of hardware and software to facilitate comparison and exchange of analyses and methods, and perhaps direct and regular communication via computer. However, there would be natural diversity among the groups with regard to the resources of particular interest, the quality of data, and the ecological and economic circumstances within which resource-use decisions are made.

At least one person within each center, probably the director or co-director will be responsible for maintaining exchange of information and ideas throughout the international network of centers. Members of the centers would come together at least once a year to present results, discuss common problems, plan joint publications, and handle administrative concerns. Some international coordinating body will be necessary; possibilities include the United Nations University, the United Nations Environmental Program, and/or the International Institute of Applied Systems Analysis.

Immediate action steps already completed or in progress to bring about this network include:

Action Step	Timeline
Study existing centers and learn from their strengths and weaknesses.	

(Centers already in existence, which now exhibit some, but not all of the ideal characteristics listed under Intended Results, which are already interacting to some extent, and which could be the core of the project, are:

The Resource Policy Center, Dartmouth College, Hanover, N.H., USA.

Winrock International Center for Agricultural Research, Morrilton, Arkansas, USA.

The Petroeconomics Group, Christian Michelson Institute, Bergen, Norway.

The Resource Policy Group, Oslo, Norway.

The All-Union Institute for System Studies, Moscow, USSR.

The Research Project on Natural Resources, Hungarian Academy of Sciences, Budapest, Hungary.)

Clarify and communicate the idea; get feedback and corrections.	1980-1982

(That is one of the purposes of this paper and of many presentations and discussions over the past two years.)

Make some small demonstration models to illustrate what the centers can produce.	1981-1982

(One such model is described in the next section of this paper.)

Find groups willing to start.	1982

(In addition to the ones listed above, I would like to add groups from the OPEC nations and from the so-called Third

World. Some interest has been expressed in Costa Rica, India, Venezuela, and Zaire.)

Hold a planning meeting. September, 1982

(The meeting will be held in Hungary, under the sponsorship of the Hungarian Ministry of Industry and UNESCO.)

Find initial funding for each group for at least the first year of operation. September, 1982

(Three groups are funded now, two more are actively seeking funding. Only one group has the possibility of endowed funding at the moment.)

Find funding for international coordination for at least the first year. September, 1982

Action steps following these will be decided by the group as a whole at the Hungary meeting.

A SAMPLE PRODUCT – THE TOTAL RESOURCE BASE DIAGRAM

Over the page is a diagram that looks relatively complicated, but in fact is a simple and incomplete representation of the many sorts of resources any human population has to work with. Six basic kinds of resources are shown – reading clockwise from the upper left-hand corner they are:

soil resources – land, forests, crops

water resources

fossil fuel resources – petroleum, coal, natural gas

mineral resources

and in the center

capital resources

human resources

Energy resources and the capital plant that accounts for energy demand are scattered throughout the diagram.

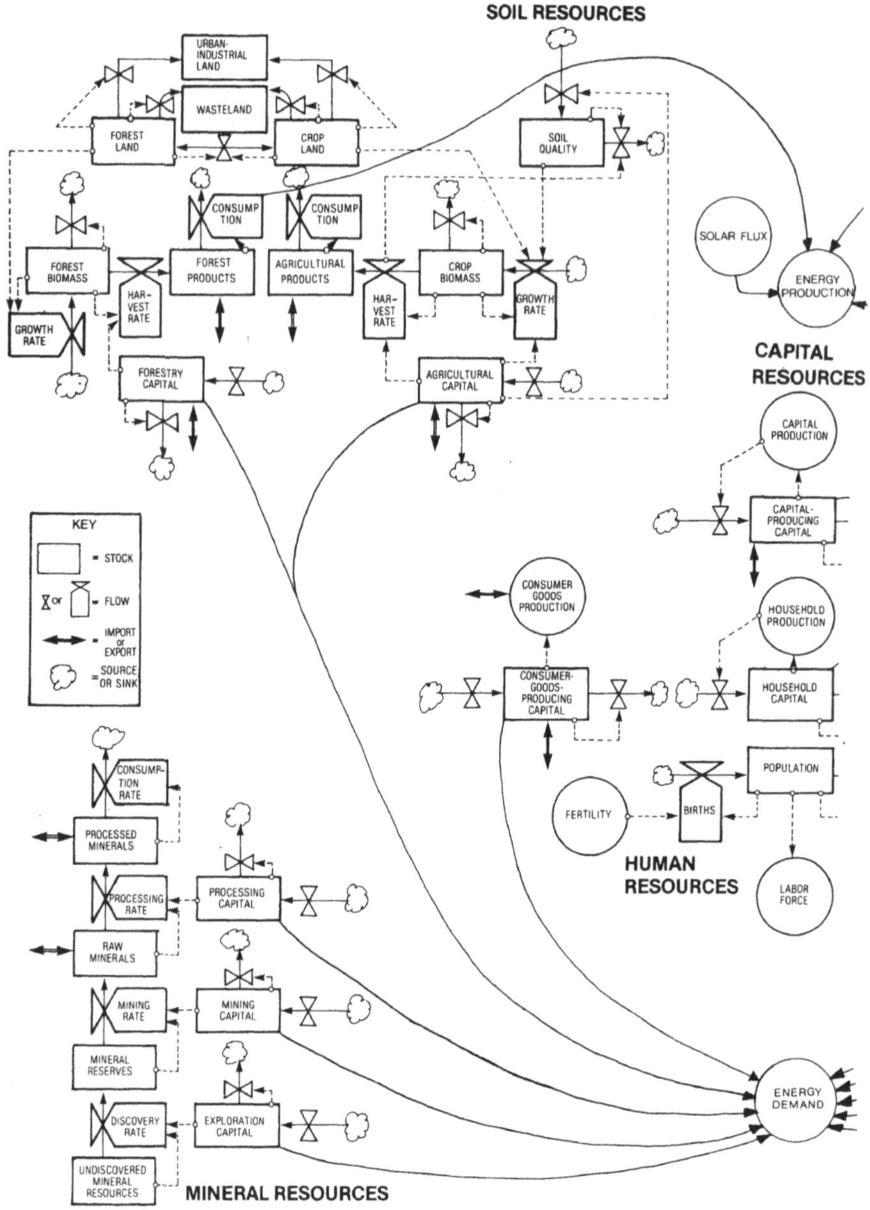

A Total Resource Management Project

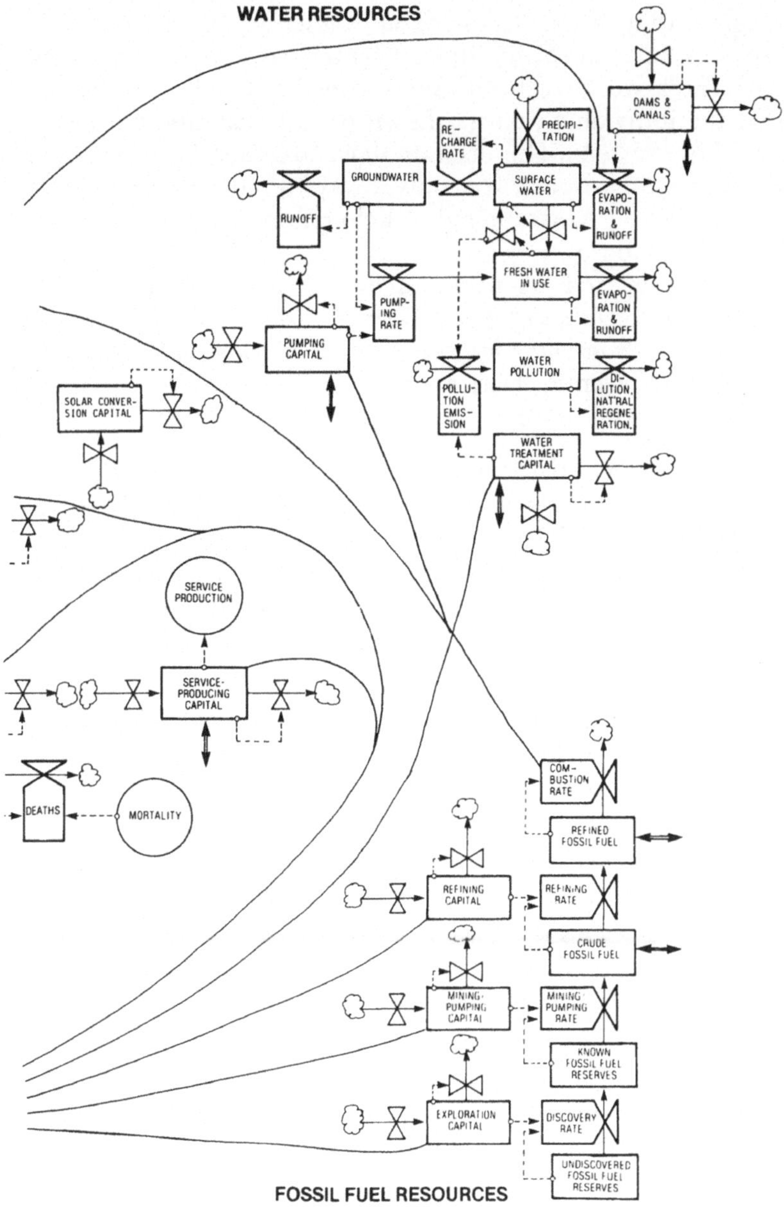
WATER RESOURCES
DAMS & CANALS
PRECIPITATION
RECHARGE RATE
GROUNDWATER
SURFACE WATER
RUNOFF
EVAPORATION & RUNOFF
FRESH WATER IN USE
PUMPING RATE
PUMPING CAPITAL
WATER POLLUTION
POLLUTION EMISSION
DILUTION, NAT'RAL REGENERATION,
WATER TREATMENT CAPITAL
SOLAR CONVERSION CAPITAL
SERVICE PRODUCTION
SERVICE-PRODUCING CAPITAL
DEATHS
MORTALITY
COMBUSTION RATE
REFINED FOSSIL FUEL
REFINING CAPITAL
REFINING RATE
CRUDE FOSSIL FUEL
MINING-PUMPING CAPITAL
MINING-PUMPING RATE
KNOWN FOSSIL FUEL RESERVES
EXPLORATION CAPITAL
DISCOVERY RATE
UNDISCOVERED FOSSIL FUEL RESERVES
FOSSIL FUEL RESOURCES

I know of no computer model or even accounting system in any region that keeps track of all these resources and their interactions simultaneously. In ten years of work at the.Resource Policy Center we have been asked to make detailed studies of each of the individual sectors shown on the diagram, but never of the whole system. The diagram does not, therefore, signify a currently running model. It is only a way to picture resources — a device to focus discussion and to help people begin to envision a total resource base.

(The diagram and its uses are described in detail in D.H. Meadows, *Understanding Long-Term Consequences of Energy Development Programs in the Third World,* DSD #412-1, Resource Policy Center, Dartmouth College).

There are several points to note about this diagram:

1) It illustrates only *physical stocks and flows* of material resources. Most of its physical flows are accompanied by monetary flows not shown on the diagram.

2) It illustrates the primary resources basic to any industrializing society of any political or economic orientation.

3) It also applies to any geographical unit, from a village to the globe as a whole.

4) It carefully distinguishes stocks of resources from flows, and therefore provides an introduction to the dynamics of resource development or exploitation over time.

5) Of the many interconnections among resources it shows only one kind — energy flows. One could also use the diagram to begin to trace out flows of labor, capital, money, and information.

6) Although the diagram is quite simple compared to an actual resource base, it is a surprisingly useful framework with which to order thoughts and discussions.

This diagram is only one example, and a very introductory one, of the sorts of systematic frameworks and pictoral communication schemes the centers could produce in order to synthesize and display information about resources.

PRODUCTS OF THE PROJECT

The international network of centers is likely to produce many results, some concrete and some not. In increasing order of importance, the products to be expected are:

1) Resource models of many types: top-down, bottom-up, aggregated, disaggregated, local, national and perhaps international.

2) International meetings, which are worthwhile in their own right in bringing together analysts from many countries interested in similar resource problems.

3) Better, more consistent, and more organized resource data, as a result of the systematic and coordinated work of the network of centers.

4) Scientific monographs describing the resource situation and management possibilities for each region in English and in the language of the region.

5) Papers in the international scientific literature comparing and contrasting resources in different regions and drawing general principles of resource management.

(The five products above are almost guaranteed from any well-run international scientific effort. They are extremely important, but if they were the only results of the project, I would not be interested in doing it. The unique products of this project are farther down this list.)

6) Educational programs, textbooks, popular communications about resources and resource management.

7) Excellent analytical centers with trained staff standing ready to meet societal needs (as defined by the policymakers, or the people, or both) and to deal with resource questions as they come up.

8) A steady stream of young professionals trained in systematic thinking and resource management, moving into private enterprises, public policymaking, and educational institutions.

(These three results are also likely to be produced if there is definite intention to produce them. However, they also are only means to the results I would most like to see, namely:)

9) Increased understanding of and respect for resource systems by all people at all levels in society.

10) Actual, permanent improvements in resource management in each region, so that all human needs are met sufficiently, reliably, equitably, sustainably, appropriately, and beautifully.

11) A network for international cooperation in using and enhancing the resources of the entire planet.

DISCUSSION POINTS

1) Dr. Spitz felt there was a need for this type of research center. The piecemeal approach followed so far has given rather discouraging results. For example, last November in the Khatmandu Valley in Nepal there were about 60 consulting fims and groups, each linked to a specific foreign agency and each doing a little bit of something such as the evaluation of a lift irrigation project, of the terracing of a hill or of post harvest losses. Similarly in Bangladesh and India the number of research agencies is quite enormous.

2) Dr. Manshard pointed to the value of learning from past experiences such as the Man and the Biosphere (MAB) program in Unesco which covers some of these aspects from a biological point of view. When Unesco tried to expend this already large intergovernmental program for a Man and His Environment project, they failed. Considering just the physical resource base listed – geology, ecology, hydrology – there are several big intergovernmental programs in these areas already. The intended effort is large, perhaps overambitious.

3) Dr. Hrabovszky asked whether the centers would use models as the only tools for analysis. Professor Meadows replied that models will be just one of many tools in an information processing center. Often of course quick, simple models answer about 90% of a question.

An overview model would provide an analytic tool and a means to disseminate ideas and stimulate a dialogue among researchers in different disciplines in different parts of the world.

4) One drawback of such a global approach encompassing all resources is that there may be an unmanageable richness of

answers and computer printout. The work may be buried in detail and complicated.

Professor Meadows was familiar with the problem of becoming buried in computer printout, indeed this is the central and continuous problem of people who try to study whole systems. The only way to deal with the problem is to be very clear on the purpose of the exercise. Unmanageable quantities of printout are usually generated when a whole system is being studied, rather than the solution to definite problems, since without a clear problem there is no reason to not include everything. It is important to not attempt anything which is not relevant to the purpose.

5) Dr. Spitz had the hope that at the end of a study it would be possible to publish books for children, because this would mean that they had come to grips with the main problems. It is easy to publish 40 books of 1000 pages each of academic standard, but an output of 10 books of 50 pages for children would be the sign of a successful project.

Professor Meadows agreed that one should aim for a wide range of communication channels, including television and even comic books.

6) Dr. Hrabovszky pointed out that the vision behind the project proposal was based on 10 years of experience at a center placed within a strongly laissez-faire society (USA) where for example government interference in land use has been minimal. This is very different from the situation in most developing countries where there are often very disjointed existing institutional systems, trying to deal with problems in a piecemeal fashion. It may not be wise to center a project on such an institution, within a university environment, particularly when there is a lot of competition for limited funds.

Professor Meadows indicated that the various centers in the network would differ widely. Some are in centrally planned economies and some are in market economies, some have better computer facilities or data availability. Each would have to work out a mode of political operation appropriate to its own region.

7) In answer to a question from Dr. Boltz, Professor Meadows stated that there was no overall budget or estimate of the total cost. The Hungarians and Russians, as those in the USA, control their own budgets. The New Hampshire center has 50 people altogether, about one half of whom are students, with a budget of $800,000 per year. Other centers will be different.

7 General Discussion

7.1 INTRODUCTION AND SUMMARY

The day's debate is not presented here in chronological order, but rather is grouped by topic. Since most speakers covered a number of points the classification is of necessity only approximate; with each contribution being presented as a coherent whole.

In answer to questions on the focus of the problem, Dr. Levy identified the agriculture-energy nexus as the topic of the proposed UNU Study. The main problem to be considered is the deprivation in food and energy of many people in the world, and one key question concerns future trends in the price of food to the increasing populations of the large cities of the developing countries. The UNU would wish to provide an analytic tool for decision makers (and this meeting considered in particular the usefulness of global models), together with a set of concepts and a number of people, educated in those concepts who could make decisions makers more aware of the wider impact of decisions.

Within the debate a number of differences of opinion appeared, such as concentration on agriculture or total food systems, emphasis on an analysis starting from the community ("bottom-up") or from a global picture ("top-down"), and the choice of either open or specific studies. There was also a difference in attitudes towards global models. Such opinions are reported here as expressed by the participants.

A summary of the debate is as follows.

The problem

- How to avoid the increase of inequalities, and how to help reduce the deep inequalities in access to food and energy between nations and within nations.

The goals

- To provide an understanding of the impact of national and international decisions at local level, and vice-versa, and
- to stimulate decision makers and appropriate groups to formulate their own visions of the future and to translate them into scenarios and action.

The general conditions

- different approaches are required at different levels (local, national, regional, global); they are to be linked in order to ensure a sufficient degree of coherence,
- modelling has to be imbedded in a larger analytical system (hybrid approach involving qualitative analysis and unquantifiable factors), and
- the food-energy interaction has to be viewed in the socio-economic context without losing sight of the goal to be reached.

The methods

- assess existing relevant knowledge within global and national (or regional) models, and in agricultural and energy production and distribution,
- encourage the formation of small teams working in separate homogenous areas (countries or sub-national regions), to examine the detailed problems of resources, socio-economic conditions, technology, development prospects, etc., possibly using existing teams,
- provide linkages among these teams based on common goals, in order to arrive at comparable results and generalizations,
- test the expected results of new technology (e.g. biotechnology) using models developed, and determine the possible socio-economic and environmental impacts,
- use local and international programmes to train people in energy and agriculture systems, analysis and management,
- disseminate knowledge through publications (monographs, text books, etc.), other communication channels, international meetings, availability of resource bases and models that can be used by others,
- connect this overall structure to other resource bases which exist or might be created, and,

• encourage the establishment of national pilot projects based on an integrated approach.

7.2 THE PROBLEM

Mr. Herrera noted that no government has used global models as instruments for decision making, but that they have had an impact on public opinion about the possible future. The global model results show that from the point of view of the Third World the future situation is hopeless. That is absolutely clear. For example, the Global 2000 Report indicates a situation at the end of the century that will be worse than now, and with the most optimistic results of other models there are only marginal improvements. The North-South gap remains, and tends to worsen.

For a historical view of the character of the gap, consider that immediately after the second world war the objectives of the governments of the central countries were more or less the same as those of the third world countries, they looked for similar styles of development. Now, the situation has changed totally. For the first time in history, a sizeable part of humanity has forgotten the problems that were every-day concern until a few decades ago. For those people in wealthy countries, to eat to-morrow, to have a decent house, education or health care are no longer a problem. The problems of these countries are now overconsumption, alienation, environmental concern and the use of free time.

One problem of the developing countries is to reap the benefits of the industrial society. But historical processes do not follow the same linear sequence, and new kinds of problems are appearing. One example is that of robotisation – with microelectronics, the whole structure of industrialised countries will change in 2-3 decades. What then will happen in Third World countries? Consider, for example, Brazil where there is about 30% unemployment, where the most important industry for the utilisation of urban manpower is the automobile industry, and where Volkswagen is now robotising production.

What does it mean to be unemployed? In advanced countries, you eat to-morrow, you have education for your children, health care, and so on. In Brazil or any developing country, you don't care about health or education, which are luxuries, you worry if you are going to eat to-morrow.

The models make sense – with the implicit assumptions of the models, there is no hope. How can we change that? What can the UNU do? The Bariloche model, for example, was a normative model. It did not consider the gap. An attempt was made rather to stimulate people in developing countries to specify their own version of the world, to try to create a new type of society.

It is important to not ask developing countries to build a model to fit with a central model, because that means that all the variables and parameters are chosen centrally. Don't start with a model, but first develop a vision of the world. The vision can *then* be compared to a model – the Third World must start the discussion on its own terms.

The Cancun meeting after Ottawa is the symbolic end of a North-South dialogue that never started. We can however have a dialogue amongst scientists, to propose solutions to global problems.

International organisations have priorities, and ask people from Third World countries to present proposals within those priorities. For many institutions, to have funds from outside is a question of survival. So they present proposals on the requested subjects. That is a waste of money, time and precious imagination. So there is a need to take a risk, to ask people to develop what they see as a possible future, and after that maybe we can put together the two parts of the world.

Mr. Matthews expressed a belief in equal pay for equal work throughout the world. The equal distribution of food and energy with which we are concerned depends on adjustments in comparative purchasing power worldwide, since the principal victims of world wide unequal opportunities and unequal pay are the hungry people who live in the world South.

World modellers might follow two lines of research that would guide and motivate leaders in the Third World towards self-reliant political action aimed at equalisation. Firstly global modelling could emphasise solidarity in the world South and assist leaders to identify initial priorities, such as the potentials and paths for maximising South-South trade, to determine what OPEC type organisations are feasible for other commodities such as coffee, rubber and bauxite, and to examine the opportunities and benefits of greater integration of continental or regional groups.

Secondly, the optimum physical levels of production and export of commodities, to maximise returns to the world South, could be considered. Possibly such models would show that in most cases a lower production by the world South of their commodities would benefit them, principally by raising prices to a fairer level and by creating better terms of trade for the South.

It could, for example, be both rational and profitable for some countries to reduce their production of oil, since:

- many wells are overpumping, and if oil were withdrawn at a slower rate, the maximum yield for the life-time of each well would be greater,
- the flaring of gases is a waste of national resources, and if oil production were limited to wells where gas could be recovered and stored, this would be an important conservation policy,
- oil left in the ground may be worth more than investments made with excess profits,
- some oil-exporting countries receive more in income than they can efficiently utilize, and
- when oil appeared, the total quantity was about 2 trillion barrels. The world North have already utilised their 25% share. If oil is as valuable an engine of development as it seems to be, there is no reason whatever why one more barrel of oil should move from South to North at any price.

Dr. Leach was of the opinion that the development process at the micro level of the village, not energy, is the leader. This is often forgotten, with bad effects on both national policies and on the poor, whom the national policies are meant to serve.

The largest single world energy problem is the severe crisis of wood scarcity in many parts of the world, so that many communities may collapse or have to migrate. Wood and dung are referred to by economists as free goods, but they are not free at all – there are very heavy costs in labor and other effects.

Also, when people move from one level of development to another and the economy becomes monetised, historically it is observed that there are often very harmful effects such as the mal-distribution of incomes – a whole nexus of poverty issues.

In the 1980's, there is a wider diversity of possible energy sources, and a transition in the development process is needed to open up the possibility of using many energy sources. The tragedy is that many well-intentioned people in the energy game (such as the green movement, the Nairobi conference, and politicians in the South who see renewable energy as an important political bandwagon) ask that we introduce renewable energy into the matrix without any understanding of the issues involved, some of the costs they are imposing on people, why it is often so difficult to introduce these technologies, and why they so often fail. The more efficient wood stove is a classic example. There are many different types around the world, they are being introduced in thousands of villages everywhere, and their success rate is very low. It is low because while the poor have many problems, the energy expert comes in with blinkers on and sees only the energy problem. If the village does not see energy as the main priority for development, the villagers will not accept what the expert tells them to do. For example, in parts of East Africa women walk 6 hours a day to collect wood, the forest cover is being depleted at 3-4% per year and there is a 3% population growth rate. The energy expert advises the planting of trees and the use of more efficient stoves, but the villagers are unconcerned because they have many other priorities, such as better houses, water supply, sanitation and rural dispensaries.

Until the process of implementing energy policies is placed within this micro-development context, we are bound to make the most dreadful mistakes. We must place a high priority on the building up of sensitivity to real issues of this kind, building up at a micro level information about village systems, maps of those transitions that have worked and of some that have not worked (and why they have not worked). These can be used as the basis of national policies that are sensitive to the real issues.

In regard to the UNU research effort, many institutions around the world are beginning to tackle these problems, to build up the experience and the data. The main priority is to link these up, not in a massive central computer base, but by making associations between these groups. A second priority is to set up an information system, across the South in particular, so that governments and agencies can learn about experiences in other countries. South-South communication is very poor on the whole.

Dr. Makhijani emphasised the seriousness of the world situation. If we connect what we read in the newspapers to the conference discussion, we must conclude that we are in the most serious crisis that humanity has ever faced, and this arises from social, political and economic history. One manifestation of this is "the silent violence of famine", the daily food problem whereby about 6 million children die every 4 months of hunger and related causes. Here is as much violence as the violence of Dachau and Auschwitz.

Models and scenarios are convenient to our present preoccupations, but modelling lacks any connection with cause and effect and cannot track history such as the great plague and imperial expansion.

There is one part of the world which is easy to study, and which illustrates key behaviour patterns. This key part of the capitalist world – third world system is South Africa. Here we may find a dynamic which will enable us to understand the history of the past 3-400 years.

Dr. Makhijani also emphasised that energy is not an end in itself, and that we need to consider demand, use and needs. In particular, needs relate to what we want to do and thus to social conflict. Social and ethical criteria should be applied to a consideration of energy needs. For example, the energy need for building nuclear weapons is not a need for everybody but may be a need for a large company wishing to make profits. This may be an anti-human need, and judgment is required in deciding how to deal with such energy needs. The whole question of what is an energy need is never a fixed question, but is always a social question, open to the arena of social, political and economic struggle.

Energy demand is primarily determined by purchasing power, which is why 80% of the world's oil is consumed in the capitalist countries. The structure of investment is controlled by those who have the purchasing power or can manipulate the population through advertising, etc. The control of agriculture is from cities by citified rulers, and there are hierarchies of metropoles.

What is this energy-agriculture problem? Consider for example bullocks, which in Nepal represent in energy terms 40 times the installed electricity generation capacity; grazing land is in many countries the most important source of energy. Since both poor and rich own bullocks, the question of power for agriculture is very complicated. Bullocks and tractors cannot be compared only from the point of view of horsepower since from many points of view (social, political) they are completely different. Energy and agriculture considerations meet around the central question of the control of land. It is important too to know where problems originated. It is not possible to understand any problem without substantial historical knowledge – the future is defined by the conflicts of the past.

Dr. Gallopin raised the issue of uncertainty. While models are based on constant dynamics, an increasing level of uncertainty is the most certain feature of the future. There is a need for both better data and for a better comprehension of the whole subsistence production sector. This requires both better models and better ways of thinking in this uncertain world. One conclusion is that in a situation of great uncertainty and complexity it is unlikely that one single approach could produce the solution. A constellation of approaches is needed.

One important item in the agenda is to try new ways of thinking of how to respond to this fluid situation. Thus in the field of models we can choose either to collect better data, statistical series and data banks which might lead to models more based on the industrialised countries, or to look for more qualitative, less data dependent approaches to try to explore the major consequences of alternative hypotheses and model structures. Also we can choose either to identify, calculate and predict an optimum, or to look for undesirable states and for policies that can move the system away from such pitfalls (such as dictatorships).

It is important to look for different approaches, capable of responding quickly and flexibly to unavoidable surprises, and maximising the capacity to handle risk.

7.3 LOCAL DEVELOPMENT

Dr. Sachs felt that development at the local level is the pertinent focus of development to the extent that in the last instance either development happens locally or there is no development at all. We need to address ourselves to the question of how this local development is being affected or is not being affected by regional, national and global policies, and to find the intended and untintended impact and lack of impact of policies. This calls for a very elaborate systems analysis, for the spelling out of what global environment means in practical terms. How, for example, are villages affected by a resolution of FAO about trade in grains, or by the lack of adequate safeguards for the ozone layer if UNEP does not do its job properly?

This will not lead to a model, but could lead to a major systems study on the articulation of the pertinent levels of decision making and on the impact of policies.

Dr. Glaeser noted that a technological system consists of techniques plus knowledge. Three agricultural techniques may be identified: traditional, capital and energy intensive, and a mixture (ecofarming). The ecological knowledge contained in socio-cultural and economic systems and in the biophysical environment should not be neglected. Consider the efficiency of food-energy systems: the worst energy efficiency is in fishing; then agriculture in so-called developed countries, with the best in the traditional agriculture systems in less-developed countries.

An important alternative is an energy-saving, environmentally sound system which gives good average yields. This exists as ecofarming or organic farming, characterised by the use of naturally converted solar energy, the manipulation of nutrient cycles, the maintenance of crop diversity, biological methods of pest, weed and erosion control, and the exploitation of the ecological succession. The quality of food is often better and the system is environmentally sound. As far as yields are concerned, a comparison of 16 organic farms and 16 traditional farms in the USA corn belt revealed little difference in yield, while the energy requirements of the organic farms were less than one third that of their conventional competitors.

Research questions for the UNU include:

- the impact of different agricultural technologies on energy efficiency, environmental hazards, food quality, food supply (including self-sufficiency and needs satisfaction),
- strategies for national food policies, to implement the optimal results, and
- a survey of the interrelationships of the development policies in developed and developing countries.

Dr. Hrabovsky felt that a great share of subsistence agriculture in the developing world satisfies these conditions, but is unable to produce enough food for a rapidly growing population, with a rapidly increasing proportion becoming urbanised. We must be careful in holding out this solution to development problems, since it requires a very slow evolution and there is not enough time to be able to rely on the dynamics of this system.

7.4 PROJECT BOUNDARIES

Mr. Maddox was of the opinion that the most tangible projects are the most likely to succeed. As an example, many genetic engineers believe that by manipulating genes of crop plants, it may be possible to enormously improve productivity and thus to reduce import requirements. Such enthusiasts believe that their dreams are certain to come true. But this may not be the case at all. Even if it were possible to fix nitrogen-fixing genes in wheat and rice, it would be possible to grow these plants only in tropical regions. Without some improvement in the photo-synthetic yield, it would not be feasible for temperate regions. It would be valuable if one could assess the feasibility of some of these genetic engineering projects on the assumption that technically they will succeed; and some modelling techniques are admirably suited to this.

The value of such a study would be to indicate the social and other impacts if such dreams do come true. Also it might feed back into the genetic engineering community a sense of realism, of how much and how little might be accomplished if their technical dreams succeed.

Dr. Hrabovsky expressed a concern over too great an emphasis on socio-economic factors. With no clear boundaries, there is the danger of getting lost, as is shown by past experience. It is important to keep to the original formulation of the food and energy problem, set within an environmental framework, but not as an amorphous mass. In the definition of problem boundaries a lot will depend on more concrete proposals.

It has been clearly agreed that there will be a three-layered analytical system – the village level, the national unit and the global setting, with important interactions amongst these. Also the roles of individual decision makers has been emphasised.

The analysis needs to consider what are the levers that can be manipulated, by whom, and at what level. There is a need for a linking up and a very careful specification of linking variables. The quantitative formal system should be surrounded by a qualitative discussion of some variables which the model could not handle, but which we know are important.

The UNU should discuss with individual global modellers how the existing models could be used as a starting point. We must look for opportunities for generalisations; for example how is it possible to generalise from village to country to world level?

7.5 SOME PROJECTS

Dr. Fergany belongs to a group of researchers from Arab countries who are interested not in building a global model, but in building an Arab model with a global environment. This would seem to be the only meaningful approach to give relevance to the question of development in Third World countries.

Basic drawbacks of present global models include a tendency to be global (and not very useful to Third World countries or regions) and technoeconomic (a major drawback if we consider development prospects in the Third World). Development in the Third World is definitely not concerned with growth in GDP, for certain regions could sustain a drop in GDP while making real accomplishments in the development sense or in real welfare. The first priority of a structural transformation of social and economic systems should concern the satisfaction of basic needs, and then a continuous increase in the level of real welfare. In this process the most important aspect is the mobilisation of the resources of the country; and most importantly the human resources.

This involves a process of socio-political change; we are really talking about the decision making institutions of the Third World, at local, national and regional levels. Decision making must relate to the nature and goals of society, and to socio-economic activity. This includes consideration of how the society is formed, what are the institutions, and how they make decisions. As an example, there is a considerable difference between studying Tanzania in socio-cultural terms (how that affects production) and in thinking of the villageisation process. In the second case, we are talking of socio-political transformations. In order to consider the development process for the Third World we must consider structural change, not stopping at the level of explaining what is happening now.

It is important that the UNU should not set up a techno-economic project. In order to have an impact on the development prospects of the Third World we should try to study aspects of socio-political organisation, in terms of not just understanding what is happening but also in terms of trying to change it. The food-energy nexus is only a subsystem. We should be studying the total system, and the most important factor in this is social organisation.

One commentator felt that we must accept that we are trying to be normative, and pointed to the Bariloche assumption of some change to the status quo. But each country and institution is in a different situation and dynamic. The mobilisation of people's perception of their own problems is part of our role.

Another commentator pointed to the dangers of outside agents (as UNU) assuming the role of agents of change.

Professor Malu wa Kalenga noted that Zaire is trying to link the prices of oil and food. They are in the process of implementing a strategy of development of the energy and agriculture sectors in such a way that the price of oil will play a diminishing and indirect role. The energy system of Zaire is based on two resources: hydroelectricity and wood. The hydroelectric potential of the country is around 120,000 Megawatts, and tropical forests cover about 55% of the country. The aim is for a hydrogen economy, with fertilizer and methanol produced using hydrogen obtained by the hydrolysis of water, and for the total electrification of the mass transit system. A 1,800 kilometer high voltage DC line has just been completed, and a large-scale railway system has been started.

There will be an extensive use of wood, and oil will play at most an indirect role in food production. This is particularly so since the problem in Zaire is not so much one of increasing the yield, but of putting more arable land into use. At the present 30% of available arable land is being used. The aim is not for mechanised agriculture, but for labor and land extensive agriculture.

Some deforestation problems are appearing, particularly around big cities like Kinshasa. One suggestion for research is for a better understanding of the ecology of tropical forests. We don't know the energy store, the share that can be safely harvested or the best practice to do so. However, much information points to an alarming conclusion with unpredictable environmented consequences, since throughout most of its extent the rain forest appears to be incapable of regeneration. Although sylviculture could be practiced carefully, continuously and responsibly, a massive blanketing of areas previously covered with multilayed species by a handful of desirable plantation trees to yield cheap and plentiful wood for fuel pulp and construction would be a dubious strategy. The local results, in parts of Zaire mainly around Kinshasa, are well known: increased storm flow and erosion.

Dr. Van Asseldonk introduced some technical remarks on a Bangladesh model. Of rural energy use, 93% is for food preparation, with 80% for cooking and the rest for processing. Oil accounts for about 7.5% (people have little buying power) and wood for about 7% (there is only a small part of the forest left).

If we look at the main components and the relevance of farm size, we find a big difference in the rural and urban consumption of food and energy, and between large and small farmers. Jute and straw waste are better provided in large farms which grow more cash crops. As a result there is less feed-fuel competition than in small farms.

There is a considerable possibility of energy saving in cooking. The heat efficiency of dung, waste and firewood is about 8%, and gas utilisation by fermentation may improve this by a factor of 5-6.

There is also a choice of several cooking practices which affect the energy efficiency. These factors are incorporated into the model.

Professor Kirshnayya provided information on two projects. It is hoped to start, with the aid of the Canadian IDRC, a documentation center for developing countries on the applications and uses of modelling techniques. This may provide information on what is happening in the modelling world for the use of a network of institutions and people in the developing world, in particular with reference to Asia and the India Ocean region.

A long term, eco-social and economic research project

studying the western hills region of India has been suggested. Field data could be collected to enable an understanding of the energy, agriculture and food actuality in that region. This would help an understanding of possibilities and a consideration of new alternatives including changes that are likely in the future. This would involve looking at employment, the social management system of villages, larger communities, migration processes and so on. A sub-model could link with other regional models into an integrating national model, making sure that the total system is as self-reliant as possible.

7.6 FORMAT FOR A UNU PROJECT

Dr. Robinson mentioned some factors relating to the problem and to the role of the UNU, to implications concerning the overall project organisation, and in particular the role of global models. We wish to look at the key elements of a particular problem – an interdisciplinary study of food and energy. Amongst the key elements that one can quickly identify as having been important in recent years are: ecological aspects including the destruction of other species and the ecosystem (remember that other parts of nature need to be considered), socio-political aspects, trade and the activities of multinationals, the development of the world economy, cultural and historical aspects. All of these are quite key in a consideration of this particular problem.

The identification of a list of key elements gives us some idea of the organisation that would be appropriate to study this problem. An appropriate organisation for an interdisciplinary study such as this is a team – which can work together, and therefore of about 5-6 people. If you get that point wrong, the whole project may fall to bits. It will take about 3 years for a study like this to really evolve.

A consideration of the elements identified will indicate that different people in different groups will provide different interpretations of what is happening. In other words there will be different paradigms to be considered. This must be remembered when setting up the teams.

Models should be put in this context, of a team which is going to look at quite a variety of different factors. Models should be appropriate to this type of problem, and left to the team to choose or develop (small models if what is most appropriate, for example). There is a great problem in moving to global models, and a danger that this may force the pattern of the research: one may study those elements which fit into the model and ignore other elements. As a person who has studied global models, Dr. Robinson believes that they are not very useful. They are primarily a propaganda tool. If you have a lot of money, are setting up a large scale project, and have a message to get over, you can often convince people more easily with the help of a global model. But most results seen coming out of global models were fed into the models in the first place.

Putting this project in the overall world context, if the UNU can operate as a university it could guarantee the academic freedom necessary for this sort of study. If you are going to study cultural, historical and socio-political aspects, and if you are going to talk about the real factors in this problem, then each of the suggested teams will surely rub somebody up the wrong way. The UNU may then be *the* appropriate international organisation for this type of effort.

The UNU should also not repeat work under way elsewhere.

The appropriate tool for this study is then first of all people; several interdisciplinary teams deliverately chosen to follow divergent paradigms. There should be contact amongst the teams, and academic freedom is absolutely essential. If you stand outside the whole modelling-planning system, and look at what happens in the real world, you find that most modelling and planning exercises are controlled by certain agents. And if a government, for example, sets up a planning department, it does not do so in order to plan that government out of existence.

The choice of models must be left to the teams; do not impose the tool and then fit the problem study around it. The process should be to identify the key elements in the problem and then look for whatever tools are appropriate. Within that focus each team should be looking for linkages, and able to spread itself as wide as it wishes. Thus, for example, if a team decides that political aspects are dominant, it should be free to consider the implications.

Dr. Levy indicated that the UNU can guarantee academic freedom, and can encourage the study of different paradigms.

Dr. Fergany developed this point. The special character of

the UNU, particularly from the point of view of academic freedom, makes the UNU vastely different from any other UN organisations, which are essentially intergovernmental bodies. This is why one would dare to make certain proposals to the UNU which one would not make to a normal UN body. This is not a call to the UNU to produce subversive literature, but to dare to be intellectually free and to propose real solutions to the problems of the world.

We should note the tale on which the title "Groping in the Dark" was built. We must not look for the key where there is light, but we must look for the key where it is – where it now is dark. To paraphrase Bernard Shaw, there are two kinds of people: reasonable and unreasonable man. The reasonable man tries to adapt himself to society, while the unreasonable man tries to adapt society to himself. Hence all progress depends on the unreasonable man. So lets all try to be a little unreasonable and creative.

The starting point for a proposal should be the lack of satisfaction of energy and food needs of large segments of the world population. The ultimate objectives are to try to understand what is happening and to decide what options are open for meeting this situation.

We should be extremely enlightened by what is happening in some Third World countries. For example, in Sri Lanka there is a tremendous development program to bring new land under cultivation, along with the necessary irrigation. However they are beginning to realise that, with the physical development under way, they are running into difficulties with the institutional structure that should go with it. We will similarly run into the strong linkages between the energy and food subsystems and economic processes.

To the three levels, local, national and global, we should add a fourth, regional, level – taking into account collective self-reliance among nations in a certain region.

There should be social scientists at all levels, and in addition those working on the technological aspects should have an understanding of and appreciation for the impact of socio-political processes on technologies. It is not model building per se that is required, but a systems analytic study, with some elements of quantitative modelling.

Dr. Norse emphasised the need to get a clear idea of international linkages – if you get too complex, you never finish. Keep it simple, within the resources of Third World countries, with models on mini-computers rather than main frame computers. Keep it transparent so you have a clear idea of your biases in your model, small so you can apply a sensitivity analysis, small so you have a reasonable chance of producing a working model in a reasonable length of time. If you don't see a concrete result for your endeavours, the team may fall apart, and there are many examples of that.

The people present at this conference represent mainly the global modelling fraternity. There is an absence of people who model small agriculture systems, and there are very good models of the Philippines, Venezuela, Nigeria, and so on, which have already started to look at the agriculture-energy nexus and which have tried to bring in some social aspects. It could be cost-effective to link into these.

There must be a hybrid approach, not an effort to link everything together; machine-machine interfaces are not always possible. We need lateral thinking; there is a lot to be taken from other disciplines.

It was pointed out that the UNU has the problem of the optimum allocation of research resources. It should not repeat what has been done, but should draw heavily on the results of other models. First then there should be a search for cooperation with other models at national, regional and global levels to find out to what extent the UNU could use these models.

Dr. Chan hoped the next UNU meeting could be in a developing country, perhaps in the ESCAP region. The majority of the population considered is in the developing countries, and there the situation is quite different. For example, in China, where a discussion of energy and agricultural development has been under way for 3 years, there is a different situation from America and Europe. The main energy source is coal, not oil. In the rural areas the main energy source is biomass from local resources, not commercial energy. Also more developing countries should be asked to participate.

The work should start with concrete examples from national or regional models, rather than from global models.

8 Synthesis View of Proceedings

JACQUES LESOURNE

In this personal synthesis, I shall first discuss briefly the use of global models and the relations between decision-makers and analysts. Then, I shall review the relations between energy and agriculture put into evidence during the seminar and present some propositions on a possible UNU programme. Finally, I shall address the central question raised at the beginning of the seminar: What do global models tell us on political implications concerning energy and agriculture?

8.1 THE USE OF GLOBAL MODELS

The first idea that should be stressed is that there has been in the last ten years, a continuous progress in global modelling. With the model of the limits to growth, the notion of global models has become accepted. With the Mesarovic and Pestel model the necessity for regional decomposition has been established. Several models, such as that of Leontiev, have shown the need for a sectoral decomposition. SARUM illustrates the importance of substitution among the different factors of production and substitution in consumption, as well as recognition of the role of prices and of barriers in international commerce. In the UNITAD model there is an improvement in the treatment of the role of financial transfers, including debts, interest and aid. At the same time there has been a differentiation between rural and urban sectors, and between different agricultural systems, as in the work of the Chinese group. In the work of Sam Cole and UNITAR there is a decomposition of the population into classes, and the relative decomposition of goods, which has permitted the treatment of a number of pertinent questions. The food and agriculture program of IIASA has developed furthest the concept of interaction among national models.

But the global models are not only perfectible tools. They have a second merit which we should recognize: Independent of the results and relationships, they imply a considerable effort at the elaboration of basic information. Therefore, they represent a progress in the availability of pertinent data at the world level. Unfortunately, the most interesting data incorporated into the models are often not easily available and this is a loss since such data may have an interest in themselves. The UNU may consider the problem of taking in charge the publication of this information, providing an easy access, with clear definitions and an attempt at a coordination of reference years.

A third remark which must be made on global modelling is that we tend to accept more and more, as was done in Interfutures, the fact that a reflexion on the future cannot be reduced to a set of calculations on a model. This too, is a step forward, for at first a scenario was considered to be represented by a computer run with a given model, and the totality of reflexion as if it were contained in the model. On the contrary, even if a model is a necessary tool, it is also vital to use other sources of information. A reflexion on the future involves forming a judgement. Hence, it is equally essential to take into account quantitative elements which are not easily included directly in the model and qualitative elements which cannot be formally modelled at least for the time being: social groups, types of conflict, political questions within countries and between countries. If the study is limited to a model, the accent is on those aspects of the problem that are most easily modelled and formalised. This problem is not unique to global modelling. In operations research, the studies which are published are often not the best, but rather those which can be developed in a mathematical formulation and by consequence are accepted by journals. The problem is also found in futures research.

I consider also a progress, and that will be my fourth remark, that we recognize how difficult it is to make some analysis of the future without a definite vision of the manner in which the world functions. There are different paradigms, and different groups differ in their understanding of the functioning of the world system. There are also differences between disciplines, and in the personalities of the research leaders, who may have come from the natural sciences, economics or (less often) sociology or political sciences. Depending on their origin, they will privilege or neglect different aspects of the world system.

After having presented these remarks, I would like to introduce a few more specific issues:

- The problem of time scale has a great importance, particularly in the consideration of technology. The longer the time scale, the more technology becomes a critical factor; in shorter time scales the problem is more one of social development and the acceptance of known technologies. Although, as in economics, there is a clear distinction between short and middle terms, there is a tendency in global modelling to treat the long term as the very long term. I think this is an error and that we should clearly distinguish the long term model covering for instance two decades and the very long term models exploring half a century.

- When the models are general, they are often badly adapted to provide a pertinent response to particular questions; but when they are partial, there is a problem of linkages, and of the manner in which concerns not included in the partial analysis are related. Hence, the need of compromises which are not always easy.

- The link between global modelling and theoretical research is very important. The models are based on output from a number of scientific disciplines, but there is not always a sufficient interaction between the relations integrated into the global models and the theoretical research. A greater effort should be made to separate the relations which are based on solid scientific research and those which are postulated to make the model complete.

- Even though it is accepted that multidisciplinarity is absolutely necessary, there are difficulties in this approach. Often scientists have no other standards than the publications of the scientific republics to which they belong. To carry out multidisciplinary research it is necessary to abandon that reference standard, with often considerable career implications. This is all the more important since the social sciences, in particular sociology and political science, are often under-represented in those studies.

• Finally, in most efforts on global modelling the work has stopped before all the possibilities of the models have been explored. There is a need for a greater permanence; so that after an enormous initial investment, the full results can be gathered.

8.2 RELATIONS BETWEEN ANALYSTS AND DECISION MAKERS

If decision makers cannot understand those who construct global models, is the only excuse a lack of intellectual capacity? I do not think so. Let us imagine a fable in which a specialist in global modelling becomes overnight a minister of foreign affairs, and finds that, due to the limits of his action, much of his modelling experience must be set aside. There has been a recent example in a European country, and we have seen a change in attitude of this person as he came from an economic institute where he had constructed models to the post of Minister of Finance of his country. We must recognise that decision makers have not the same objectives as model-builders. We may agree as scientists that organizations throughout the world should be effective, viable and sustainable. But this is not the objective of the decision makers. Most of the time, they must be concerned with the reactions of various social groups; for example, when the price of wheat is fixed in the EEC, the agriculture minister of France must consider what he will have to say to the wheat producers the following week. Even if he places great importance on long term effects, he has to care, in a democratic country, about short-term goals.

We must also realise that the people constructing global models are not only scientific experts. They constitute also an ideological lobby. Although I agree with many of the aspects of this ideology, I have also a reaction, almost a sentiment of rejection, to the idea that a group may set out to make decisions on a global scale, and to the idea that politicians may base decisions only on the logic of the global modellers.

Analysts and decision makers do not share the same perception of time. Often a decision maker must define a point of view at an international conference or in a bilateral discussion with another government, when there is not enough time to bring together the modelling groups who study the problem.

The modellers often estimate that it will take three years to carry out a study but in three years the problem may no longer be of interest, and the minister may no longer be in power. As a consequence the image of the relationship between analysts and decision makers as a sort of dialogue is a false conception.

It is impossible to escape from a hierarchy of relationships. At the base are fundamental studies, which take often several years. Then there are intermediate studies which examine more limited issues and may last between a few months and one year. Finally there is a need for translators who are capable, as problems arise, to present propositions to the ministers. It is only by such a hierarchy that one can assume that basic work such as global modelling has some effect on decisions. Otherwise the concept of time is too distant.

Analysts and decision makers do not refer to control instruments in the same sense. Thus in models, variables and control functions are aggregated and relatively abstract, and thus do not correspond directly to the instruments with which a decision maker is familiar and on which he can act.

Decision makers are of course not the only public. In industrial societies, particularly in the West, the number of decision makers is in any case very large. We need to consider not only ministers and high functionaries, but also a lot of other persons for it is difficult for a minister for example to take an initiative without some backing from a collection of groups or from the individuals who make up the society.

The questions has been raised whether we should present an optimistic or a pessimistic vision of the future. We should be neither optimistic or pessimistic, for in both cases there is a risk of a demobilising effect. If results are centered on otpimism, the reaction may be that the situation is not too bad and there is no need to worry. If the results are centered on pessimism, on the menace of catastrophe, the reaction may be one of complete powerlessness. Above all the goal of the global modellers and of the long-term analysts must be to make the world intelligible.

Analysts have in common a systemic view of the world. Hence, the major message they should convey is the sense of systemic behaviour. With a great deal of humility we must put into evidence the existence of feedback loops and what we learn from them. As for the modelling results of different teams, they will be at times convergent and at other times divergent. This raises a difficult issue, since in the expression of results we face the basic problem that everything which is simple is false, while everything which is complex is useless. But if simple and false messages are given to the media, other members of the modelling community will stress the errors and the result will be confusion. It is then extremely important to place the accent on areas of agreement. As for the points of disagreement, it is important to show that the debate is open, that attitudes depend on systems of values and on the balancing of risks of different possibilities and that it is normal and expected that a consensus should not appear.

8.3 RELATIONS BETWEEN ENERGY AND AGRICULTURE

The seminar has stressed the multiple linkages between energy and agriculture, such as the choice of the use of land, of investment finance and equipment, the use of each as input into the other, interactions with the environment, the necessity of a balance of payments equilibrium (as, for example, deficits in energy trade affect the ability to import foodstuffs), the increasing importation of food by petrol exporters and the greater problems of non-oil countries, and the complex issues of political pressures between societies. Some linkages are specific to the Third World and the poorest countries, but there are differences between countries and regions, as well as between regions within countries, and between urban and rural areas. The non-commercial aspects of economies are often fundamental. The participants have also underlined the diversity of power structures (for historical and social reasons, inherited from the past).

We must be particularly sensitive to the problems of viability and discontinuity, and devote our attention to the risks of fundamental collapses of both ecological and social systems. Conflict occurs in a world in which hopes are raised by communications and whose reality cannot evolve at the same speed as anticipations.

Problems of transportation and distribution must be considered. Distances are important in some Third World countries, and distribution problems often lead to a loss of food stuffs.

As many other participants, I think essential to study micro-societies, considering education and organization, and obstacles to the use of new technology. The question of food reserves at the various levels, and over various time scales must be included into the analysis.

8.4 A PROGRAM FOR THE UNU

Let us consider some of the issues which will have to be faced in the preparation of an "agriculture and energy program" for the UNU. These issues have been raised during the discussion:

- A first problem is to fix the object of the intended multi-disciplinary study, but there is not a simple link between the process and the product, as, for example, in chemistry, and the process itself will to some extent define the product or rather the products, since in many important projects, there is a multiplicity of products.
- A second decision concerns definition of the studies which should be carried out at local, national and global levels as well as at regional levels in two senses: sub-national ecologically homogenous regions, and complementary groups of geographically close countries.
- An important issue may be to explore how global and national politics impact on the food and energy situation of local poor populations in poor countries. It would be interesting to examine how a particular decision taken by a particular government has an impact on a particular environment, as on such fundamental aspects as energy and food supplies.
- At the local level, it is important to remember that, as well as global models, many models of local systems exist, and these will be pertinent to the study. The UNU should also consider the question of the creativity of small groups and the question of generalisation. Possibly it is at a local level that one can best study the problems of discontinuities, of risks, of environment, of political and institutional structures, and of barriers to certain technologies.
- At a national level, a number of models exist, usually with economic bases. One solution is to use such existing models and

to try to enrich parts of the models in order to progress more quickly and to profit from past research and existing information bases. It will be important to introduce into the models the instruments which are significant for decision makers and for the public power. Some aspects concerning, for example, customs, fiscal structures, budgetary spending and the placing of state investments should be introduced into the models if they are not already present.

- Globally, the study should also take advantage of existing models, but it could also try to improve them according to its aims.

- In any study of this kind, the planning process is essential. The formation of the central group is therefore very funadamental. It must be permanent and the quality of the staff will determine the quality of the work. The terms of reference should give to this group a significant amount of freedom and the ability to invent a process in which it can believe since this group will in any case carry out the program of research that it wishes to do. For example, in France although the government directs research projects, for a long time research groups have learnt how to actually do what they want while relating apparently the work to the government wishes.

- As pluridisciplinarity is fundamental, a degree of decentralization must be included. Different groups in different countries of the Third world will develop their own research. However the links between the central and other groups must include a common vision, within which each group may define its own point of view not in a constrained homogeneity imposed by the central group.

- The question of duration of the project is important. If sufficient funding is available, it is best to spread it over a long period since there will then be time to change the approach, to contribute to the formation of institutes and to eliminate errors. Too many global modelling groups have been like abstract painters who must use only one stroke and not make the least error. But since in world studies, mistakes are unavoidable, enough time is needed to correct these and to improve on the basic set of options.

8.5 GLOBAL MODELS AND POLITICAL IMPLICATIONS FOR ENERGY AND AGRICULTURE

This conference has examined the political implications of global models for the problems relating energy and agriculture.

At first, the result is disappointing: Thus far, there have been some striking divergences in model forecasts. For example while one study expects no long-term change in the international real price of cereals, another suggests a price doubling by the end of the century. Many other conclusions are vague and not very meaningful for decision-makers.

However, the judgment should be much more balanced: Much can be learnt from past model studies: an emphasis on the importance of this problem both today and in the world of tomorrow; the recognition of the enormous differences between regions and the identification of particularly critical areas (sub-Saharian Africa, certain regions of South Asia, parts of Latin America); an understanding of important linkages. The disagreement on results and on the fundamental links is often due to the different levels of aggregation. This concerns for example the problem of aid and the consequences on national revenues. This should encourage us to intensify our research effort.

Index